煤炭高等教育“十四五”规划教材

矿物加工导论双语教程

Bilingual Course on Mineral Processing Technology

主　编　吕宪俊

副主编　朱向楠　任阳光

中国矿业大学出版社

·徐州·

内容提要

本书以矿物加工工程的工艺过程和主要工艺方法为主线，全面介绍了矿物加工工程的基本原理、工艺方法和主要设备。本书内容覆盖面广，涵盖了矿物学基础(绪论)、破碎与筛分、磨矿与分级、重力分选、重介质分选、磁电分选、浮选、湿法冶金、产品脱水与尾矿处理、矿石分拣与电子垃圾回收等内容。本书的课文主要选自英语原文著作和公开发表的科技文献，语言规范，词汇丰富。全书共分10个单元、36个小节，每个小节正文后附有注释和练习题，书后附有生词和短语。

本书可作为矿物加工工程专业研究生的专业英语参考教材，也可作为矿物加工工程领域工程技术人员、外贸人员的参考资料。

图书在版编目(CIP)数据

矿物加工导论双语教程 ：汉、英 / 吕宪俊主编. —徐州 ：中国矿业大学出版社，2023.8

ISBN 978-7-5646-5753-6

Ⅰ. ①矿… Ⅱ. ①吕… Ⅲ. ①选矿—教材—汉、英 Ⅳ. ①TD9

中国国家版本馆CIP数据核字(2023)第037373号

书　　名　矿物加工导论双语教程
Kuangwu Jiagong Daolun Shuangyu Jiaocheng

主　　编　吕宪俊

责任编辑　褚建萍

出版发行　中国矿业大学出版社有限责任公司
(江苏省徐州市解放南路　邮编221008)

营销热线　(0516)83885370　83884103

出版服务　(0516)83995789　83884920

网　　址　http://www.cumtp.com　**E-mail**:cumtpvip@cumtp.com

印　　刷　江苏淮阴新华印务有限公司

开　　本　787 mm×1092 mm　1/16　**印张** 19　**字数** 632千字

版次印次　2023年8月第1版　2023年8月第1次印刷

定　　价　46.00元

(图书出现印装质量问题，本社负责调换)

前　言

随着科学技术的发展和对外经济技术交流的不断增加，对大学生、科技人员和外贸人员英语能力的要求不断提高。为适应这一要求，我们编写了这部《矿物加工导论双语教程》。该书旨在培养矿物加工工程专业学生阅读和翻译英文专业科技文献的能力，使学生掌握必要的专业词汇，熟悉科技英语文献的写作特点和翻译技巧。通过课后练习等环节，该书还旨在培养学生的专业英语阅读、翻译和表达能力。全书共分10个单元、36个小节，适用于64学时左右的课堂教学安排。

本书由吕宪俊担任主编、朱向楠和任阳光担任副主编，吕宪俊、朱向楠负责全书的整理、定稿，任阳光负责课后练习题和注释的撰写以及全书的校对。各章节具体编写分工如下：吕宪俊负责1.1、1.4、2.3、2.4、3.1、3.2、4.3章节的编写，石常省负责1.2、1.3章节的编写，王秀兰负责2.1章节的编写，白春华负责2.2章节的编写，张乾负责3.3章节的编写，朱向楠负责3.4、4.4、6.3、7.4、7.5、8.1、8.2章节的编写，郑钢丰负责4.1、7.3章节的编写，张开永负责4.2、8.3章节的编写，张鸿波负责5.1、5.2章节的编写，杨宏丽负责6.1、6.2章节的编写，陶有俊负责7.1、7.2章节的编写，任阳光负责8.4、9.3、9.4、10.1、10.2章节的编写，李志红负责9.1、9.2章节的编写。

在编写本书过程中，我们参阅和引用了许多国内外同行的资料，得到了中国煤炭教育协会的大力帮助和指导，在此一并表示衷心的感谢！

由于编者水平有限，书中不当之处在所难免，恳请广大读者批评指正。

编　者

2023年1月

前言

CONTENTS

1 INTRODUCTION(绪论)

1.1 Minerals, Rocks and Ores(矿物、岩石与矿石)

1. Minerals

(1) Definition

Minerals by definition are natural inorganic substances possessing definite chemical compositions and crystal structures. Some flexibility, however, is allowed in this definition. Many minerals exhibit isomorphism, where substitution of atoms within the crystal structure by similar atoms takes place without affecting the structure. The mineral olivine, for example, has the chemical composition $(Mg, Fe)_2SiO_4$, but the ratio of Mg atoms to Fe atoms varies in different olivines. The total number of Mg and Fe atoms in all olivines, however, has the same ratio to that of the Si and O atoms. Minerals can also exhibit polymorphism, different minerals having the same chemical composition, but markedly different physical properties due to a difference in crystal structure. Thus, graphite and diamond have exactly the same composition, being composed entirely of carbon atoms, but have widely different properties due to the different arrangement of the carbon atoms within the crystal lattice.

(2) Occurrence

The forms in which metals are found in the crust of the earth and as seabed deposits depend on their reactivity with their environment, particularly with oxygen, sulphur, and carbon dioxide. Gold and the platinum are found principally in the native and metallic form. Silver, copper, and mercury are found native, as well as in the form of sulphides, carbonates, and chlorides. The more reactive metals are always in compound form, such as the oxides and sulphides of iron and the oxides and silicates of aluminum and beryllium. The naturally occurring compounds are known as minerals, most of which have been given names according to their composition (e.g., galena-lead sulphide, PbS; sphalerite-zinc sulphide, ZnS; cassiterite-tin oxide, SnO_2).

(3) Properties

A mineral is a naturally occurring chemical compound, usually of crystalline form (Figure 1.1). A mineral has one specific chemical composition, whereas a rock can be an aggregate of different minerals. The study of minerals is called mineralogy.

There are over 5,300 known mineral species, while the silicate minerals compose over 90% of the earth's crust. Silicon and oxygen constitute approximately 75% of the earth's crust,

Figure 1.1 Common sulphide mineral crystals

which translates directly into the predominance of silicate minerals.

Minerals can be described by their various physical properties, which are related to their chemical composition and structure. Common distinguishing characteristics include crystal structure and habit, hardness, lustre, diaphaneity, colour, streak, tenacity, cleavage, fracture, parting, and specific gravity. More specific tests for describing minerals include magnetism, taste or smell, radioactivity and reaction to acid.

The term "mineral" is often used in a much more extended sense to include anything of economic value which is extracted from the earth. Thus coal, chalk, clay, and granite do not come within the definition of a mineral. Granite, for instance, which is the most abundant igneous rock, i. e., a rock formed by cooling of molten material, or magma within the earth's crust, is composed of three main mineral constituents, feldspar, quartz, and mica. These three homogeneous mineral components occur in varying proportions in different granites, and even in different parts of the same granite mass.

2. Rocks

A rock is composed of one or more of the minerals forming the earth's crust (Figure 1.2). The scientific study of rocks is called petrology. Rocks are commonly divided, according to their origin, into three major classes—igneous, sedimentary, and metamorphic.

Figure 1.2 Granite with different mineral compositions

(1) Igneous rocks

Igneous rock originates from the cooling and solidification of molten matter from the earth's interior. If the rock is formed on the earth's surface (i. e. , from the solidification of lava), it is called extrusive rock; igneous rock that has cooled and solidified slowly beneath the earth's surface is intrusive rock. Igneous rocks are commonly divided into classes by texture. Some rocks are markedly granular (e. g. , granite, syenite, diorite, gabbro, peridotite, and pyroxenite), while others (e. g. , basalt, trachite, dacite, and andesite) are composed of grains visible only under a microscope. Both fine-grained and coarse-grained igneous rocks frequently contain grains called phenocrysts that are larger than the surrounding grains; such rocks are said to be porphyritic in texture. Rocks with grains of uniform size are called equigranular.

(2) Sedimentary rocks

Sedimentary rocks originate from the consolidation of sediments derived in part from living organisms but chiefly from older rocks of all classes. The sediments of inorganic origin are chiefly removed from older rocks by erosion and transported to the place of deposition; chemical precipitation from solution is a secondary cause of deposition of inorganic matter. Sedimentary rocks are commonly distinguished, according to their place of deposition, such as continental, marine, littoral, estuarine, lacustrine, and fluvial.

The characteristic feature of sedimentary rocks is their stratification, they are frequently called stratified rocks. Sedimentary rocks made up of angular particles derived from other rocks are said to have a clastic texture, in contrast to pyroclastic sediments, which are particles of volcanic origin. Among the important varieties of sedimentary rock, distinguished both by texture and by chemical composition, are conglomerate, sandstone, tillite, sedimentary breccia, shale, marl, chalk, limestone, coal, lignite, gypsum, and rock salt. Characteristic occurrences in sedimentary rocks are fossils, footprints, raindrop impressions, concretions, oolites, ripple marks, rill marks, and crossbedding. Some of these features are useful in determining the antiquity of sedimentary formations and in interpreting geologic history.

(3) Metamorphic rocks

Metamorphic rocks originate from the alteration of the texture and mineral constituents of igneous, sedimentary, and older metamorphic rocks under extreme heat and pressure deep within the earth. Some (e. g. , marble and quartzite) are massive in structure; others, and particularly those which have been subject to the more extreme forms of metamorphism, are characterized by foliation (i. e. , the arrangement of their minerals in roughly parallel planes, giving them a banded appearance). A distinguishing characteristic of many metamorphic rocks is their slaty cleavage. Among the common metamorphic rocks are schist (e. g. , mica schist and hornblende schist), gneiss, quartzite, slate, and marble.

3. Ores

(1) Definition

An ore is a naturally occurring mineral aggregate that contains valuable constituents (such as metals or minerals) and can be profitably mined. This establishes the market price of the valuable metal or mineral as a critical criterion in the definition, which will vary according to the commercial demands. Nearly all rock deposits contain some metallic minerals, but in many cases the concentration of metal is too low to justify mining the ore. With the passage of time and depletion of richer ores, a mineral deposit may be updated to an ore. Improvement in metallurgical extraction and the introduction of new methods also become factors in making available deposits. Thus, the introduction of the flotation process in mineral processing permitted the extraction of copper from material containing less than 0.5% of the metal and formerly regarded as waste.

Minerals with no commercial value, called gangue minerals, are usually found mixed with the ore. Some ores are simple chemical compounds, while others are chemically complex minerals. Important ores of aluminum, iron, manganese, and tin are oxides; important ores of antimony, copper, lead, mercury, nickel, silver, and zinc are sulfides. Some metals, called native metals, occur uncombined in nature, e.g., copper, gold, platinum, and silver.

The minimum metal content required for a deposit to qualify as an ore varies from metal to metal. Many nonferrous ores contain, as mined, as little as 1% metal, and often much less. Gold may be recovered profitably in ores containing only 5 parts per million (ppm) of the metal, whereas iron ores containing less than about 20% metal are regarded as low grade.

There are many plants where minerals are recovered in secondary circuits, treating tailings, where the feed grades are much lower than would be economic on a mined ore. Typical ore grades for tungsten ores are in the range 0.5%~1.5% WO_3, but the Climax Molybdenum plant in the United States treats 45,000 tonnes per day of tailings, containing less than 0.1% WO_3, and is one of the two major producers of tungsten concentrate in the United States.

(2) Classification

Ores are frequently classed according to the nature of the valuable mineral. Thus, in native ores the metal is presented in the elementary form; sulphide ores contain the metal in the form of a sulphide, and in oxidised ores the valuable mineral may be present as oxide, sulphate, silicate, carbonate, or some hydrated from of these. Complex ores are those containing profitable amounts of more than one valuable mineral. Metallic minerals are often found in certain associations, within which they may occur as mixtures of a wide range of particle sizes or as single-phase solid solutions or compounds. Galena and sphalerite, for example, commonly associate, as do copper sulphide minerals and sphalerite to a lesser extent. Pyrite (FeS_2) is very often associated with these minerals.

Ores are also classified by the nature of their gangues, such as calcareous or basic (lime rich), or siliceous, or acidic (silica rich).

(3) Metallic or non-metallic ores

Ores of economic value can be classed as metallic or non-metallic, according to the use of the mineral (Figure 1.3). Certain minerals may be mined and processed for more than one purpose. In one category the mineral may be a metal ore, i. e., when it is used to prepare the metal, as when bauxite is used to make aluminum. The alternative is for the compound to be classified as a non-metallic ore, i. e., when bauxite or natural aluminum oxide is used to make material for refractory bricks or abrasives. Many non-metallic ore minerals associate with metallic ore minerals and are mined and processed together, e. g., galena, the main source of lead often associates with fluorite (CaF_2) and barytes ($BaSO_4$), both important non-metallic minerals.

Figure 1.3 Metallic and non-metallic ores

Notes

1. Many minerals exhibit isomorphism, where substitution of atoms within the crystal structure by similar atoms takes place without affecting the structure.

许多矿物呈现类质同象,即晶体结构中的原子被相似的原子取代而不影响其晶体结构。

2. Thus, graphite and diamond have exactly the same composition, being composed entirely of carbon atoms, but have widely different properties due to the different arrangement of the carbon atoms within the crystal lattice.

因此,石墨和金刚石这两种矿物的成分完全相同,都由碳原子组成,但由于碳原子在晶格中的排列方式不同,它们的性质却大相径庭。

3. Minerals can be described by their various physical properties, which are related to their chemical composition and structure.

矿物可以用与其化学组成和结构有关的各种物理性质来描述。

4. Rocks are commonly divided, according to their origin, into three major classes—igneous, sedimentary, and metamorphic.

岩石按其成因一般分为火成岩、沉积岩和变质岩三大类。

5. Galena and sphalerite, for example, commonly associate, as do copper sulphide

minerals and sphalerite to a lesser extent.

例如，方铅矿和闪锌矿通常共生在一起，而硫化铜矿物与闪锌矿的共生现象相对较少。

6. In one category the mineral may be a metal ore, i. e. , when it is used to prepare the metal, as when bauxite is used to make aluminum. The alternative is for the compound to be classified as a non-metallic ore, i. e. , when bauxite or natural aluminum oxide is used to make material for refractory bricks or abrasives.

当矿物用于制备金属时，它可能作为金属矿石一类，如铝土矿用于炼铝时就是金属矿石。另一种分类是将其归为非金属矿石，即当铝土矿或者天然氧化铝作为制造耐火砖或磨料的原料时就属于非金属矿石。

Exercises

1. Answer the following questions according to the text.

(1) What is the definition of minerals?

(2) What is isomorphism?

(3) How to classify rocks?

(4) How many types of ores can be classified according to the nature of the valuable mineral?

(5) What are complex ores?

2. Translate the following phrases into English.

(1) 无机化合物　(2) 化学组成　(3) 共生　(4) 有用矿物

(5) 脉石矿物　(6) 火成岩　(7) 非金属矿石　(8) 解理

(9) 沉积岩　(10) 变质岩

3. Translate the following paragraphs into Chinese.

Paragraph 4 in section 1 (3) “The term ‘mineral’ is often used in a much more ... ”.

4. Translate the following sentences into English.

(1) 金属在地壳中的存在形式取决于它们与环境的反应性。

(2) 矿物是一种天然形成的化合物，通常为晶体形式。

(3) 通常根据有用矿物的性质对矿石进行分类。

(4) 根据有用矿物的用途，矿石可分为金属矿石和非金属矿石。

(5) 没有商业价值的矿物称为脉石矿物。

1.2 Types and Applications of Coal(煤的类型及应用)

1. Origin of coal

(1) Coal-forming materials

Coal is a fossil fuel formed in swamp ecosystems where plant remains were saved by water and mud from oxidization and biodegradation. Coal is a readily combustible black or brownish-black rock. It is a sedimentary rock, but the harder forms, such as anthracite coal, can be regarded as metamorphic rocks because of later exposure to elevated temperature and pressure. It is composed primarily of carbon along with assorted other elements, chiefly hydrogen, sulfur, oxygen, and nitrogen (Figure 1.4).

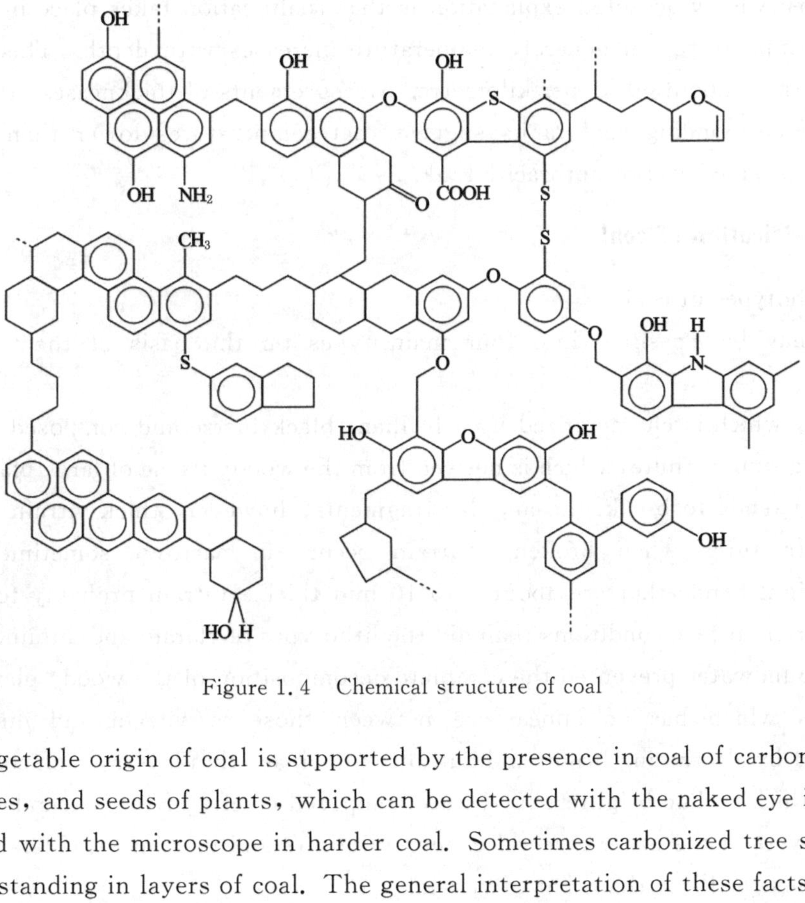

Figure 1.4 Chemical structure of coal

The vegetable origin of coal is supported by the presence in coal of carbonized fibers, stems, leaves, and seeds of plants, which can be detected with the naked eye in the softer varieties and with the microscope in harder coal. Sometimes carbonized tree stumps have been found standing in layers of coal. The general interpretation of these facts is that coal originated in swamps similar to present-day peat bogs and in lagoons, probably partly from plants growing in the area and partly from plant material carried in by water and wind.

(2) Formation processes

The general sequence of coalification is from lignite to subbituminous to bituminous to anthracite. Since microbial activity ceases within a few metres of earth's surface, the coalification process must be controlled primarily by changes in physical conditions that take place with depth.

Three major physical factors—duration, temperature, and pressure—may influence the coalification process. In laboratory experiments artificially prepared coals are influenced by the duration of the experiment, but in nature the length of time is substantially longer and the overall effect of time remains undetermined. Low-rank coal (i. e., brown coal) in the Moscow Basin was deposited during Carboniferous time but was not buried deeply and never reached a higher rank.

The effect of increasing pressure due to depth of burial is not considered to cause coalification. In fact, increasing overburden pressure might have the opposite effect if volatile compounds such as methane that must escape during coalification are retained. Pressure may influence the porosity and moisture content of coal.

The most widely accepted explanation is that coalification takes place in response to increasing temperature. In general, temperature increases with depth. This geothermal gradient averages about 30 ℃ per kilometre. Measurements of thicknesses of sedimentary cover and corresponding coal ranks suggest that temperatures lower than 200 ℃ are sufficient to produce coal of anthracite rank.

2. Classification of coal

(1) Lithotypes of coal

Coals may be classified into four main types on the basis of their macroscopic appearance.

Vitrain, which is characterized by a brilliant black lustre and composed primarily of the maceral group vitrinite, which is derived from the woody tissue of large plants. Vitrain is brittle and tends to break into angular fragments; however, thick vitrain layers show conchoidal fractures when broken. Vitrain occurs in narrow, sometimes markedly uniform, bright bands that are about 3 to 10 mm thick. Vitrain probably formed under somewhat drier surface conditions than did the lithotypes of clarain and durain. On burial, stagnant groundwater prevented the complete decomposition of the woody plant tissues.

Clarain, which has an appearance between those of vitrain and durain and is characterized by alternating bright and dull black laminae (thin layers, each commonly less than 1 mm thick). The brightest layers are composed chiefly of the maceral vitrinite and the duller layers of the other maceral groups (liptinite and inertinite). Clarain exhibits a silky lustre less brilliant than that of vitrain.

Durain, which is characterized by a hard granular texture and composed of the maceral groups of liptinite and inertinite as well as relatively large amounts of inorganic minerals. Durain occurs in layers more than 3 to 10 mm thick, although layers more than 10 cm thick have been recognized. Durains are usually dull black to dark gray in colour. Durain is thought to have formed in peat deposits below water level, where only liptinite and inertinite components resisted decomposition and where inorganic minerals accumulated from sedimentation.

Fusain, which is commonly found in silky and fibrous lenses that are only millimetres thick and centimetres long. Most fusain is extremely soft and crumbles readily into a fine, soot like powder that soils the hands. Fusain is composed mainly of fusinite (carbonized woody plant tissue) and semifusinite from the maceral group inertinite, which is rich in carbon and highly reflective. It closely resembles charcoal, both chemically and physically, and is believed to have been formed in peat deposits swept by forest fires, by fungal activity that generated intense heat, or by subsurface oxidation of coal.

(2) Ranking by coalification

As geological processes apply pressure to peat over time, it is transformed successively into the following types of coal.

Lignite is the lowest rank of coal and used almost exclusively as fuel.

Sub-bituminous coal whose properties range from those of lignite to those of bituminous coal and is used primarily as fuel for steam-electric power generation.

Bituminous coal a dense coal, usually black, sometimes dark brown, often with well defined bands of bright and dull material, used primarily as fuel in steam-electric power generation, with substantial quantities also used for heat and power applications in manufacturing and to make coke.

Anthracite the highest rank, a harder, glossy, black coal used primarily for residential and commercial space heating.

In addition to the major ranks (lignite, subbituminous, bituminous, and anthracite), each rank may be divided into coal groups such as high-volatile bituminous coal. These categories differ slightly between countries.

Virtually all classification systems use the percentage of volatile matter present to distinguish coal ranks. In the ASTM classification, high-volatile bituminous (and higher ranks) are classified on the basis of their volatile matter content. Coals of lower rank are classified primarily on the basis of their heat values, because of their wide ranges in volatile matter content (including moisture). According to the German classification, coal is classified as follows (Table 1. 1).

Table 1. 1 German classification of coal

Designation	Volatiles /%	C/%	H/%	O/%	S/%	Heat content /(kJ/kg)
Lignite	45~65	60~75	6. 0~5. 8	34~17	0. 5~3	<28,470
Flame coal	40~45	75~82	6. 0~5. 8	>9. 8	~1	<32,870
Gas flame coal	35~40	82~85	5. 8~5. 6	9. 8~7. 3	~1	<33,910
Gas coal	28~35	85~88	5. 6~5. 0	7. 3~4. 5	~1	<34,960
Fat coal	19~28	88~90	5. 0~4. 5	4. 5~3. 2	~1	<35,380

Table 1.1 (continued)

Designation	Volatiles /%	C/%	H/%	O/%	S/%	Heat content /(kJ/kg)
Forge coal(Meagre coal)	14~19	90~91	4.5~4.0	3.2~2.8	~1	<35,380
Nonbaking coal	10~14	91~92	4.0~3.8	2.8~3.5	~1	<35,380
Anthracite	7~12	>92	<3.75	<2.5	~1	<35,300

3. Application of coal

(1) Coal as fuel

Coal is extracted from the ground by coal mining. Since 1983, the world's top coal producer has been China. In 2015 China produced 3,747 million tonnes of coal —48% of 7,861 million tonnes world coal production. In 2015 other large producers were United States [813 million tonnes (Mt)], India (678 Mt), European Union (539 Mt) and Australia (503 Mt). In 2010 the largest exporters were Australia with 328 million tonnes (27% of world coal export) and Indonesia with 316 million tonnes (26%), while the largest importers were Japan with 207 million tonnes (18% of world coal import), China with 195 million tonnes (17%) and South Korea with 126 million tonnes (11%).

Coal is primarily used as a solid fuel to produce electricity and heat through combustion. Coal is the largest source of energy for the generation of electricity worldwide, as well as one of the largest worldwide anthropogenic sources of carbon dioxide releases. When coal is used for electricity generation, it is usually pulverized and then burned in a furnace with a boiler. The furnace heat converts boiler water to steam, which is then used to spin turbines which turn generators and create electricity. An alternative approach of using coal for electricity generation with improved efficiency is the integrated gasification combined cycle (IGCC) power plant. Instead of pulverizing the coal and burning it directly as fuel in the steam-generating boiler, the coal is gasified to create syngas, which is burned in a gas turbine to produce electricity. Another alternative form of coal combustion is coal-water slurry fuel (CWS), which was developed in the Soviet Union.

At least 40% of the world's electricity comes from coal, and in 2016, 30% of the United States' electricity came from coal, down from approximately 49% in 2008. As of 2012 in the United States, use of coal to generate electricity was declining, as plentiful supplies of natural gas became available at low prices.

(2) Coking and use as coke

Coke is a solid carbonaceous residue derived from low-ash, low-sulfur bituminous coal from which the volatile constituents are driven off by baking in an oven without oxygen at temperatures as high as 1,000 ℃ so that the fixed carbon and residual ash are fused together. Metallurgic coke is used as a fuel and as a reducing agent in smelting iron ore in a blast furnace. Coke from coal is grey, hard, and porous and has a heating value of 29.6 MJ/kg. Byproducts of this conversion of coal to coke include coal tar, ammonia, light oils, and "coal gas".

Petroleum coke is the solid residue obtained in oil refining, which resembles coke but contains too many impurities to be useful in metallurgical applications.

(3) Gasification

Coal gasification breaks down the coal into its components, usually by subjecting it to high temperature and pressure, using steam and measured amounts of oxygen. This leads to the production of syngas, a mixture mainly consisting of carbon monoxide (CO) and hydrogen (H_2).

In the past, coal was converted to make coal gas, which was piped to customers to burn for illumination, heating, and cooking. At present, the safer natural gas is used instead. South Africa still uses gasification of coal for much of its petrochemical needs.

Gasification is also a possibility for future energy use, as the produced syngas can be cleaned-up relatively easily leading to cleaner burning than burning coal directly. The cleanliness of the cleaned-up syngas is comparable to natural gas enabling to burn it in a more efficient gas turbine rather than in a boiler used to drive a steam turbine.

Coal is also an important feedstock in production of a wide range of chemical products. The main route to these products is coal gasification to produce syngas. Primary chemicals that are produced directly from the syngas include methanol, hydrogen and carbon monoxide, which are the chemical building blocks from which a whole spectrum of derivative chemicals are manufactured, including olefins, acetic acid, formaldehyde, ammonia, urea and others.

(4) Liquefaction

Coal can also be converted into liquid fuels like gasoline or diesel by several different processes.

The Fischer-Tropsch process of indirect synthesis of liquid was used in Nazi Germany for many years and is today used by SASOL in South Africa. Coal would be gasified to make syngas (a mixture of CO and H_2 gas) and the syngas can be condensed using Fischer-Tropsch catalysts to make light hydrocarbons which are further processed into gasoline and diesel.

A direct liquefaction process, Bergius process (liquefaction by hydrogenation), is also available but has not been used outside Germany, where such processes were operated both during World War Ⅰ and World War Ⅱ. Several other direct liquefaction processes have been developed, among these being the SRC-Ⅰ and SRC-Ⅱ (Solvent Refined Coal) processes developed by Gulf Oil and implemented as pilot plants in the United States in the 1960s and 1970s.

Another direct hydrogenation process was explored by the NUS Corporation in 1976 and patented by Wilburn C. Schroeder. The process involved dried, pulverized coal mixed with roughly 1% molybdenum catalysis. Hydrogenation occurred by use of high temperature and pressure synthesis gas produced in a separate gasifier. The process ultimately yielded a synthetic crude product, naphtha, a limited amount of C_3/C_4 gas, light-medium weight liquids (C_5—C_{10}) suitable for use as fuels, small amounts of NH_3 and significant amounts of CO_2.

Yet another process (Karrick process) to manufacture liquid hydrocarbons from coal is low temperature carbonization (LTC), which was developed by Lewis C. Karrick, an oil shale technologist at the U. S. Bureau of Mines in the 1920s. Coal is coked at temperatures between 450 ℃ and 700 ℃ compared to 800 ℃ to 1,000 ℃ for metallurgical coke. These temperatures optimize the production of coal tars richer in lighter hydrocarbons than normal coal tar. The coal tar is then further processed into fuels.

Notes

1. Coal is a fossil fuel formed in swamp ecosystems where plant remains were saved by water and mud from oxidization and biodegradation.

煤是在沼泽生态环境中形成的化石燃料，在这种环境中由于水和泥的保护作用，植物遗骸不受氧化和生物降解的影响。

2. It is composed primarily of carbon along with assorted other elements, chiefly hydrogen, sulfur, oxygen, and nitrogen.

它主要由碳和其他元素组成，主要为氢、硫、氧和氮。

3. An alternative approach of using coal for electricity generation with improved efficiency is the integrated gasification combined cycle (IGCC) power plant.

提高煤炭发电效率的另一种方法是电厂使用整体煤气化联合循环(IGCC)技术。

4. Coke is a solid carbonaceous residue derived from low-ash, low-sulfur bituminous coal from which the volatile constituents are driven off by baking in an oven without oxygen at temperatures as high as 1,000 ℃ so that the fixed carbon and residual ash are fused together.

焦炭是烟煤在隔绝氧气的条件下和 1 000 摄氏度的高温下焙烧脱除挥发分后，固定碳和残留灰分熔融在一起的混合物。

5. Coal gasification breaks down the coal into its components, usually by subjecting it to high temperature and pressure, using steam and measured amounts of oxygen.

煤气化是将煤分解为不同的组分，通常使用蒸汽和一定量的氧气气氛，在高温、高压条件下进行。

6. Gasification is also a possibility for future energy use, as the produced syngas can be cleaned-up relatively easily leading to cleaner burning than burning coal directly.

煤气化的产物也可能作为一种未来的能源使用，因为煤气化生成的合成气较容易净化，从而比直接燃烧煤炭更为清洁。

Exercises

1. Answer the following questions according to the text.

(1) How to classify coal according to its coalification?

(2) What are the major applications of coal?

(3) What is coal gasification?

(4) How many processes are there for coal liquefaction?

(5) What are the physical factors that may influence the coalification process of coal?

2. Translate the following phrases into English.

(1) 沉积岩　　(2) 蒸汽发电　　(3) 煤化过程　　(4) 低灰分烟煤

(5) 挥发分　　(6) 煤气化　　(7) 直接液化　　(8) 间接液化

(9) 煤岩组分　　(10) 褐煤

3. Translate the following paragraphs into Chinese.

Paragraph 2 in section 3 (1) "Coal is primarily used as a solid fuel to produce electricity ..."

4. Translate the following sentences into English.

(1) 煤是一种容易燃烧的黑色或棕黑色的岩石。

(2) 煤化程度的顺序一般是从褐煤到亚烟煤到烟煤再到无烟煤。

(3) 影响煤化过程的主要物理因素有三个:持续时间、温度和压力。

(4) 根据煤的宏观形貌,煤可分为四种主要类型:镜煤、亮煤、暗煤和丝炭。

1.3 Coal Assay(煤质分析)

1. Chemical analysis

Coal is a heterogeneous mixture of organic compounds, together with a certain amount of inorganic material in the form of moisture and mineral impurities. The general nature of these materials can be evaluated by a combination of two sets of analytical data: Proximate analysis and Ultimate analysis.

(1) Proximate analysis

Proximate analysis gives a measure of the relative amount of volatile and non-volatile organic compounds in the coal as well as the percentage of moisture and non-combustible mineral materials.

Mositure

The moisture in a coal may occur in four possible forms: ① Surface moisture: water held on the surface of coal particles or macerals. ② Hydroscopic moisture: water held by capillary action within the microfractures of the coal. ③ Decomposition moisture: water held within the coal's decomposed organic compounds. ④ Mineral moisture: water which comprises part of the crystal structure of hydrous silicates such as clays.

Determination of total moisture is based on the loss in mass between the sealed sample as received in the laboratory and that of the same sample fully dried. This is achieved by any of the following methods: ① Heating the coal within a solution of toluene. ② Drying in a minimum free-space oven at 150 ℃ within a nitrogen atmosphere. ③ Drying in air at 100～105 ℃ and relative loss of mass determined. Methods ① and ② are suitable with low-rank coals but method ③ is only suitable for high-rank coals as free air drying low-rank coals may promote oxidation.

The inherent or air-dried moisture content is represented by relative loss of mass when the air-dried coal is heated, without oxidation, to a little over 100 ℃. Inherent moisture is analysed similarly, though it may be done in a vacuum.

Volatile matter

The material referred to as "volatile matter" in a coal sample represents the components of the coal, except for the moisture content, that are liberated at high temperature in absence of air. This is usually a mixture of short and long chain hydrocarbons, aromatic hydrocarbons and some sulphur. This material may be release from the organic compounds or from the mineral impurities, and the significance of the latter should not be overlooked. If the percentage of volatile matter is to be used to classify the coal, correction for that derived from inorganic sources (carbonates, clays etc.) must be made.

The volatile matter content is determined by heating the coal under rigidly specified conditions. British and Australian Standards require the coal to be heated in cylindrical silica crucible in a muffle furnace of specified dimensions at (900 ± 5) ℃ for 7 min. American (ASTM) procedures involve heating at (950±25) ℃ using a platinum crucible in a vertical electric furnace. These differences in procedure give slightly different results for the same coal, and the method used must therefore be stated with the analytical results.

Ash

The ash of a coal is the non-combustible inorganic residue that remains when coal is burned out. Determination of ash content requires heating the coal in a ventilated furnace, and expressing the mass of the residue remaining as percentage of the original coal. The heating rate is normally slow at first, to prevent fixation of sulphur or other compounds in the ash. In British and Australian practice the sample is heated slowly from ambient temperatures to 815 ℃, where it is held until constant mass is attained.

American procedures depend on the nature of the minerals present. In normal circumstance, the coal is heated slowly to between 700 ℃ and 750 ℃ and held at that temperature until constant mass is developed. For coals that have a high proportion of calcite and pyrite, a two-stage process is required to prevent fixation of sulphur in the ash. The coal is heated to 500 ℃ over 1 h and 750 ℃ over 2 h, and held at 750 ℃ until its mass becomes constant.

Fixed carbon

The fixed carbon content of a coal is the carbon found in the material that remains after the volatile matter has been expelled. It represents the decomposition residue of the organic components in coal, and carries with small amounts of nitrogen, sulphur, hydrogen and possibly oxygen as absorbed or chemically combined material.

Fixed carbon is not determined directly, but is simply the difference, in an air-dried coal, between the sum of the other components (moisture, volatile matter and ash) and 100%. The fixed carbon content is used as an index of the yield of coke expected from a coal on carbonization, or as a measure of the solid combustible material that remains in coal-burning equipment after the volatile fraction has been liberated. If corrections are applied for ash or mineral matter, it may be used as an index of coal rank and parameter in coal classification.

(2) Ultimate analysis

Ultimate analysis determines the total amounts of each of the principal chemical elements in the coal such as carbon, hydrogen, oxygen, nitrogen and sulphur. In some studies, the data may be recalculated to remove the percentage of these elements contained in the moisture and mineral impurities, in order to obtain the chemical composition of the organic fraction alone.

Carbon, hydrogen and oxygen are of great commercial significance in assessing the coking, gasification and liquefaction properties of the coal, while nitrogen and sulphur represent possible sources of pollution in a coal to be used for these or for combustion applications. The percentages of carbon, hydrogen and oxygen in the organic fraction of the coal can also be used as indices of rank and as parameters in coal classifications.

Carbon and hydrogen

Carbon and hydrogen occur mainly as complex hydrocarbon compounds. They are liberated as CO_2 and H_2O when coal is burned, and as a result are most readily determined together. However, the combustion process also liberates CO_2 from any carbonate minerals that may be present, and additional H_2O is derived from water in clays or other hydrous minerals. Hydrogen is also found in the inherent moisture of the are-dried coal, and thus any measure of "total" carbon and hydrogen includes a certain amount of material derived from inorganic sources. Corrections and separate analyses can be made to allow for the inorganic material.

Most techniques for the determination of carbon and hydrogen are based on heating the air-dried coal in a stream of dry oxygen, and collecting the CO_2 and H_2O produced in a series of absorption tubes. However, oxides of nitrogen and sulphur, as well as possibly small amounts of chlorine are also produced by the combustion process, and provision must be made in the apparatus to ensure they do not interfere with selective absorption system.

Two methods of determination are covered by standard procedures in Britain and Australian. These are the Liebig method, where the coal is heated to 800 ℃, and the high temperature method, where it is heated to 1,350 ℃. The high temperature method requires fewer additional processes to allow for sulphur, nitrogen etc. and is favoured by many commercial laboratories. American (ASTM) standards provide only for the Liebig method to be used.

Sulphur

Sulphur may occur in coal in various ways: ① Organic sulphur, where it is incorporated into the hydrocarbon compounds of the coal substance. ② As sulphide minerals, such as pyrite, in the inorganic fraction (pyritic sulphur). ③ As sulphate minerals, mostly hydrous iron or calcium sulphates, usually produced by atmospheric oxidation for these sulphides.

The total sulphur content of a coal may be determined by different methods. One of these is the "high temperature method" where the coal is burned in oxygen at 1,350 ℃, converting all the sulphur present to SO_2. The SO_2 is then converted to sulphuric acid for titrimetric determination.

Sulphur is also analysed to determine whether it is a sulfide mineral or in a sulfate form. This is achieved by dissolution of the sulfates in hydrochloric acid and precipitation as barium sulphate. Sulfide content is determined by measurement of iron content, as this will determine the amount of sulphur present as iron pyrite.

Oxygen

Oxygen is a component of many of the organic compounds in a coal. It is also found in the moisture, and in many of the mineral species, such as clays and carbonates. The oxygen content of a coal is traditionally determined by subtracting the amount of other chemical components (C, H, N and S) form 100%. Techniques involving direct determination are also available, which avoid the effect of cumulative error in the analysis by difference. These methods involve heating the coal in an inert atmosphere at temperatures of at least 900 ℃ or up to nearly 2,000 ℃, liberating the oxygen as carbon monoxide. The monoxide is then converted catalytically to CO_2, the amount of which is determined in absorption train to give the coal's oxygen content.

2. Special tests for combustion

(1) Specific energy

The energy liberated by a coal on combustion is chiefly due to the exothermic interaction of its hydrocarbon compounds with oxygen. In an operating furnace, the combustion is unconfined. Water vapors and other compounds (acid forming gases) escape directly to the atmosphere without condensation. The heat given off per unit mass of coal in unconfined conditions is called the "net" specific energy. In the laboratory, however, combustion takes place under confined conditions, and most of the products are allowed to condense after ignition. This causes the total amount of energy given off by the coal to be increased due to the latent heat given off in the process, and the result, known as the "gross" specific energy, usually represents an over estimation of the amount of heat actually liberated from the coal under operating conditions.

The energy output of a coal is determined using a bomb calorimeter which measures the specific energy output of a coal during complete combustion. The results of specific energy determinations are expressed in the SI system of units as megajoules per kilogram (MJ/kg). Other units also used for specific energy or calorific value are calories per gram, kilocalories per kilogram and British thermal units per pound.

(2) Ash fusion temperatures

The behavior of the coal's ash residue at high temperatures can be a critical factor in selection of coals for combustion applications. The behaviour of the ash at high temperature is related to its chemical composition. Material with a high proportion of aluminum generally appeals to be the most refractory. Small amounts of iron, calcium and potassium, however, if present, may act as fluxing agents and reduce the temperature at which fusion occurs.

Ash fusion temperatures are determined by viewing a moulded specimen of finely ground ash through an observation window in a high-temperature furnace. The moulded ash, usually in the form of a cube or a pyramid, is heated steadily from 1,000 ℃ to as high temperature as possible, preferably around 1,600 ℃. And, the following temperatures are recorded: ① Deformation temperature: This is reached when the corners of the mould first become rounded. ② Softening (sphere) temperature: This is reached when the top of the mould takes on a spherical shape. ③ Hemisphere temperature: This is reached when the entire mould takes on a hemisphere shape. ④ Flow (fluid) temperature: This is reached when the molten ash collapses to a flattened button on the furnace floor.

(3) Crucible swelling index (Free Swelling Index)

The simplest test to evaluate whether a coal is suitable for production of coke is the Free Swelling Index test. This involves heating a small sample of coal in a standardised crucible to around 800 ℃. After heating for a specified time, or until all volatiles are driven off, a small coke button remains in the crucible. The cross sectional profile of this coke button compared to a set of standardised profiles determines the Free Swelling Index.

3. Physical and mechanical properties

(1) Relative density

Relative density or specific gravity of the coal depends on the rank of the coal and degree of mineral impurity. Knowledge of the density of each coal ply is necessary to determine the properties of composites and blends. The density of the coal seam is necessary for conversion of resources into reserves.

Relative density is normally determined by the loss of a sample's weight in water. This is best achieved using finely ground coal, as bulk samples are quite porous.

(2) Particle size distribution

The particle size distribution of milled coal depends partly on the rank of the coal, which determines its brittleness, and on the handling, crushing and milling it has undergone. Generally coal is utilized in furnaces and coking ovens at a certain size, so the crushability of the coal must be determined. It is necessary to know this data before coal is mined, so that suitable crushing machinery can be designed to optimise the particle size for transport and use.

(3) Abrasion testing

Abrasion is the property of the coal which describes its propensity and ability to wear away machinery and undergo autonomous grinding. While carbonaceous matter in coal is relatively soft, quartz and other mineral constituents in coal are quite abrasive. This is tested in a calibrated mill, containing four blades of known mass. The coal is agitated in the mill for 12,000 revolutions at a rate of 1,500 revolutions per minute. The abrasion index is determined by measuring the loss of mass of the four metal blades.

(4) Hardgrove grindability index

Most modern coal burning equipment requires the coal to be ground to a fine powder before it is fed into the combustion chamber. The ease with which depends on the strength or tenacity of the coal, and is measured by the Hardguove grindability index.

In this test, a given mass of coal with a fixed particle size distribution is ground in a specified way in a standardized ring ball mill (Hardgrove mill). The resulting powdered coal is screened and the Hargrove grindability index (H. G. I.) calculated from the mass of fine material produced. Coals with a high Hardgrove index are relatively soft and easy to grind, those with a low value (less than 50) are hard and much more difficult to make into pulverized fuel.

Notes

1. Coal is a heterogeneous mixture of organic compounds, together with a certain amount of inorganic material in the form of moisture and mineral impurities. The general nature of these materials can be evaluated by a combination of two sets of analytical data: Proximate analysis and Ultimate analysis.

煤是由有机化合物和以水分、矿物杂质形式存在的无机物质组成的非均匀混合物。这些物质的性能可以通过两组分析数据的组合来确定:工业分析和元素分析。

2. The material referred to as "volatile matter" in a coal sample represents the components of the coal, except for the moisture content, that are liberated at high temperature in absence of air.

煤样中称作挥发分的物质是煤中除水分以外的成分,它是在高温、隔绝空气的条件下释放出来的。

3. The ash of a coal is the non-combustible inorganic residue that remains when coal is burned out. Determination of ash content requires heating the coal in a ventilated furnace, and expressing the mass of the residue remaining as percentage of the original coal.

煤的灰分是煤燃烧后残留的不可燃无机物。测定灰分含量需要在通风炉中进行。通常用残余物的质量占原煤质量的百分比表示灰分含量。

4. Carbon, hydrogen and oxygen are of great commercial significance in assessing the coking, gasification and liquefaction properties of the coal, while nitrogen and sulphur represent possible sources of pollution in a coal to be used for these or for combustion applications.

碳、氢、氧的含量对于评价煤炭的焦化、气化、液化性质具有重要的意义。氮和硫则代表煤炭用于焦化、气化和液化或燃烧时煤中可能的污染源。

5. The energy output of a coal is determined using a bomb calorimeter which measures the specific energy output of a coal during complete combustion.

煤释放的能量是用弹式量热计测定的,该量热计可测量煤在完全燃烧时释放的比能。

6. Coals with a high Hardgrove index are relatively soft and easy to grind, those with a low value (less than 50) are hard and much more difficult to make into pulverized fuel.

哈氏可磨性指数高的煤相对较软、易于研磨,可磨性指数低(小于 50)的煤较硬、难以制成粉状燃料。

Exercises

1. Answer the following questions according to the text.

(1) What's the function of proximate analysis of coal?

(2) How many forms of moisture may occur within coal samples?

(3) What is the critical factor in selection of coals for combustion applications?

(4) How to analyze the volatile matter in coal?

(5) How to determine the crucible swelling index?

2. Translate the following phrases into English.

(1) 煤质分析　(2) 热值　(3) 相对密度　(4) 固定碳

(5) 粒度分布　(6) 水分含量　(7) 工业分析　(8) 元素分析

(9) 燃烧试验　(10) 灰分含量

3. Translate the following paragraphs into Chinese.

Paragraph 1 in section 1 "Coal is a heterogeneous mixture of organic compounds ...", paragraph 1 in section 1 (1) "Proximate analysis gives a measure of ..." and paragraphs 1 and 2 in section 1 (2) "Ultimate analysis determines the ...".

4. Translate the following sentences into English.

(1) 工业分析给出了煤中挥发性和非挥发性有机化合物的相对含量以及水分和不可燃矿物物料的百分含量。

(2) 元素分析确定了煤中碳、氢、氧、氮和硫等主要化学元素的含量。

(3) 煤的相对密度取决于煤化程度和矿物杂质含量。

1.4 Mineral Processing Methods(矿物加工方法)

1. Concept of mineral processing

Mineral processing is a process of separating the grains of valuable minerals from the gangue minerals, to produce a concentrate, containing most of the valuable minerals and a tailing, containing predominantly the gangue minerals. There are two fundamental operations in mineral processing, namely the liberation of the valuable minerals from their waste gangue minerals, and separation of these values from the gangue, this latter process being known as concentration.

It has been predicted that the importance of mineral processing of metallic ores may decline as the physical process utilized are replaced by the hydrometallurgical routes, because higher recoveries are obtained by some chemical methods. This may certainly apply when the useful mineral is very finely disseminated in the ore and adequate liberation from the gangue is not possible, in which case a combination of chemical and mineral processing techniques may be advantageous. However, in the majority of cases the energy consumed in direct smelting or leaching of low-grade ores would be so enormous as to make the cost prohibitive. Compared with these processes, mineral processing methods are inexpensive, and their use is readily justified on economic grounds. Mineral processing reduces the bulk and weight of material which must be transported to the smelter, thus reducing shipping and handling costs; and smelting costs are greatly reduced due to the reduced tonnage which must be treated.

Liberation of the valuable minerals from the gangue is accomplished by comminution, which involves crushing, and, if necessary, grinding, to such a particle size that the product is a mixture of relatively clean particles of mineral and gangue.

After the minerals have been liberated from the gangue, the ore is subjected to some process of concentration, which separates the ore into two or more products. Separation is usually achieved by utilizing some specific difference in physical or chemical properties between the valuable mineral and gangue minerals in the ore.

2. Mineral separation methods

The most important physical methods which are used to concentrate ores are as follows:

(1) Separation dependent on optical and radioactive properties, etc. this is often called sorting, which commonly included hand selection of high-grade ores.

(2) Separation dependent on specific gravity differences. This utilizes the differential movement of minerals due to mass effects, usually in hydraulic currents.

(3) Separation utilizing the different surface properties of the minerals. Froth

flotation, which is undoubtedly the most important method of concentration, is affected by the degree of affinity of the minerals for rising air-bubbles within the agitated pulp. By adjusting the "climate" of the pulp by various reagents, it is possible to make the valuable minerals air-avid and the gangue minerals water-avid. This results in separation by transfer of the valuable minerals to the air-bubbles which form the froth floating on the surface of the pulp.

(4) Separation dependent on magnetic properties. Low intensity magnetic separators can be used to concentrate ferromagnetic minerals such as magnetite (Fe_3O_4), while high-intensity separators are used to separate paramagnetic minerals from their gangue. Magnetic separation is an important process in the beneficiation of iron ores, but also finds application in the treatment of paramagnetic non-ferrous minerals. It is widely used to remove paramagnetic wolframite [$(Fe, Mn)WO_4$] and hematite (Fe_2O_3) from tin ores, and has found considerable application in the processing of non-metallic minerals.

(5) Separation dependent on electrical conductivity properties. High-tension separation can be used to separate conducting from non-conducting minerals.

Comminution and concentration are the two primary operations in mineral processing, but many other important steps are involved, among which are sizing of the ore at different stages in treatment, by the use of screens and classifiers, and dewatering of the mineral pulps, using thickeners, filters, and driers.

3. Mineral liberation

(1) Degree of liberation

One of the major objects of comminution is the liberation of the valuable minerals from the associated gangue minerals at the coarsest possible particle size. If such an aim is achieved, then not only is energy saved by the reduction of the amount of fines produced, but any subsequent separation stages become easier and cheaper to operate. If high-grade solid products are required, then good liberation is essential; however, for subsequent hydrometallurgical processes, such as leaching, it may only be necessary to expose the required mineral.

In practice complete liberation is seldom achieved, even if the ore is ground down to the grain size of the desired mineral particles. This is illustrated by Figure 1. 5, which shows a lump of ore which has been reduced to a number of cubes of identical volume and of a size below that of the grains of mineral observed in the original ore sample. It can be seen that each particle produced containing mineral also contains a portion of gangue; complete liberation has not been attained; the bulk of the major mineral—the gangue—has, however, been liberated from the minor mineral—the value.

The particles of "locked" mineral and gangue are known as middlings, and further liberation from this fraction can only be achieved by further comminution.

The "degree of liberation" refers to the percentage of the mineral occurring as free

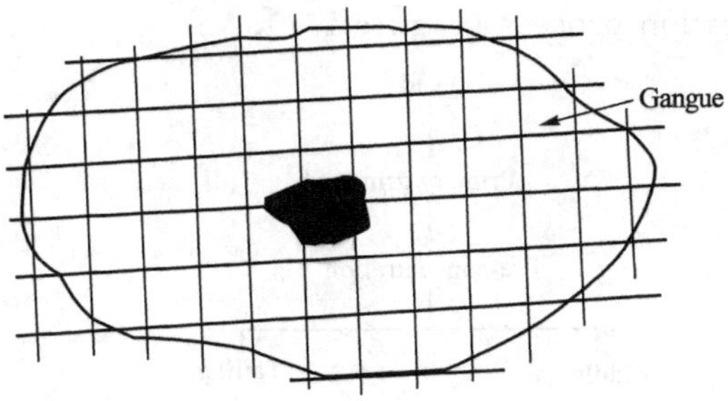

Figure 1.5 "Locking" of mineral and gangue

particles in the ore in relation to the total content. This can be high if there are weak boundaries between mineral and gangue particles, which are often the case with ores composed mainly of rock-forming minerals, particularly sedimentary materials. Usually, however, the adhesion between mineral and gangue is strong and, during comminution, the various constituents are cleft across.

Figure 1.6 is a cross-section through a typical ore particle and illustrates effectively the liberation dilemma often facing the mineral processor. Regions A represent valuable mineral, and region AA is rich in valuable mineral, but is highly intergrown with the gangue mineral. Comminution produces a range of fragments, ranging from fully liberated mineral and gangue particles, to those illustrated. Particles of type 1 are rich in mineral, and are classed as concentrate as they have an acceptable degree of locking with the gangue which limits the concentrate grade. Particle of type 4 would likewise be classed as tailings, the small amount of mineral present reducing the recovery of mineral into the concentrate. Particle of type 2 and 3, however, would probably be classed as middlings, although the degree of regrinding needed to promote economic liberation of mineral from particle of type 3 would be greater than that in particle of type 2.

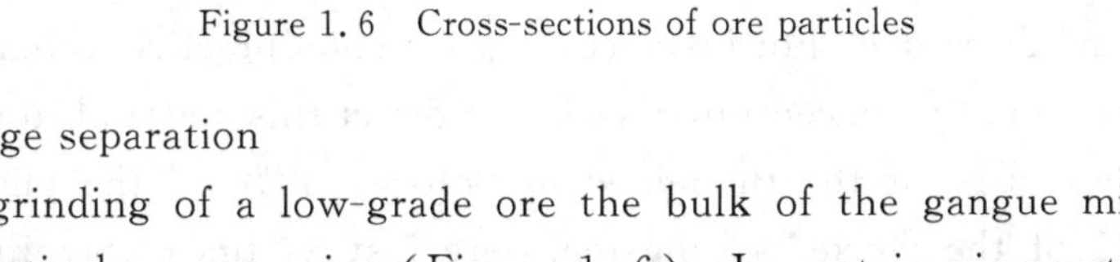

Figure 1.6 Cross-sections of ore particles

(2) Two-stage separation

During the grinding of a low-grade ore the bulk of the gangue minerals are often liberated at a relatively coarse size (Figure 1.6). In certain circumstances it may be economic to grind to a size much coarser than the optimum in order to produce in the subsequent concentration process a large middlings fraction and tailings which can be discarded at a coarse grain size. The middlings fraction can then be reground to produce a feed to the final concentration process (Figure 1.7).

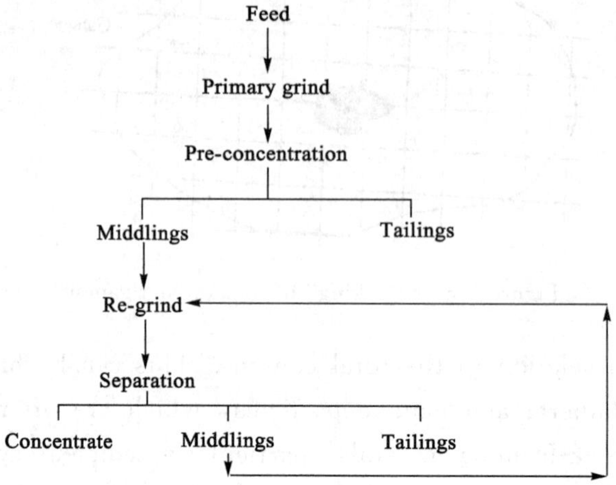

Figure 1.7 Flowsheet for process utilizing two-stage separation

This method discards most of the coarse gangue early in the process, thus considerably reducing grinding costs, as needless comminution of liberated gangue is avoided. It is often used on minerals which can easily be separated from the free gangue, even though they are themselves locked to some extent with gangue. It is the basis of the heavy medium process of preconcentration.

4. Mineral concentration

(1) Major factors affecting mineral concentration

The object of mineral processing, regardless of the method used, is always the same, i. e., to separate the minerals into two or more products with the values in the concentrates, the gangue in the tailings, and the "locked" particles in the middlings. Such separations are, of course, never perfect, so that much of the middlings produced are, in fact, misplaced particles, i. e., those particles which ideally should have reported to the concentrate or the tailings. This is often particularly serious when treating ultra-fine particles, where the efficiency of separation is usually low. In such cases, fine liberated valuable mineral particles often report in the middlings and tailings. The technology for treating fine-sized minerals is, as yet, poorly developed, and, in some cases, very large amounts of fines are discarded. For instance, it is common practice to remove material less than 10 μm in size from tin concentrator feeds and direct this material to the tailings, and, in the early 1970s, 50% of the tin mined in Bolivia, 30% of the phosphate mined in Florida, and 20% of the world's tungsten were lost as fines. Significant amounts of copper, uranium, fluorspar, bauxite, zinc, and iron were also similarly lost.

It should be pointed out that the process is also limited by the mineralogical nature of the ore. For example, in an ore containing native copper it is theoretically possible to produce a concentrate containing 100% Cu, but, if the ore mineral was chalcopyrite

($CuFeS_2$), the best concentrate would contain only 34.5% Cu.

(2) Mineral concentration indexes

The recovery, in the case of the concentration of a metallic ore, is the percentage of the total metal contained in the ore that is recovered in the concentrate; a recovery of 90% means that 90% of the metal in the ore is recovered in the concentrate and 10% is lost in the tailings. It is possible to speak of the recovery of metal in tailings, middlings, and other products, but normally the term refers to the amount of metal recovered in the concentrates—the valuable products. The recovery, when dealing with non-metallic ores, refers to the percentage of the total mineral contained in the ore that is recovered into the concentrate, i. e., recovery is usually expressed in terms of the valuable end product.

The ratio of concentration is the ratio of the weight of the feed to the weight of the concentrates. It is a measure of the efficiency of the concentration process, and it is closely related to the grade or assay of the concentrate; the value of the ratio of concentration will generally increase with the grade of concentrate.

The grade, or assay, usually refers to the content of the marketable end product in the material. Thus, in metallic ores, the per cent metal is often quoted, although in the case of very low-grade ores, such as gold, the metal content may be expressed as parts per million (ppm). Some metals are sold in oxide form, and hence the grade may be quoted in terms of the marketable oxide content, e. g., % WO_3, % U_3O_8, etc. In non-metallic operations, grade usually refers to the mineral content, e. g., % CaF_2 in fluorite ores; diamond ores are usually graded in carats per 100 tonnes (t), where 1 carat is 0.2 g. Coal is graded according to its ash content, i. e., the amount of incombustible mineral present within the coal; most of the coal produced in Britain is consumed in power stations, which require a feed with an ash content of between 15% and 20%.

The enrichment ratio is the ratio of the grade of the concentrate to the grade of the heads, and again is related to the efficiency of the process.

Ratio of concentration and recovery are essentially independent of each other, and in order to evaluate a given operation it is necessary to know both. For example, it is possible to obtain a very high grade of concentrate and ratio of concentration by simply picking a few lumps of pure galena from a lead ore, but the recovery would be very low. On the other hand, a concentrating process might show a recovery of 99% of the metal, but it might also put 60% of the gangue minerals in the concentrate. It is, of course, possible to obtain 100% recovery by not concentrating the ore at all.

There is an approximately inverse relationship between recovery and grade of concentrate in all concentrating processes. If an attempt is made to attain a very high-grade concentrate, the tailings assays are higher and the recovery is low. If a high recovery of metal is aimed for, there will be more gangue in the concentrate and the grade of concentrate and ratio of concentration will both decrease.

Notes

1. Mineral processing is a process of separating the grains of valuable minerals from the gangue minerals, to produce a concentrate, containing most of the valuable minerals and a tailing, containing predominantly the gangue minerals.

选矿是将有用矿物颗粒与脉石矿物分离，以生产含有大部分有用矿物的精矿和含有脉石矿物尾矿的过程。

2. Liberation of the valuable minerals from the gangue is accomplished by comminution, which involves crushing, and, if necessary, grinding, to such a particle size that the product is a mixture of relatively clean particles of mineral and gangue.

从脉石中解离有用矿物是通过粉碎完成的，粉碎包括破碎，如果有必要，也包括使产品的粒度达到相对纯净的矿物和脉石的混合物的程度。

3. Separation dependent on specific gravity differences. This utilizes the differential movement of minerals due to mass effects, usually in hydraulic currents.

取决于密度差异的分选。该方法利用了矿物由于质量效应引起的差动运动，通常在水流中进行。

4. Froth flotation, which is undoubtedly the most important method of concentration, is affected by the degree of affinity of the minerals for rising air-bubbles within the agitated pulp.

泡沫浮选法确实是最重要的一种分选方法，它受到各种矿物在搅拌矿浆中对上升气泡亲和程度的影响。

5. Comminution and concentration are the two primary operations in mineral processing, but many other important steps are involved, among which are sizing of the ore at different stages in treatment, by the use of screens and classifiers, and dewatering of the mineral pulps, using thickeners, filters, and driers.

粉碎和富集是矿物加工中的两个主要操作，但矿物加工还涉及许多其他重要步骤，其中包括使用筛子和分级机在处理的不同阶段对矿石进行分级，以及使用浓密机、过滤机和烘干机对矿浆进行脱水。

6. The recovery, in the case of the concentration of a metallic ore, is the percentage of the total metal contained in the ore that is recovered in the concentrate.

以金属矿石的选矿为例，回收率是指在精矿中回收的金属占矿石中所含总金属的百分比。

7. The ratio of concentration is the ratio of the weight of the feed to the weight of the concentrates. It is a measure of the efficiency of the concentration process, and it is closely related to the grade or assay of the concentrate; the value of the ratio of concentration will generally increase with the grade of concentrate.

选矿比是给矿重量与精矿重量的比值。它是衡量选矿工艺效率的指标，与精矿的品位密切相关；选矿比一般会随着精矿品位的增加而增加。

8. The grade, or assay, usually refers to the content of the marketable end product in

the material. Thus, in metallic ores, the per cent metal is often quoted, although in the case of very low-grade ores, such as gold, the metal content may be expressed as parts per million (ppm).

品位通常指物料中可销售的最终产品的含量。因此,在分选金属矿石时,往往采用金属百分含量表示品位,而当矿石品位极低时,如金矿,金属品位可以用百万分之几(ppm)来表示。

Exercises

1. Answer the following questions according to the text.

(1) What is the object of mineral processing?

(2) What does the recovery mean?

(3) How to calculate the enrichment ratio?

(4) How to understand the reverse relationship between recovery and grade of concentrate in all concentrating processes.

(5) What is the ratio of concentration and its function?

(6) What are the two fundamental operations in mineral processing?

2. Translate the following phrases into English.

(1) 矿物加工　(2) 矿物解离　(3) 细粒嵌布　(4) 低品位矿石

(5) 解离度　(6) 光学性质　(7) 磁性　(8) 导电性

(9) 连生颗粒　(10) 矿物富集　(11) 中矿　(12) 再磨

(13) 粗粒脉石　(14) 单体颗粒

3. Translate the following paragraphs into Chinese.

Paragraphs 1, 2 and 3 in section 2 “The most important physical methods which are used ...”.

4. Translate the following sentences into English.

(1) 在低品位矿石的磨矿过程中,大部分脉石矿物通常在相对粗的粒度下解离出来。

(2) 粉碎的主要目的之一是在尽可能粗的粒度下从伴生脉石矿物中解离出有用矿物。

(3) “解离度”是指矿石中以单体颗粒形式存在的矿物占矿物总量的百分比。

(4) 选矿的目标始终是相同的,即将矿物分离成两种或多种产品,有用矿物在精矿中、脉石在尾矿中、连生颗粒在中矿中。

2 CRUSHING AND SCREENING(破碎与筛分)

2.1 Comminution(粉碎)

1. Introduction

Because most minerals are finely disseminated and intimately associated with the gangue, size reduction of ores is normally done in order to liberate the valuable minerals from the host rock before separation can be undertaken. This is achieved by comminution in which the particle size of the ore is progressively reduced until the clean particles of mineral can be separated by such methods as are available. Comminution in its earliest stages is carried out in order to make the freshly excavated material easier to handle by scrapers, conveyors, and ore carriers, and in the case of quarry products to produce material of controlled particle size.

Explosives are used in mining to remove ores from their natural beds, and blasting can be regarded as the first stage in comminution. Comminution in the mineral processing plant, or "mill", takes place as a sequence of crushing and grinding processes. Crushing reduces the particle size of run-of-mine ore to such a level that grinding can be carried out until the mineral and gangue is substantially produced as separate particles.

Crushing is accomplished by compression of the ore against rigid surfaces, or by impact against surfaces in a rigidly constrained motion path. This is contrasted with grinding which is accomplished by abrasion and impact of the ore by the free motion of unconnected media such as rods, balls, or pebbles.

Crushing is usually a dry process, and is performed in several stages, reduction ratios being small, ranging from three to six in each stage. The reduction ratio of a crushing stage can be defined as the ratio of maximum particle size entering to maximum particle size leaving the crusher, although other definitions are sometimes used.

Grinding is usually performed "wet" to provide a slurry feed to the concentration process, although dry grinding has limited applications. Tumbling mills with either steel rods or balls, or sized ore as the grinding media, are used in the last stages of comminution. There is an overlapping size area where it is possible to crush or grind the ore.

Stirred mills are now commonly used in mineral processing, though they have been present in other industries for many years. They represent the broad category of mills which use a stirrer to provide motion to the steel, ceramic, or rock media. Both vertical and horizontal configurations exist, and since they can operate with smaller media sizes, they are far more suitable for fine grinding applications than ball mills.

A relatively new comminution device, the high-pressure grinding rolls(HPGR), utilizes compression breakage of a particle bed, in which energy efficient inter-particle breakage occurs. The reduction ratio obtained in a single pass through the HPGR is substantially higher than that obtained in conventional rolls crushers. Some evidence has also been reported for downstream benefits such as reduced grinding strength and improved leachability due to microcracking. The HPGR offers a realistic potential to markedly reduce the comminution energy requirements needed by tumbling mills. Reports have suggested the HPGR to be between 20% and 50% more efficient than conventional crushers and mills.

2. Principles of comminution

(1) Breaking stress

Minerals being crystals have a tendency to break into endless numbers of sizes and shapes every time they are introduced to energy. The difficulty in size reduction lays in the art of limiting the number of over and under sizes produced during the reduction.

Most minerals are crystalline materials in which the atoms are regularly arranged in three-dimensional arrays. The configuration of atoms is determined by the size and types of physical and chemical bonds holding them together. In the crystalline lattice of minerals, these inter-atomic bonds are effective only over small distances, and can be broken if extended by a tensile stress. Such stresses may be generated by tensile or compressive loading (Figure 2.1).

Even when rocks are uniformly loaded, the internal stresses are not evenly distributed, as the rock consists of a variety of minerals dispersed as grains of various sizes. The distribution of stress depends upon the mechanical properties of the individual minerals, but more importantly upon the presence of cracks or flaws in the matrix, which act as sites for stress concentration (Figure 2.2).

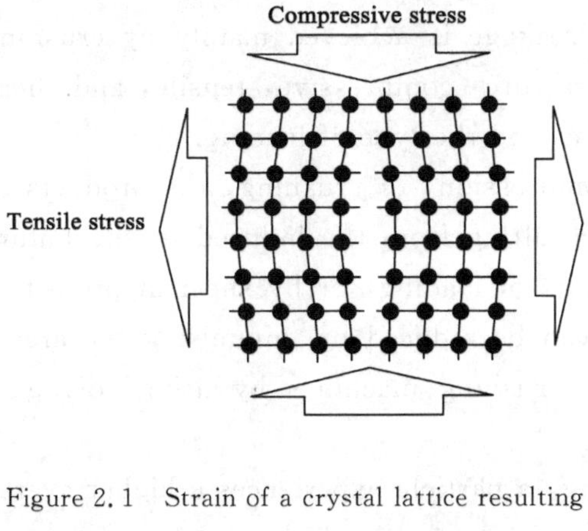

Figure 2.1 Strain of a crystal lattice resulting from tensile or compressive stresses

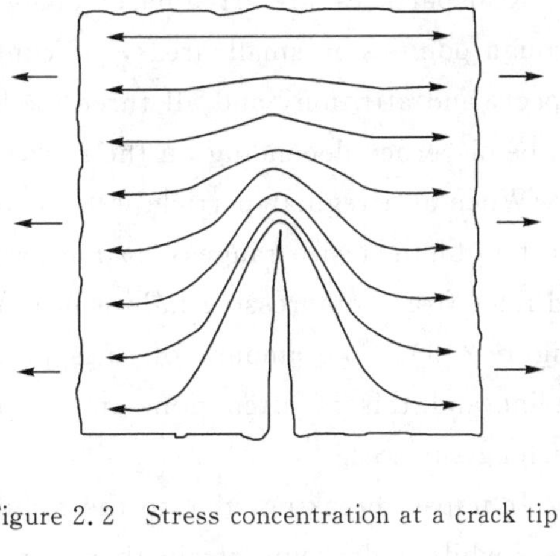

Figure 2.2 Stress concentration at a crack tip

It has been shown that the increase in stress at such a site is proportional to the square root of the crack length perpendicular to the stress direction. Therefore, there is a critical value for the crack length at any particular level of stress at which the increased stress level at the crack tip is sufficient to break the atomic bond at that point. Such rupture of the bond will increase the crack length, thus increasing the stress concentration and causing a rapid propagation of the crack through the matrix, thus causing fracture.

Although the theories of comminution assume that the material is brittle, crystals can, in fact, store energy without breaking, and release this energy when the stress is removed. Such behaviour is known as elastic. When fracture does occur, some of the stored energy is transformed into free surface energy, which is the potential energy of atoms at the newly produced surfaces. Due to this increase in surface energy, newly formed surfaces are often more chemically active, and are more amenable to the action of flotation reagents, etc. , as well as oxidizing more readily.

Griffith(1921) showed that materials fail by crack propagation when this is energetically feasible, i. e. , when the energy released by relaxing the strain energy is greater than the energy of the new surface produced. Brittle materials relieve the strain energy mainly by crack propagation, whereas "tough" materials can relax strain energy without crack propagation by the mechanism of plastic flow, where the atoms or molecules slide over each other and energy is consumed in distorting the shape of the material. Crack propagation can also be inhibited by encounters with other cracks or by meeting crystal boundaries. Fine-grained rocks, such as taconites, are therefore usually tougher than coarse-grained rocks.

The energy required for comminution is reduced in the presence of water, and can be further reduced by chemical additives which adsorb onto the solid. This may be due to the lowering of the surface energy on adsorption providing that the surfactant can penetrate into a crack and reduce the bond strength at the crack tip before rupture.

(2) Modes of fracture

Real particles are irregularly shaped, and loading is not uniform but is achieved through points, or small areas, of contact. Breakage is achieved mainly by crushing, impact, and attrition, and all three modes of fracture (compressive, tensile, and shear) can be discerned depending on the rock mechanics and the type of loading.

When an irregular particle is broken by compression, or crushing, the products fall into two distinct size ranges: coarse particles resulting from the induced tensile failure, and fines from compressive failure near the points of loading, or by shear at projections (Figure 2. 3). The amount of fines produced can be reduced by minimising the area of loading and this is often done in compressive crushing machines by using corrugated crushing surfaces.

In impact breaking, due to the rapid loading, a particle experiences a higher average stress while undergoing strain than is necessary to achieve simple fracture, and tends to

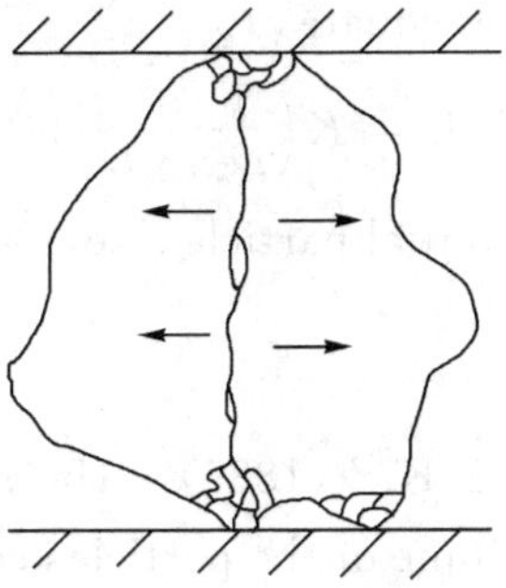

Figure 2.3 Fracture by crushing

break apart rapidly, mainly by tensile failure. The products are often very similar in size and shape.

Attrition (shear failure) produces much fine material, and may be undesirable depending on the comminution stage and industry sector. Attrition occurs mainly in practice due to particle particle interaction (inter-particle comminution), which may occur if a crusher is fed too fast, contacting particles thus increasing the degree of compressive stress and hence shear failure.

3. Comminution theory

Comminution theory is concerned with the relationship between energy input and the product particle size made from a given feed size. Various theories have been expounded, none of which is entirely satisfactory.

The greatest problem lies in the fact that most of the energy input to a crushing or grinding machine is absorbed by the machine itself, and only a small fraction of the total energy is available for breaking the material. It is to be expected that there is a relationship between the energy required to break the material and the new surface produced in the process, but this relationship can only be made manifest if the energy consumed in creating new surface can be separately measured.

In a ball mill, for instance, it has been shown that less than 1% of the total energy input is available for actual size reduction, the bulk of the energy being utilized in the production of heat.

Another factor is that a material which is plastic will consume energy in changing shape, a shape which it will retain without creating significant new surface. All the theories of comminution assume that the material is brittle, so that no energy is absorbed in processes such as elongation or contraction which is not finally utilized in breakage.

(1) Von Rittinger's theory

The oldest theory is that of Von Rittinger(1867), which states that the energy consumed in the size reduction is proportional to the area of new surface produced. The surface area of a known weight of particles of uniform diameter is inversely proportional to the diameter, hence Rittinger's law equates to

$$E = K\left(\frac{1}{D_2} - \frac{1}{D_1}\right) \tag{2.1}$$

where E is energy input, D_1 is the initial particle size, D_2 is the final particle size, and K is a constant.

(2) Kick's theory

The second theory is that of Kick(1885). He stated that the work required is proportional to the reduction in volume of the particles concerned. Where f is the diameter of the feed particles and p the diameter of the product particles, the reduction ratio R is f/p. According to Kick's law, the energy required for comminution is proportional to $\lg R/\lg 2$.

(3) Bond's theory

Bond(1952) developed an equation which is based on the theory that the work input is proportional to the new crack tip length produced in particle breakage, and equals the work represented by the product minus that represented by the feed. In particles of similar shape, the surface area of unit volume of material is inversely proportional to the diameter. The crack length in unit volume is considered to be proportional to one side of that area and therefore inversely proportional to the square root of the diameter.

For practical calculations the size in microns which 80% passes is selected as the criterion of particle size. The diameter in microns which 80% of the product passes is designated as P, the size which 80% of the feed passes is designated as F, and the work input in kilowatt hours per short ton is W, Bond's third theory equation is

$$W = \frac{10W_i}{\sqrt{P}} - \frac{10W_i}{\sqrt{F}} \tag{2.2}$$

where W_i is the work index. The work index is the comminution parameter which expresses the resistance of the material to crushing and grinding; numerically it is the kilowatt hours per short ton required to reduce the material from theoretically infinite feed size to 80% passing 100 μm.

4. Crushing circuits

A typical flowsheet for a crushing plant producing ball mill feed is shown in Figure 2.4. The circuit is typical of current practice in that the secondary product is screened and conveyed to a storage bin, rather than feeding the tertiary crushers directly. The intermediate bins allow good mixing of the secondary screen oversize with the circulating load, and regulation of the tertiary crusher feed, providing more efficient crushing.

Another "Classical" three-stage ore crushing prior to rod mill is shown in Figure 2.5.

Vibrating screens are sometimes placed ahead of the secondary crushers to remove undersize material, or scalp the feed, and thereby increase the capacity of the secondary crushing plant. Undersize material tends to pack the voids between the large particles in

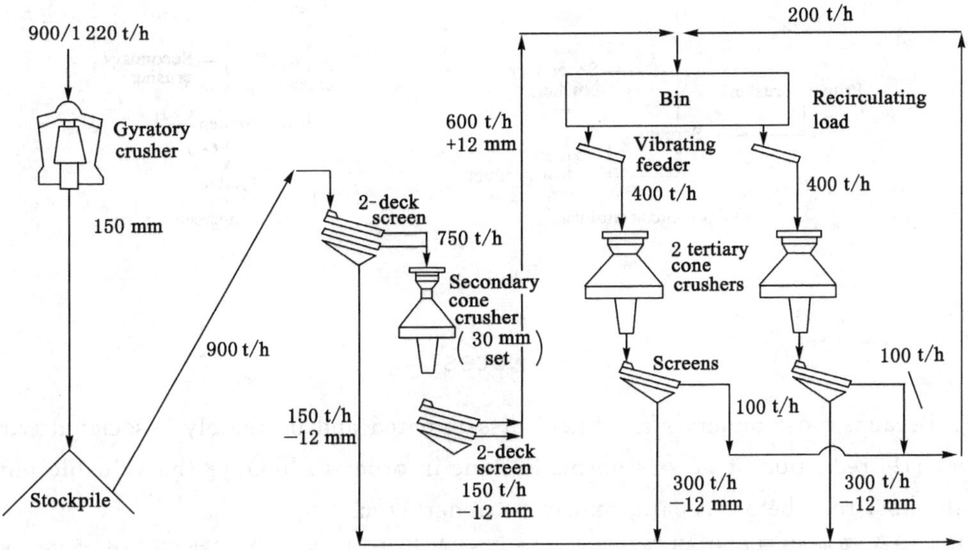

Figure 2.4 Three-stage crushing circuit for ball mill feed (after Motz, 1978)

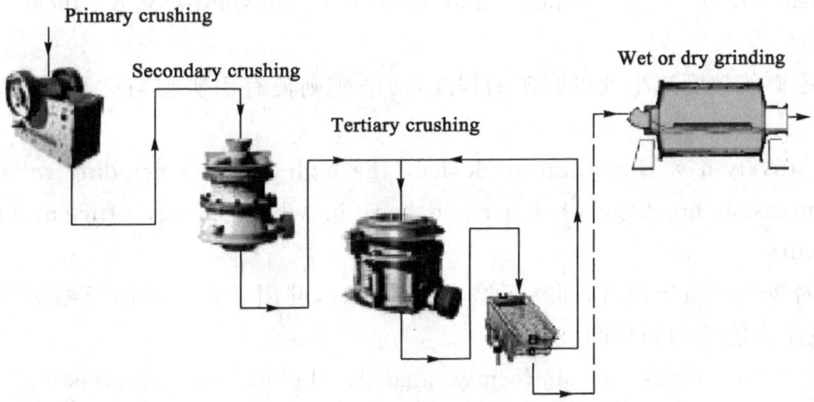

Figure 2.5 Three-stage ore crushing prior to rod mill

the crushing chamber, and can choke the crusher, causing damage, because the packed mass of rock is unable to swell in volume as it is broken.

Crushing may be in open or closed circuit depending on product size (Figure 2.6). In open-circuit crushing, undersize material from the screen is combined with the crusher product and is then routed to the next operation. Open-circuit crushing is often used in intermediate crushing stages, or when the secondary crushing plant is producing a rod mill feed. If the crusher is producing ball-mill feed it is good practice to use closed-circuit crushing in which the undersize from the screen is the finished product. The crusher product is returned to the screen so that any oversize material will be recirculated.

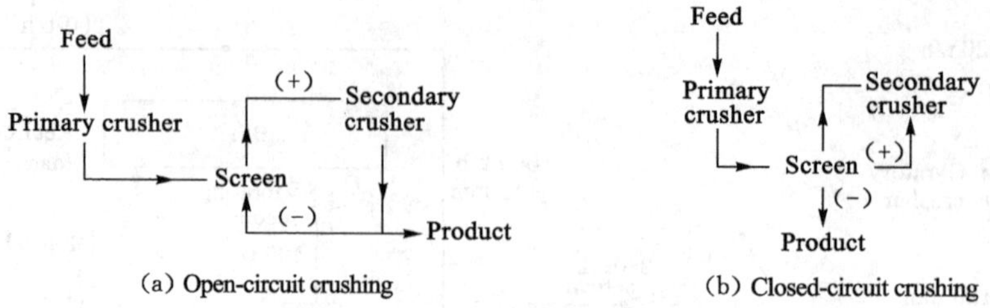

Figure 2.6

Notes

1. Because most minerals are finely disseminated and intimately associated with the gangue, size reduction of ores is normally done in order to liberate the valuable minerals from the host rock before separation can be undertaken.

由于大多数矿物呈微细浸染状，且与脉石紧密共生，因此通常在进行分离之前，需要对矿石破碎以便将有用矿物从母岩中解离出来。

2. Crushing reduces the particle size of run-of-mine ore to such a level that grinding can be carried out until the mineral and gangue is substantially produced as separate particles.

碎矿使原矿的粒度减小到可磨矿的程度，而磨矿则使有用矿物与脉石矿物基本上成为单体颗粒。

3. A relatively new comminution device, the high-pressure grinding rolls (HPGR), utilizes compression breakage of a particle bed, in which energy efficient inter-particle breakage occurs.

一种相对较新的粉碎设备，即高压磨辊(HPGR)，利用了颗粒层的压缩破碎，在这种情况下，可以发生高能效的颗粒间破碎。

4. Even when rocks are uniformly loaded, the internal stresses are not evenly distributed, as the rock consists of a variety of minerals dispersed as grains of various sizes.

即使岩石受力均匀，内应力也不会均匀分布，因为岩石是由各种矿物组成的，而矿物又是呈各种不同粒度分布的。

5. Therefore, there is a critical value for the crack length at any particular level of stress at which the increased stress level at the crack tip is sufficient to break the atomic bond at that point.

因此在任一特定应力水平上，裂隙长度均有一临界值，此时在裂隙顶端提高应力水平足以使该点的原子键断裂。

6. Although the theories of comminution assume that the material is brittle, crystals can, in fact, store energy without breaking, and release this energy when the stress is removed.

尽管粉碎理论假定物料是脆性的，但实际上晶体可以贮存能量而不致碎裂，当应力取消后,能量被释放。

7. Attrition occurs mainly in practice due to particle particle interaction (inter-particle comminution), which may occur if a crusher is fed too fast, contacting particles thus increasing the degree of compressive stress and hence shear failure.

在实践中研磨主要是由于颗粒与颗粒间的相互作用(粒间破碎)而产生的,如果破碎机给矿的速度过快,研磨就可能发生,因为相互接触的颗粒会增大压应力,从而产生剪切破碎。

8. In a ball mill, for instance, it has been shown that less than 1% of the total energy input is available for actual size reduction, the bulk of the energy being utilized in the production of heat.

例如,在球磨机中,只有不到1%的总能量输入可用于实际磨矿,大部分能量用于产生热量。

Exercises

1. Answer the following questions according to the text.

(1) How to accomplish crushing?

(2) What are the factors that determine the configuration of atoms?

(3) What factors does the comminution theory concern?

(4) How to choose open-circuit crushing or closed-circuit crushing?

(5) How to achieve breakage?

2. Translate the following phrases into English.

(1) 剪切　(2) 破碎应力　(3) 粉碎设备　(4) 晶格

(5) 三段破碎机　(6) 振动磨机　(7) 冲击破碎　(8) 破碎回路

3. Translate the following paragraphs into Chinese.

Paragraphs 1 and 2 in section 1 "Because most minerals are finely disseminated and intimately ...".

4. Translate the following sentences into English.

(1) 矿石中有用矿物与脉石矿物紧密共生。

(2) 破碎一般为干式作业,采用分段方法完成。

(3) 在矿物晶体的晶格内,原子间键只在很短的距离内有效。

(4) 破碎主要是通过压碎、冲击和研磨实现的。

(5) 如果破碎机生产球磨机给料,最好使用闭路破碎。

2.2 Crushers(破碎机)

1. Introduction

Crushing is the first mechanical stage in the process of comminution in which the main objective is the liberation of the valuable minerals from the gangue. It is generally a dry operation and is usually performed in two or three stages. Lumps of run-of-mine ore can be as large as 1.5 m across and these are reduced in the primary crushing stage to 10～20 cm in heavy-duty machines.

In most operations, the primary crushing schedule is the same as the mining schedule. When primary crushing is performed underground, this operation is normally a responsibility of the mining department; when primary crushing is on the surface, it is customary for the mining department to deliver the ore to the crusher and for the mineral processing department to crush and handle the ore from this point through the successive ore-processing unit operations. Primary crushers are commonly designed to operate 75% of the available time, mainly because of interruptions caused by insufficient crusher feed and by mechanical delays in the crushers.

Secondary crushing includes all operations for reclaiming the primary crusher product from ore storage to the disposal of the final crusher product, which is usually between 0.5 cm and 2 cm in diameter. The primary crusher product from most metalliferous ores can be crushed and screened satisfactorily, and the secondary plant generally consists of one or two size-reduction stages with appropriate crushers and screens. If, however, the ore tends to be slippery and tough, the tertiary crushing stage may be substituted by coarse grinding in rod mills. On the other hand, more than two size-reduction stages may be used in secondary crushing if the ore is extra-hard, or in special cases where it is important to minimize the production of fines.

2. Primary crushers

Primary crushers are heavy-duty machines, used to reduce the run-of-mine ore down to a size suitable for transport and for feeding the secondary crushers or autogenous/semi-autogenous mills. They are always operated in open circuit, with or without heavy-duty scalping screens (grizzlies). There are two main types of primary crusher in metalliferous operations-jaw and gyratory crushers.

(1) Jaw crushers

The distinctive feature of this class of crusher is the two plates which open and shut like animal jaws. The jaws are set at an acute angle to each other, and one jaw is pivoted so that it swings relative to the other fixed jaw. Material fed into the jaws is alternately nipped and released to fall further into the crushing chamber. Eventually it falls from the discharge aperture.

Jaw crushers are classified by the method of pivoting the swing jaw (Figure 2.7). In the Blake crusher the jaw is pivoted at the top and thus has a fixed receiving area and a variable discharge opening. In the Dodge crusher the jaw is pivoted at the bottom, giving it a variable feed area but fixed delivery area. The Dodge crusher is restricted to laboratory use, where close sizing is required, and is never used for heavy-duty crushing as it chokes very easily. The Universal crusher is pivoted in an intermediate position, and thus has a variable delivery and receiving area.

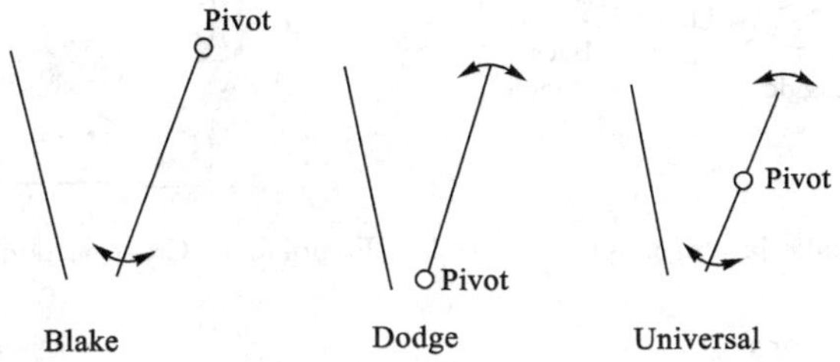

Figure 2.7 Jaw-crusher types

There are two forms of the Blake crusher—double toggle and single toggle.

Double-toggle Blake crushers

In this model (Figure 2.8), the oscillating movement of the swinging jaw is effected by vertical movement of the pitman. This moves up and down under the influence of the eccentric. The back toggle plate causes the pitman to move sideways as it is pushed upward. This motion is transferred to the front toggle plate and this in turn causes the swing jaw to close on the fixed jaw. Similarly, downward movement of the pitman allows the swing jaw to open.

A feature of all jaw crushers is the heavy flywheel attached to the drive, which is necessary to store energy on the idling half of the stroke and deliver it on the crushing half. Since the jaw crusher works on half-cycle only, it is limited in capacity for its weight and size. Due to its alternate loading and release of stress, it must be very rugged and needs strong foundations to accommodate the vibrations.

Single-toggle jaw crushers

In this type of crusher (Figure 2.9) the swing jaw is suspended on the eccentric shaft, which allows a lighter, more compact design than with the double-toggle machine. The motion of the swing jaw also differs from that of the double-toggle design. Not only does the swing jaw move towards the fixed jaw, under the action of the toggle plate, but it also moves vertically as the eccentric rotates. This elliptical jaw motion assists in pushing rock through the crushing chamber. The single-toggle machine therefore has a somewhat higher capacity than the double-toggle machine of the same gape. The eccentric movement, however, increases the rate of wear on the jaw plates. Direct attachment of the swing jaw to the eccentric imposes a high degree of strain on the drive shaft, and so maintenance costs tend to be higher than with the double-toggle machine.

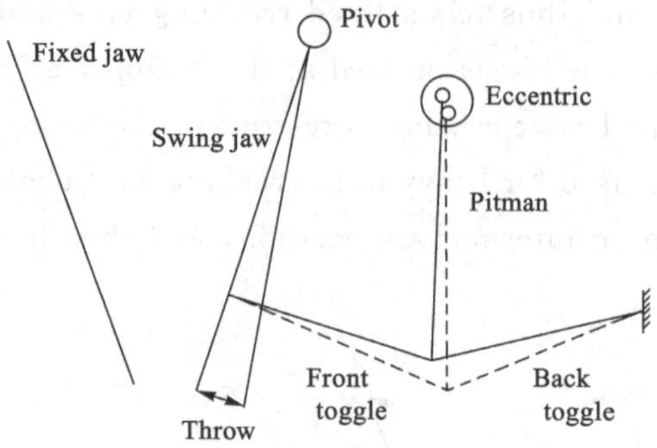

Figure 2.8 Blake jaw crusher

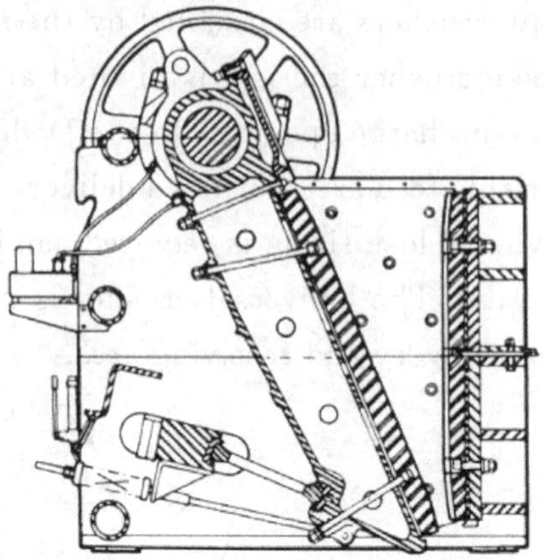

Figure 2.9 Cross-section of single-toggle jaw crusher

(2) Gyratory crushers

Gyratory crushers are principally used in surface-crushing plants, although a few currently operate underground. The gyratory crusher (Figure 2.10) consists essentially of a long spindle, carrying a hard steel conical grinding element, the head, seated in an eccentric sleeve. The spindle is suspended from a "spider" and, as it rotates, normally between 85 r/min and 150 r/min, it sweeps out a conical path within the fixed crushing chamber, or shell, due to the gyratory action of the eccentric. As in the jaw crusher, maximum movement of the head occurs near the discharge. This tends to relieve the choking due to swelling. The spindle is free to turn on its axis in the eccentric sleeve, so that during crushing the lumps are compressed between the rotating head and the top shell segments, and abrasive action in a horizontal direction is negligible.

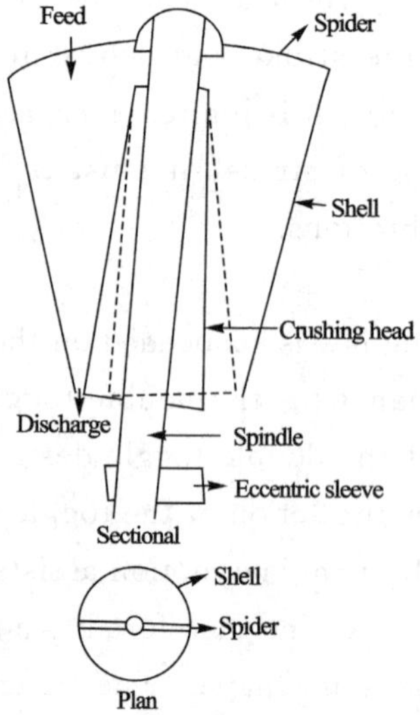

(a) Functional diagram

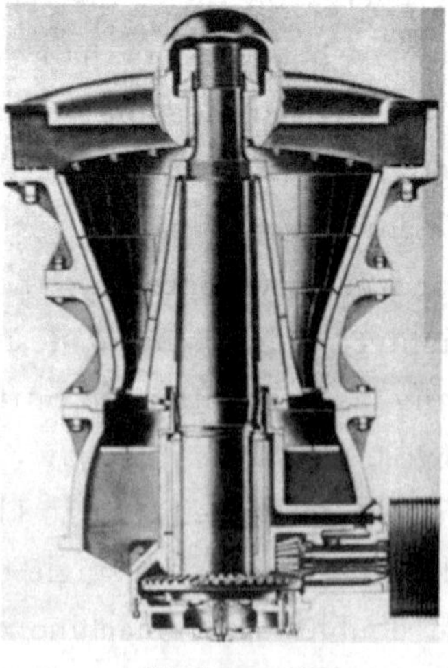

(b) Cross-section

Figure 2.10 Gyratory crusher

In fact, the gyratory crusher can be regarded as an infinitely large number of jaw crushers each of infinitely small width. Since the gyratory, unlike the jaw crusher, crushes on full cycle, it has a much higher capacity than a jaw crusher of the same gape, and is usually favored in plants handling very large throughputs.

3. Secondary crushers

Secondary crushers are much lighter and much easier to handle than the heavy-duty, rugged primary machines. Secondary crushers also operate with dry feeds, and their purpose is to reduce the ore to a size suitable for grinding. In those cases where size reduction can be more efficiently carried out by crushing, there may be a tertiary stage before the material is passed to the grinding mills. Tertiary crushers are, to all intents and purposes, of the same design as secondaries, except that they have a closer set. The bulk of secondary crushing of metalliferous ores is performed by cone crushers, although crushing rolls and hammer mills are used for some applications.

(1) Cone crusher

The cone crusher is a modified gyratory. The essential difference is that the shorter spindle of the cone crusher is not suspended, as in the gyratory, but is supported in a curved, universal bearing below the gyratory head or cone.

Since a large gape is not required the crushing shell of "bowl" flares outwards which allows for the swell of broken ore by providing an increasing cross-sectional area towards the discharge end. The flare of the bowl allows a much greater head angle than in the gyratory crusher, while retaining the same angle between the crushing members (Figure 2.11). This gives the cone crusher a high capacity, since the capacity of gyratory crushers is roughly proportional to the diameter of the head.

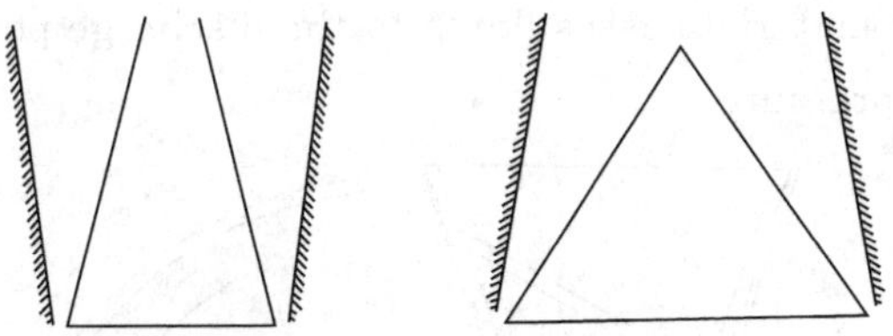
(a) Gyratory (b) Cone crushers

Figure 2.11 Head and shell shapes

Unlike a gyratory crusher, which is identified by the dimensions of the feed opening and the mantle diameter, a cone crusher is rated by the diameter of the cone lining. Cone crushers range in size from 559 mm to 3.1 m and have capacities up to 1,100 t/h with a discharge setting of 19 mm, although two 3.1 m Symons cone crushers, each with capacities of 3,000 t/h, have been installed in a South African iron-ore plant.

The Symons cone crusher is the most widely used type of cone crusher. It is produced in two forms: the Standard for normal secondary crushing and the Short-head for fine, or tertiary duty (Figures 2.12). They differ mainly in the shape of their crushing chambers.

The Standard cone has "stepped" liners which allow a coarser feed than in the Short-head. They deliver a product varying from 0.5 cm to 6 cm. The Short-head has a steeper head angle than the Standard, which helps to prevent choking from the much finer material being handled. It also has a narrower feed opening and a longer parallel section at the discharge, and delivers a product of 0.3～2.0 cm.

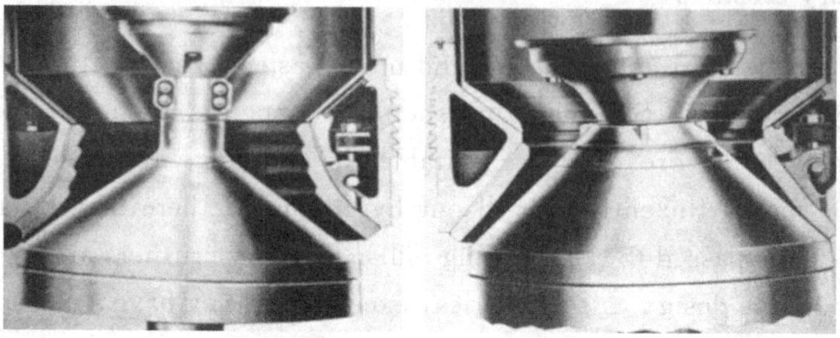

(a) Standard cone crusher (b) Short-head cone crusher

Figure 2.12

(2) Roll crushers

Roll crushers, or crushing rolls, have a useful application in handling friable, sticky, frozen, and less abrasive feeds, such as limestone, coal, chalk, gypsum, phosphate, and soft iron ores. Jaw and gyratory crushers have a tendency to choke near the discharge when crushing friable rock with a large proportion of maximum size pieces in the feed.

The mode of operation of roll crushers is extremely simple, the standard spring rolls (Figure 2.13) consisting of two horizontal cylinders which revolve towards each other. The set is determined by shims which cause the spring-loaded roll to be held back from the solidly mounted roll. Unlike jaw and gyratory crushers, where reduction is progressive by repeated pressure as the material passes down to the discharge point, the crushing process in rolls is one of single pressure.

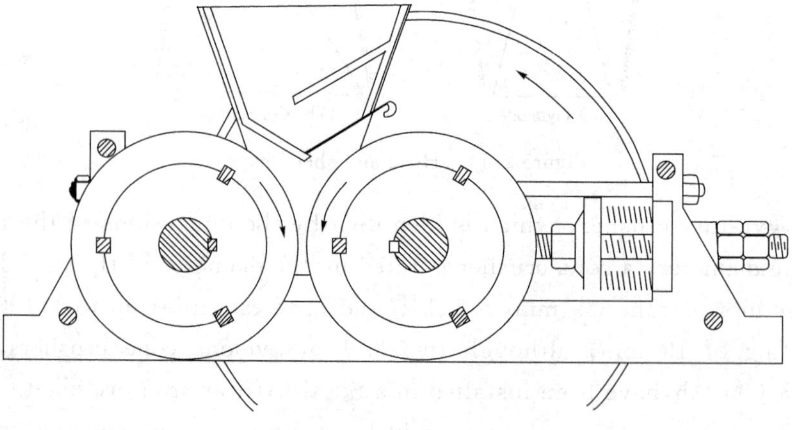

Figure 2.13 Crushing rolls

Roll crushers are also manufactured with only one rotating cylinder, which revolves towards a fixed plate. Other roll crushers use three, four, or six cylinders. In some crushers the diameters and speeds of the rolls may differ. The rolls may be gear driven, but this limits the distance adjustment between the rolls; and modern rolls are driven by V-belts from separate motors.

Smooth-surfaced rolls are usually used for fine crushing, whereas coarse crushing is often performed in rolls having corrugated surfaces, or with stub teeth arranged to present a chequered surface pattern.

The pressure exerted on the feed particles in conventional roll crushers is in the range 10～30 MPa, but work in Germany led to the development in the mid-1980s of the *High-Compression Roller Mill*, which utilizes forces in excess of 50 MPa, by the action of a hydraulic pressures system acting on a piston which presses the movable roller against the material bed (Figure 2.14) (density >70% solids by volume). Under such high forces, the product is a compacted cake containing a high proportion of fines and grains with microcracks. The compacted cake is subsequently deagglomerated, releasing the fines, and it has been shown that the specific energy consumption for compression and ball mill-deagglomeration is considerably less than that of ball mill grinding alone. The typical comminution energy in an high pressure grinding roll (HPGR) is 2.5～3.5 kW・h/t, compared to 15～25 kW・h/t in ball mill grinding. These mills are now being utilized in the cement, diamond, and limestone industries, and there is some evidence to show that mineral liberation can be improved by using these devices, so that they may be useful in the comminution of industrial and metalliferous ores. The mills were originally designed to be operated with smooth rolls, but studded rolls have become standard in the new designs, because of their improved wear-resistant characteristics.

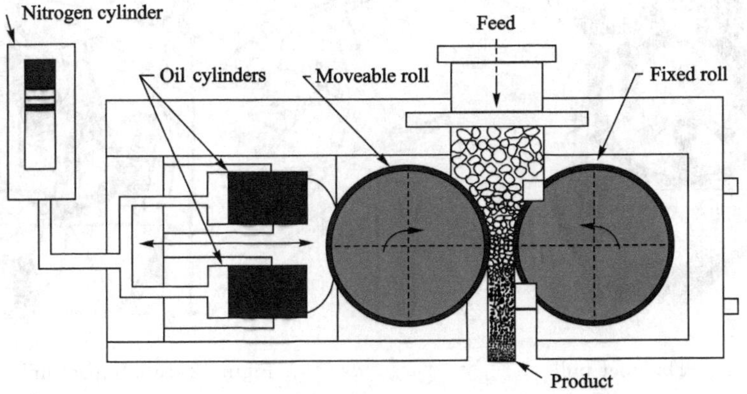

Figure 2.14 High pressure grinding rolls (Courtesy JKMRC, The University of Queensland)

(3) Impact crushers

In this class of crusher, comminution is by impact rather than compression, by sharp blows applied at high speed to free-falling rock. The moving parts are beaters, which transfer some of their kinetic energy to the ore particles on contacting them. The internal stresses created in the particles are often large enough to cause them to shatter.

Figure 2.15 shows a cross-section through a typical hammer mill. The hammers are pivoted so that they can move out of the path of oversize material, or tramp metal, entering the crushing chamber. Pivoted hammers exert less force than they would if rigidly attached, so they tend to be used on smaller impact crushers or for crushing soft material. The exit from the mill is perforated, so that material which is not broken to the required size is retained and swept up again by the rotor for further impacting.

This type of machine is designed to give the particles velocities of the order of that of the hammers. Fracture is either due to the severity of impact with the hammers or to the subsequent impact with the casing or grid. Since the particles are given very high velocities, much of the size reduction is by attrition, i. e., breaking of particle on particle, and this leads to little control on product size and a much higher proportion of fines than with compressive crushers.

The hammers can weigh over 100 kg and can work on feed up to 20 cm. The speed of the rotor varies between 500 r/min and 3,000 r/min. Due to the high rate of wear on these machines, they are limited in use to relatively non-abrasive materials. They have extensive use in limestone quarrying and in the crushing of coal. A great advantage in quarrying is in the fact that they produce a very good cubic product.

For much coarser crushing, the fixed hammer impact mill is often used (Figure 2.16). In these machines the material falls tangentially on to a rotor, running at 250～500 r/min, receiving a glancing impulse, which sends it spinning towards the impact plates.

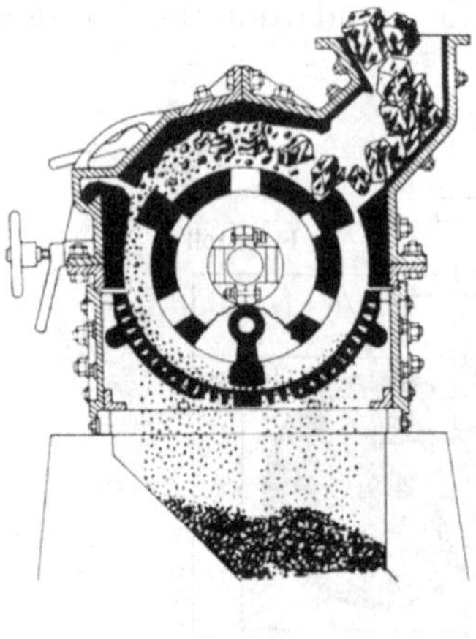

Figure 2.15 Hammer mill

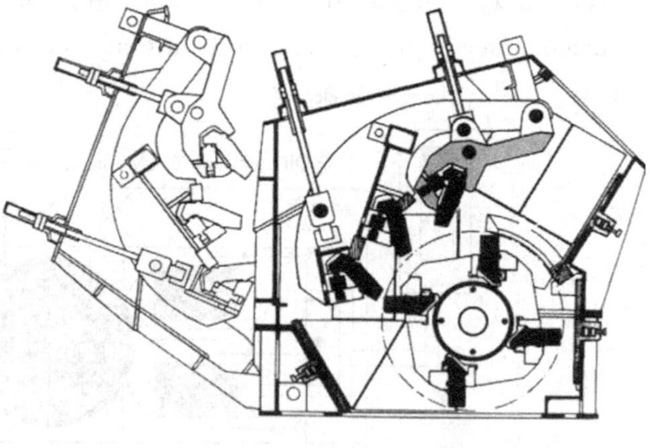

Figure 2.16 Impact mill

The fractured pieces which can pass between the clearances of the rotor and breaker plate enter a second chamber created by another breaker plate, where the clearance is smaller, and then into a third smaller chamber. This is the grinding path which is designed to reduce flakiness and gives very good cubic particles.

The rotary impact mill gives a much better control of product size than does the hammer mill, since there is less attrition. The product shape is much more easily controlled and energy is saved by the removal of particles once they have reached the size required.

Notes

1. Crushing is the first mechanical stage in the process of comminution in which the main objective is the liberation of the valuable minerals from the gangue.

破碎是粉碎过程的第一段机械作业,粉碎的主要目的是将有用矿物从脉石中解离出来。

2. The primary crusher product from most metalliferous ores can be crushed and screened satisfactorily, and the secondary plant generally consists of one or two size-reduction stages with appropriate crushers and screens.

绝大多数金属矿的粗碎产品都能够达到令人满意的破碎和筛分效果,中碎车间一般使用适宜的破碎机和筛分机进行一段或两段碎矿。

3. The jaws are set at an acute angle to each other, and one jaw is pivoted so that it swings relative to the other fixed jaw.

两块颚板呈锐角配置,一块颚板装在枢轴上,相对于另一块固定颚板摆动。

4. The gyratory crusher (Figure 2.10) consists essentially of a long spindle, carrying a hard steel conical grinding element, the head, seated in an eccentric sleeve.

旋回式破碎机(图 2.10)有一个安装在偏心套筒上的长主轴,主轴上带有坚硬的钢质锥形研磨体,即破碎头(运动圆锥)。

5. The spindle is suspended from a "spider" and, as it rotates, normally between 85 r/min and 150 r/min, it sweeps out a conical path within the fixed crushing chamber, or shell, due to the gyratory action of the eccentric.

竖轴悬吊在"横梁"上,由于偏心轴的回旋作用,其在固定破碎腔内做圆锥轨迹运动,竖轴旋转时,速度一般介于 85 r/min 和 150 r/min 之间。

6. Since the gyratory, unlike the jaw crusher, crushes on full cycle, it has a much higher capacity than a jaw crusher of the same gape, and is usually favored in plants handling very large throughputs.

因为旋回式破碎机是全周期破碎,不像颚式破碎机是半周期工作,所以旋回式破碎机的处理能力比给料口相同的颚式破碎机大得多。处理量大的选矿厂往往优先选用旋回式破碎机。

7. Tertiary crushers are, to all intents and purposes, of the same design as secondaries, except that they have a closer set.

除了排料口更小以外,实质上细碎机与中碎机的设计相同。

8. Roll crushers, or crushing rolls, have a useful application in handling friable, sticky, frozen, and less abrasive feeds, such as limestone, coal, chalk, gypsum, phosphate, and soft iron ores.

辊式破碎机或破碎辊主要用于处理易碎、黏性、冷冻和磨蚀性小的给料,例如石灰石、煤、白垩、石膏、磷酸盐和软的铁矿石。

9. Pivoted hammers exert less force than they would if rigidly attached, so they tend to be used on smaller impact crushers or for crushing soft material.

与刚性连接的锤头相比,置于销轴上的锤头施加的力更小,因而多用于规格较小的破碎机或用于破碎软质物料。

Exercises

1. Answer the following questions according to the text.

(1) What is the main objective of crushing?

(2) What are the main types of primary crusher in metalliferous operations?

(3) How does the swinging jaw move in the double-toggle Blake crusher?

(4) What is the structure of the gyratory crusher?

(5) What is the essential difference between cone crusher and gyratory crusher?

2. Translate the following phrases into English.

(1) 旋回式破碎机　(2) 粗磨　(3) 动颚　(4) 中碎

(5) 开路破碎　(6) 产品粒度　(7) 过粉碎　(8) 辊式破碎机

(9) 颚式破碎机　(10) 圆锥破碎机

3. Translate the following paragraphs into Chinese.

Paragraphs 1 and 2 in section 1 "Crushing is the first mechanical stage in the process of ...".

4. Translate the following sentences into English.

(1) 如果矿石较硬且光滑,那么可由棒磨机粗磨取代细碎。

(2) 这类破碎机的显著特点是其由两块如动物上下颚一样可以张开和闭合的颚板组成。

(3) 大部分金属矿石的中碎是由圆锥破碎机完成的。

(4) 旋回式破碎机可以被认为是宽度无限小而数量无限多的颚式破碎机。

(5) 中碎破碎机也采取干式给矿,其目的是将矿石破碎至适于磨矿的粒度。

2.3 Particle Size Analysis(粒度分析)

1. Introduction

(1) Objective of size analysis

Size analysis of the various products of a concentrator constitutes a fundamental part of laboratory testing procedure. It is of great importance in determining the quality of grinding and in establishing the degree of liberation of the values from the gangue at various particle sizes. In the separation stage, size analysis of the products is used to determine the optimum size of the feed to the process for maximum efficiency and to determine the size range at which any losses are occurring in the plant, so that they may be reduced. It is essential, therefore, that methods of size analysis must be accurate and reliable, as important changes in plant operation may be made on the results of the laboratory tests.

(2) Particle size and shape

The primary function of precision particle analysis is to obtain quantitative data about the size and size distribution of particles in the material. However, the exact size of an irregular particle cannot be measured. The terms "length" "breadth" "thickness" or "diameter" have little meaning because so many different values of these quantities can be determined. The size of a spherical particle is uniquely defined by its diameter. For a cube, the length along one edge is characteristic, and for other regular shapes there are equally appropriate dimensions. For irregular particles, it is desirable to quote the size of a particle in terms of a single quantity, and the expression most often used is the "equivalent diameter". This refers to the diameter of a sphere that would behave in the same manner as the particle when submitted to some specified operation. The assigned equivalent diameter usually depends on the method of measurement, hence the particle-sizing technique should, wherever possible, duplicate the process one wishes to control.

Several equivalent diameters are commonly encountered. For example, the Stokes' diameter is measured by sedimentation and elutriation techniques; the projected area diameter is measured microscopically and the sieve-aperture diameter is measured by means of sieving. The latter refers to the diameter of a sphere equal to the width of the aperture through which the particle just passes. If the particles under test are not true spheres, and they rarely are in practice, this equivalent diameter refers only to their second largest dimension.

There is a wide range of instrumental and other methods of particle size analysis available. A short list of some of the more common methods is given in Table 2. 1, together with their effective size ranges (these can vary greatly depending on the technology used), whether they can be used wet or dry and whether fractionated samples are available for later analysis.

Table 2.1 Common methods of particle size analysis

Method	Wet or dry	Fractionated sample?	Approx. useful size range/μm
Test sieving	Both	Yes	5～100,000
Laser diffraction	Both	No	0.1～2,000
Optical microscopy	Dry	No	0.2～50
Electron microscopy	Dry	No	0.005～100
Elutriation (cyclosizer)	Wet	Yes	5～45
Sedimentation (gravity)	Wet	Yes	1～40
Sedimentation (centrifuge)	Wet	Yes	0.05～5

2. Sieve analysis

Sieve analysis is one of the oldest methods of size analysis and is accomplished by passing a known weight of sample material successively through finer sieves and weighing the amount collected on each sieve to determine the percentage weight in each size fraction. Sieving, when applied to irregularly shaped particles, is complicated by the fact that a particle with a size near that of the nominal aperture of the test sieve may pass only when presented in a favourable position. As there is inevitably a variation in the size of sieve apertures, due to irregularity of weaving, prolonged sieving will cause the larger apertures to exert an unduly large effect on the sieve analysis.

(1) Test sieves

Test sieves are designated by the nominal aperture size, which is the nominal central separation of opposite sides of a square aperture or the nominal diameter of a round aperture. A variety of sieve aperture ranges are currently used, the most popular being the German Standard, DIN 4188; ASTM standard, E11; the American Tyler series; the French series, AFNOR; and the British Standard, BS 1796. Woven-wire sieves were originally designated by a mesh number, which referred to the number of wires per inch, which is the same as the number of square apertures per square inch. This has the serious disadvantage that the same mesh number on the various standard ranges corresponds to different aperture sizes, depending on the thickness of wire used in the woven-wire cloth. Sieves are now designated by aperture size, which gives the user directly the information needed.

Since some workers and the older literature still refer to sieve sizes in terms of mesh number, Table 2.2 lists mesh numbers for the British Standards series against nominal aperture size. Wire cloth screens are woven to produce nominally uniform square apertures within required tolerances. Wire cloth in sieves with a nominal aperture of 75 μm and greater are plain woven, while those in cloths with apertures below 63 μm may be twilled.

Table 2.2 BS 1796 wire-mesh sieves

Mesh number	Nominal aperture size/μm	Mesh number	Nominal aperture size/μm
3	5,600	36	425
3.5	4,750	44	355
4	4,000	52	300
5	3,350	60	250
6	2,800	72	212
7	2,360	85	180
8	2,000	100	150
10	1,700	120	125
12	1,400	150	106
14	1,180	170	90
16	1,000	200	75
18	850	240	63
22	710	300	53
25	600	350	45
30	500	400	38

(2) Testing methods

The general procedures for test sieving are comprehensively covered in BS 1796. Machine sieving is almost universally used, as hand sieving is long and tedious, and its accuracy and precision depends to a large extent on the operator.

Sieves can be procured in a range of diameters, depending on the particle size and mass of material to be sieved. A common diameter for laboratory sieves is 200 mm.

The sieves chosen for the test are arranged in a stack, or nest, with the coarsest sieve on the top and the finest at the bottom. A tight-fitting pan or receiver is placed below the bottom sieve to receive the final undersize, and a lid is placed on top of the coarsest sieve to prevent escape of the sample.

The material to be tested is placed in the uppermost, coarsest sieve, and the nest is then placed in a sieve shaker which vibrates the material in a vertical plane (Figure 2.17), and, on some models, a horizontal plane. The duration of screening can be controlled by an automatic timer. During the shaking, the undersize material falls through successive sieves until it is retained on a sieve having apertures which are slightly smaller than the diameter of the particles. In this way the sample is separated into size fractions. After the

Figure 2.17 Vibrating sieve shaker

required time, the nest is taken apart and the amount of material retained on each sieve weighed. Most of the near-mesh particles, which block the openings, can be removed by inverting the sieve and tapping the frame gently. Failing this, the underside of the gauze may be brushed gently with a soft brass wire or nylon brush. Blinding becomes more of a problem for the finer the aperture, and brushing, even with a soft hair brush, of sieves finer than about 150 μm aperture tends to distort the individual meshes.

Wet sieving can be used on material already in the form of a slurry, or it may be necessary for powders which form aggregates when dry-sieved. A full description of the techniques is given in BS 1796. Water is the liquid most frequently used in wet sieving, although for materials which are water repellent, such as coal or sulphide ores, a wetting agent may be necessary. The test sample may be washed down through a nest of sieves, with the finest sieve at the bottom. At the completion of the test the sieves, together with the retained oversize material, are dried at a suitable low temperature and weighed.

(3) Presentation of results

There are several ways in which the results of a sieve test can be tabulated. The three most convenient methods such as the weight percentage of material in each size range, the cumulative percentage of material passing through the sieves, and the cumulative percentage of material retained on the sieves, are shown in Table 2.3. Table 2.3 shows:

① The sieve size ranges used in the test.

② The weight of material in each size range, e.g., 1.32 g of material passed through the 250 μm sieve, but was retained on the 180 μm sieve: the material therefore is in the size range −250 μm to +180 μm.

③ The weight of material in each size range expressed as a percentage of the total weight.

④ The nominal aperture sizes of the sieves used in the test.

⑤ The cumulative percentage of material passing through the sieves, e.g., 87.5% of the material is less than 125 μm in size.

⑥ The cumulative percentage of material retained on the sieves.

Table 2.3 Results of typical sieve test

Sieve size range /μm	Sieve fractions		Nominal apertrue size /μm	Cumulative/%	
	g	%		undersize	oversize
+250	0.04	0.1	250	99.9	0.1
−250～+180	1.32	2.9	180	97.0	3.0
−180～+125	4.23	9.5	125	87.5	12.5
−125～+90	9.44	21.2	90	66.3	33.7
−90～+63	13.10	29.4	63	36.9	63.1
−63～+45	11.56	26.0	45	10.9	89.1
−45	4.87	10.9			

The results of a sieving test should always be plotted graphically in order to assess their full significance. There are many different ways of recording the results, the most common being that of plotting cumulative undersize (or oversize) against particle size. Although arithmetic graph paper can be used, it suffers from the disadvantage that points in the region of the finer aperture sizes tend to become congested. A semi-logarithmic plot avoids this, with a linear ordinate for percentage oversize or undersize and a logarithmic abscissa for particle size. Figure 2. 18 shows graphically the results of the sieve test tabulated in Table 2. 3.

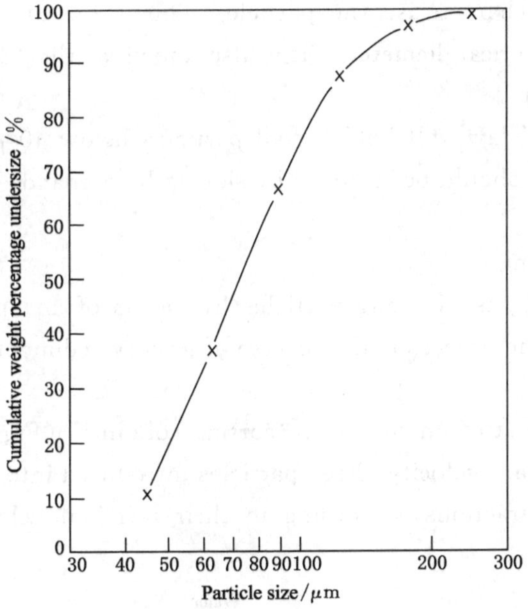

Figure 2. 18 Screen analysis graph

A valuable quantity which can be determined from such curves is the "median size" of the sample. This refers to the mid-point in the size distribution —50% of the particles are smaller than this size and 50% are larger.

3. Elutriation techniques

(1) Stokes' equivalent diameter

In sedimentation techniques, the material to be sized is dispersed in a fluid and allowed to settle under carefully controlled conditions; in elutriation techniques, samples are sized by allowing the dispersed material to settle against a rising fluid velocity. Both techniques separate the particles on the basis of resistance to motion in a fluid.

This resistance to motion determines the terminal velocity which the particle attains as it is allowed to fall in a fluid under the influence of gravity. For particles within the sub-sieve range, the terminal velocity is given by the equation derived by Stokes(1891), namely:

$$v = d^2 g(D_s - D_f)/18\eta \tag{2.3}$$

where v is the terminal velocity of the particle, m/s; d is the particle diameter, m; g is the acceleration due to gravity, m/s^2; D_s is the particle density, kg/m^3; D_f is the fluid density, kg/m^3; and η is the fluid viscosity, $N \cdot s/m^2$, $\eta = 0.001\ N \cdot s/m^2$ for water at 20 ℃.

Stokes' law is derived for spherical particles; non-spherical particles will also attain a terminal velocity, but this velocity will be influenced by the shape of the particles. Nevertheless, this velocity can be substituted in the Stokes' equation to give a value of d, which can be used to characterise the particle. This value of d is referred to as the "Stokes' equivalent spherical diameter". It is also known as the "Stokes' diameter" or the "sedimentation diameter".

In general, Stokes' law will hold for all particles below 40 μm dispersed in water; particles above this size should be removed by sieving beforehand. The lower limit may be taken as 1 μm.

(2) Elutriation sizing

Elutriation is a process of sizing particles by means of an upward current of fluid, usually water or air. The process is the reverse of gravity sedimentation, and Stokes' law applies.

All elutriators consist of one or more "sorting columns" (Figure 2.19) in which the fluid is rising at a constant velocity. Feed particles introduced into the sorting column will be separated into two fractions, according to their terminal velocities, calculated from Stokes' law.

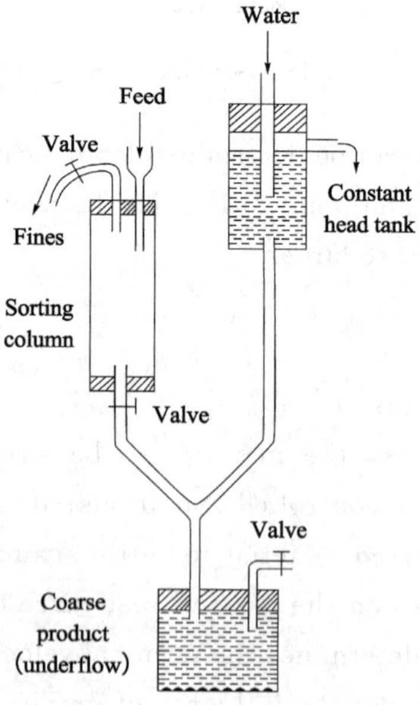

Figure 2.19 Simple elutriator

Those particles having a terminal velocity less than that of the velocity of the fluid will report to the overflow, while those particles having a greater terminal velocity than the fluid velocity will sink to the underflow. Elutriation is carried out until there are no visible signs of further classification taking place or the rate of change in weights of the products is negligible.

This involves the use of much water, involving dilution of the undersize fraction, but it can be shown that this is not as serious as in beaker decantation. Consider a sorting column of depth h, sorting material at a separating size of d. If the upward velocity of water flow is v, then by Stokes' law, $v \propto d^2$.

Particles smaller than the separating size d will move upwards in the water flow at a velocity dependent on their size.

Thus, particles of size d_1, where $d_1 < d$, will move upwards in the sorting column at a velocity v_1, where $v_1 \propto (d^2 - d_1^2)$.

The time required for a complete volume change in the sorting column is h/v, and the time required for particles of size d_1 to move from the bottom to the top of the sorting column is h/v_1.

Therefore the number of volume changes required to remove all particles of size d_1 from the sorting column $= (h/v_1)/(h/v) = d^2/(d^2 - d_1^2) = 1/[1-(d_1/d)^2]$. The number of volume changes required for various values of d_1/d are: 5.3, 1.3 and 1.0 for d_1/d of 0.9, 0.5 and 0.1.

Notes

1. It is of great importance in determining the quality of grinding and in establishing the degree of liberation of the values from the gangue at various particle sizes.

筛分分析对于确定磨矿效果和确定不同粒度下有用矿物与脉石矿物的解离度具有重要意义。

2. The primary function of precision particle analysis is to obtain quantitative data about the size and size distribution of particles in the material.

精确颗粒分析的主要作用是获得有关物料中颗粒大小和粒度分布的定量数据。

3. This refers to the diameter of a sphere that would behave in the same manner as the particle when submitted to some specified operation.

等效直径是指在某一特定操作中具有与该颗粒相同行为的球体的直径。

4. Woven-wire sieves were originally designated by a mesh number, which referred to the number of wires per inch, which is the same as the number of square apertures per square inch.

编织线筛最初由网目号表示，该网目号是指每英寸的线数，该数量与每平方英寸的方形孔数相同。

5. Wire cloth in sieves with a nominal aperture of 75 μm and greater are plain woven, while those in cloths with apertures below 63 μm may be twilled.

标称孔径为 75 μm 或更大的筛网中的金属丝网是平纹编织的，而标称孔径小于 63 μm 的筛网中的金属丝网可能会采用斜纹编织。

6. A tight-fitting pan or receiver is placed below the bottom sieve to receive the final undersize, and a lid is placed on top of the coarsest sieve to prevent escape of the sample.

在底部筛子的下方放置一个紧贴的托盘或接收装置，以收集最终的筛下产品，并在最粗的筛子的顶部放置一个盖子，以防止样品漏出。

7. Wet sieving can be used on material already in the form of a slurry, or it may be necessary for powders which form aggregates when dry-sieved.

湿筛可用于已成浆状的物料，也可用于干法筛分时形成聚团的粉末。

8. A valuable quantity which can be determined from such curves is the "median size" of the sample. This refers to the mid-point in the size distribution —50% of the particles are smaller than this size and 50% are larger.

可以从这些曲线中确定的重要数据是样品的"中位粒径"。这是指有 50%的物料粒径小于粒度分布的中位粒径，而 50%的物料大于这个粒径。

9. Elutriation is a process of sizing particles by means of an upward current of fluid, usually water or air. The process is the reverse of gravity sedimentation, and Stokes' law applies.

淘析是一种通过向上流动的流体(通常是水或空气)来实现颗粒分级的方法。该过程与重力沉降相反，适用斯托克斯定律。

Exercises

1. Answer the following questions according to the text.

(1) What is the function and importance of size analysis?

(2) How to determine the size of a particle?

(3) What are the common methods of particle size analysis?

(4) What are the general procedures for sieving test?

(5) What are the three most convenient methods to tabulate the results of a sieve test?

2. Translate the following phrases into English.

(1) 粒度分析	(2) 标准筛	(3) 等效直径	(4) 重力沉降
(5) 中粒径	(6) 筛分粒度分布	(7) 淘析	(8) 标称孔径
(9) 上升流	(10) 运动阻力		

3. Translate the following paragraphs into Chinese.

Paragraph 1 in section 1 (2) "The primary function of precision particle analysis is ...".

4. Translate the following sentences into English.

(1) 筛分分析是最古老的粒度分析方法之一，通过将已知质量的样品物料依次通过越来越细的筛网并称量每个筛网上收集的物料的质量以确定各粒级的质量百分比来完成。

(2) 有许多不同的方法记录结果，最常见的是绘制累积筛下物(或筛上物)与粒径的关系曲线。

(3) 粒度分析对于评估磨矿回路的性能非常重要。

2.4 Screening(筛分)

1. Efficiency of screening

(1) Performance of screens

Industrial sizing of materials is extensively used, and the types of equipment are many and varied. Screening is generally carried out on relatively coarse material, as the efficiency decreases rapidly with fineness. Fine screens are very fragile and expensive and tend to become blocked rather easily with retained particles. Dry screening is generally limited to material above about 5 mm in size, while wet screening down to around 250 μm is common. Although there are screen types that are capable of efficient size separations down to 40 μm, sizing below 250 μm is also undertaken by classification.

There is, of course, a wide range of purposes for screening. The main purposes in the minerals industry are: ① to prevent the entry of undersize into crushing machines, so increasing their capacity and efficiency; ② to prevent oversize material from passing to the next stage in closed-circuit fine crushing and grinding operations; ③ to prepare a closely sized end product. This is important in quarrying, where the final product size is an important part of the specification.

In its simplest form, the screen is a surface having many apertures, or holes, usually with uniform dimensions. Particles presented to that surface will either pass through or be retained, according to whether the particles are smaller or larger than the governing dimensions of the aperture. The efficiency of screening is determined by the degree of perfection of separation of the material into size fractions above or below the aperture size. There has been no universally accepted method of defining screen performance and a number of methods are employed. The most common screen performance criteria are those which define an efficiency based on the recovery of material at a given size, or on the mass of misplaced material in each product. This immediately leads to a range of possibilities, such as undersize in the overscreen product, oversize in the through-screen product, or a combination of the two.

An efficiency equation can be calculated from a mass balance across a screen as follows:

Consider a screen (Figure 2. 20) the feed to which is F t/h. Two products are generated. A coarse product of C t/h overflows from the screen, and a fine product of U t/h passes through the screen.

Let f be the fraction of material above the cut point size in the feed; c be the fraction of material above the cut point size in the overflow; and u be the fraction of material above the cut point size in the underflow. f, c, and u can be determined by sieving a representative sample of each of the fractions on a laboratory screen of the same aperture

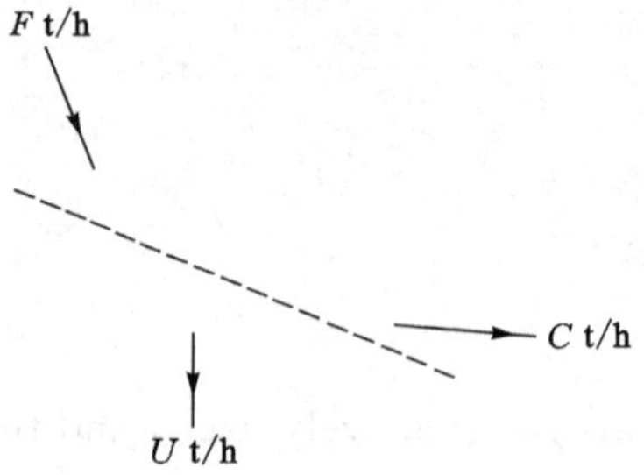

Figure 2.20 Mass balance on a screen

size as the industrial screen and assuming this to be 100% efficient.

The mass balance on the screen is: $F=C+U$, the mass balance of the oversize material is: $Ff=Cc+Uu$, and the mass balance of the undersize material is: $F(1-f)=C(1-c)+U(1-u)$.

Hence, $C/F=(f-u)/(c-u)$, and $U/F=(c-f)/(c-u)$.

The recovery of oversize material into the screen overflow is:

$$Cc/Ff=c(f-u)/f(c-u) \tag{2.4}$$

and the corresponding recovery of undersize material in the screen underflow is:

$$U(1-u)/F(1-f)=(1-u)(c-f)/(1-f)(c-u) \tag{2.5}$$

These two Equations (2.4) and (2.5), measure the effectiveness of the screen in separating the coarse material from the underflow and the fine material from the overflow.

A combined effectiveness, or overall efficiency, E, is then obtained by multiplying the two equations together:

$$E=c(f-u)(1-u)(c-f)/f(c-u)^2(1-f) \tag{2.6}$$

For screens where the aperture and the cut point are similar (and if there are no broken or deformed apertures), the amount of coarse material in the underflow is usually very low. A simplification of Equation (2.6) can be obtained by assuming that it is, in fact, zero (i.e., $u=0$), in which case the formula for fines recovery and that for the overall efficiency both reduce to:

$$E=(c-f)/c(1-f) \tag{2.7}$$

This formula is widely used and implies that recovery of the coarse material in the overflow is 100%.

(2) Factors affecting screen performance

Feed rate

The principle of sieve sizing analysis is to use a low feed rate and a very long screening time to effect an almost complete separation. In industrial screening practice, economics dictate that relatively high feed rates and short particle dwell times on the screen should be used. At these high feed rates, a thick bed of material is presented to the screen, and fines must travel to the bottom of the particle bed before they have an opportunity to pass through the screen surface. The net effect is reduced efficiency. High capacity and high efficiency are often opposing requirements for any given separation, and a compromise is necessary to achieve the optimum result.

Screen angle

If a particle approaches the aperture at a shallow angle, it will "see" a narrower effective aperture dimension and near mesh particles are less likely to pass. The slope of the screening surface affects the angle at which particles are presented to the screen apertures. Some screens utilize this effect to achieve separations significantly finer than the screen aperture. For example, sieve bends cut at approximately half the aperture size. Where screening efficiency is important, horizontal screens are selected. The screen angle also affects the speed at which particles are conveyed along the screen, and therefore the dwell time on the screen and the number of opportunities particles have of passing the screen surface.

Particle shape

Most granular materials processed on screens are non-spherical. While spherical particles pass with equal probability in any orientation, irregular-shaped near-mesh particles must orient themselves in an attitude that permits them to pass. Elongated and slabby particles will present a small cross-section for passage in some orientations and a large cross-section in others. The extreme particle shapes therefore have a low screening efficiency. Mica, for instance, screens poorly on square aperture screens, its flat, plate-like crystals tending to "ride" over the screen apertures.

Open area

The chance of passing through the aperture is proportional to the percentage of open area in the screen material, which is defined as the ratio of the net area of the apertures to the whole area of the screening surface. The smaller the area occupied by the screen deck construction material, the greater the chance of a particle reaching an aperture. Open area generally decreases with the fineness of the screen aperture. In order to increase the open area of a fine screen, very thin and fragile wires or deck construction must be used. This fragility and the low throughput capacity are the main reasons for classifiers replacing screens at fine aperture sizes.

Vibration

Screens are vibrated in order to throw particles off the screening surface so that they can again be presented to the screen, and to convey the particles along the screen. The right type of vibration also induces stratification of the feed material, which allows the fines to walk through the layer of particles to the screen surface while causing larger particles to rise to the top. Stratification tends to increase the rate of passage in the middle section of the screen.

Moisture

The amount of surface moisture present in the feed has a marked effect on screening efficiency, as does the presence of clays and other sticky materials. Damp feeds screen very poorly as they tend to agglomerate and "blind" the screen apertures. As a rule of thumb, screening at less than around 5 mm aperture size must be performed on perfectly dry or wet material, unless special measures are taken to prevent blinding. Wet screening allows finer sizes to be processed efficiently down to 250 μm and finer. Adherent fines are washed off large particles, and the screen is cleaned by the flow of pulp and additional water sprays.

2. Stationary screens

There are many different types of industrial screens, which may be classified as either stationary or moving screens.

(1) Grizzly screens

Very coarse material is usually screened on a grizzly screen, which, in its simplest form, consists of a series heavy parallel bars set in a frame (Figure 2.21). Some grizzlies employ chains instead of bars and some are shaken or vibrated mechanically to help the sizing and to aid in the removal of the oversize. The most common use of grizzlies in mineral processing is for sizing the feed to primary and secondary crushers. If a crusher has a 100 mm setting, then feed can be passed over a grizzly with a 100 mm gap in order to reduce the load on the crusher.

Figure 2.21 Grizzly screen (Courtesy of DOVE equipment and Metso Minerals)

The bars of a grizzly are inclined usually at an angle of between 20°～50°, the greater the incline, the greater is the throughput, but the lower is the efficiency. Feed flows down the grizzly in the direction of the bars to assist flow and to reduce clogging. Clogging is a major problem, especially on the cross-members which hold the bars together. The bars are usually tapered in cross-section to minimize clogging once particles have entered between them. The size of particle screened on grizzlies can be as large as 300 mm, or as small as 20 mm.

(2) Sieve bends

Wedge or profile wire or slotted polyurethane panels are used in *sieve bends* and *inclined flat screens* for dewatering and very fine screening applications. The sieve bend has a curved screen composed of horizontal wedge bars (Figure 2.22), whereas flat screens are installed on a slope of between 45°and 60°. Feed slurry enters the upper surface of the screen tangentially and flows down the surface in a direction perpendicular to the openings between the wedge bars. As the stream of slurry passes each opening a thin layer is peeled off and directed to the underside of the screen. According to Fontein(1954), particles roughly twice the thickness of this layer are dragged along with the undersize fraction; particles larger than this size pass across the openings as their greatest part projects into the liquid flowing over the slot. In general, a separation is produced at a size roughly equivalent to half the bar spacing and so very little plugging of the apertures should take place. Separation can be undertaken down to 50 μm and screen capacities are up to 180 m^3/h.

Figure 2.22 Sieve bends

Sieve bends and inclined wedge-wire screens are sometimes installed with mechanical devices to periodically vibrate or rap the screen surface in order to removed blinded particles.

3. Moving screens

(1) Vibrating screens

Vibrating screens are the most important and versatile screening machines for mineral processing applications. The success of the vibrating screen has made many older screen types obsolete in the minerals industry including shaking and reciprocating screens. Vibrating screens have a rectangular screening surface with feed and oversize discharge at opposite ends. They perform size separations from 300 mm in size down to 45 μm and they are used in a variety of sizing, grading, scalping, dewatering, wet screening, and washing applications. Vibrating screens of most types can be manufactured with more than one screening deck. On multiple-deck systems, the feed is introduced to the top coarse screen;

the undersize falling through to the lower screen decks, thus producing a range of sized fractions from a single screen.

Inclined screens

Inclined or circular motion screens (Figure 2.23) are widely used as sizing screens. A vertical circular or elliptical vibration is induced mechanically by the rotation of unbalanced weights or flywheels attached usually to a single drive shaft. The amplitude of throw can be adjusted by adding or removing weight elements bolted to the flywheels. Single-shaft screens must be installed on a slope, usually between 15° and 28° to permit flow of material along the screen.

Horizontal screens

Horizontal vibrating screens have a horizontal or near-horizontal screening surface (Figure 2.24), and therefore need less headroom than inclined screens. Horizontal screens must be vibrated with a linear or an elliptical vibration produced by a double or triple-shaft vibrator. The accuracy of particle sizing on horizontal screens is superior to that on inclined screens; however because gravity does not assist the transport of material along the screen they have lower capacity than inclined screens. Horizontal screens are used in sizing applications where screening efficiency is critical, and in drain-and-rinse screens in heavy medium circuits.

Figure 2.23 DOVE vibrating screen
(Courtesy of DOVE equipment)

Figure 2.24 Horizontal screen
(Courtesy Schenk Australia)

Resonance screens

They are a type of horizontal screen consisting of a screen frame connected by rubber buffers to a dynamically balanced frame having a natural resonance frequency which is the same as that of the vibrating screen body. The vibration energy imparted to the screen frame is stored up in the balancing frame, and re-imparted to the screen frame on the return stroke. The energy losses are reduced to a minimum, and the sharp return motion produced by the resonant action imparts a lively action to the deck and promotes good screening.

Banana screens

Banana or Multi-slope screens have become widely used in high-tonnage sizing applications where both efficiency and capacity are important. Banana screens (Figure 2. 25) typically have a variable slope of around 40°～30° at the feed end of the screen, reducing to around 0°～15° in increments of 3. 5°～5°. Banana screens are usually designed with a linear-stroke vibrator. The steep sections of the screen cause the feed material to flow rapidly at the feed end of the screen. The resulting thin bed of particles stratifies more quickly and therefore has a faster screening rate for the very fine material than would be possible on a slower moving thick bed. Towards the discharge end of the screen, the slope decreases to slow down the remaining material, enabling more efficient screening of the near-size material. The capacity of banana screens is significantly greater and is reported to be up to three or four times that of conventional vibrating screens.

Figure 2. 25 Banana screen (Courtesy Schenk Australia)

Dewatering screens

They are a type of vibrating screen that are fed a thick slurry and produce a drained sand product. Dewatering screens are often installed with a slight up-hill incline to ensure that water does not flow over with the product. A thick bed of particles forms, trapping particles finer than the screen aperture.

(2) High frequency screens

Efficient screening of fine particles requires a vibration with small amplitude and high frequency. Frequencies up to 3,600 r/min are used to separate down to 100 μm compared with vibrating screens for coarser applications that are vibrated at around 700～1,200 r/min. The vibration of the screening surface can be created by electric motors or with electrical solenoids.

High-frequency wet screens such as the Derrick stack sizer (Figure 2. 26) permit screening down to 45 microns. Screening efficiency decreases rapidly once the free water has passed through the screen, therefore these screens incorporate water-sprays to periodically re-pulp the screen oversize to ensure good washing.

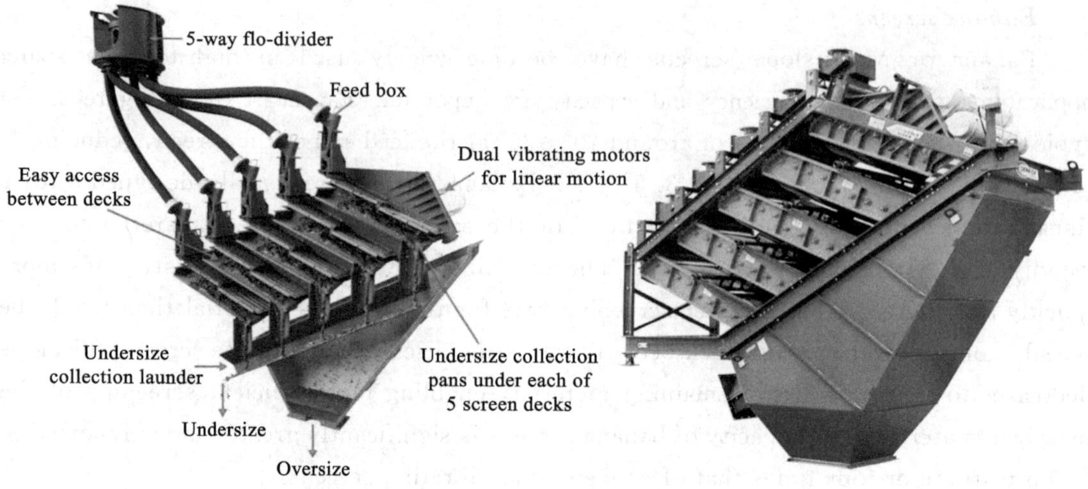

Figure 2.26 Derrick stack sizer (Courtesy of Derrick Corporation)

Notes

1. In its simplest form, the screen is a surface having many apertures, or holes, usually with uniform dimensions. Particles presented to that surface will either pass through or be retained, according to whether the particles are smaller or larger than the governing dimensions of the aperture.

筛子最简单的形式是一个具有许多孔的平面，这些孔洞通常具有统一的尺寸。经过筛面的颗粒要么透过筛孔，要么被截留在筛面上。

2. The most common screen performance criteria are those which define an efficiency based on the recovery of material at a given size, or on the mass of misplaced material in each product.

通常用筛分效率来评价筛子的筛分性能。筛分效率是指给定粒度物料的回收率或每个产品中错配颗粒的含量。

3. The bars are usually tapered in cross-section to minimize clogging once particles have entered between them.

棒条的横截面通常是锥形的，以便在颗粒进入棒条之间时尽量减少堵塞。

4. According to Fontein(1954), particles roughly twice the thickness of this layer are dragged along with the undersize fraction; particles larger than this size pass across the openings as their greatest part projects into the liquid flowing over the slot.

根据 Fontein(1954)的研究，粒度约为薄层厚度两倍的颗粒被带入筛下，大于该尺寸的颗粒会从筛面缝隙流过，因为这些颗粒会随流经筛面的介质流走。

5. Vibrating screens are the most important and versatile screening machines for mineral processing applications. The success of the vibrating screen has made many older screen types obsolete in the minerals industry including shaking and reciprocating screens.

振动筛是矿物加工应用中最重要、用途最广的筛分设备。振动筛的成功应用使得包括摇动筛和往复筛等在内的许多老式筛分设备在矿物加工行业已被淘汰。

6. They perform size separations from 300 mm in size down to 45 μm and they are used in a variety of sizing, grading, scalping, dewatering, wet screening, and washing applications.

振动筛适用的粒度范围是 300 mm 到 45 μm,可用于各种粗粒筛分、分级、脱介、脱水、细粒湿法筛分和洗涤。

7. They are a type of horizontal screen consisting of a screen frame connected by rubber buffers to a dynamically balanced frame having a natural resonance frequency which is the same as that of the vibrating screen body.

共振筛是一种水平筛,由通过橡胶减振器与动态平衡机架相连接的筛框组成,机架具有与振动筛本体相同的自然共振频率。

8. Frequencies up to 3,600 r/min are used to separate down to 100 μm compared with vibrating screens for coarser applications that are vibrated at around 700～1,200 r/min.

频率为 3 600 r/min 的振动筛用于筛分 100 μm 以下的物料,而应用于粗粒物料的振动筛的振动频率一般为 700～1 200 r/min。

Exercises

1. Answer the following questions according to the text.

(1) What are the purposes for screening?

(2) What determines the efficiency of screening?

(3) What are the factors that affect screen performance?

(4) What are the commonly used stationary sieves in industry?

(5) What are the commonly used moving screens in industry?

2. Translate the following phrases into English.

(1) 筛分性能　(2) 筛上产品　(3) 弧形筛　(4) 动筛

(5) 固定筛　(6) 香蕉筛　(7) 共振筛　(8) 振动筛

(9) 格筛　(10) 筛分效率

3. Translate the following paragraphs into Chinese.

Paragraph 1 in section 3 (1)"Vibrating screens are the most important and versatile ...".

4. Translate the following sentences into English.

(1) 格条筛可以筛分大到 300 mm、小到 20 mm 的物料。

(2) 振动筛有一个矩形筛面,筛上产品从筛子另一端排出。

(3) 一般对较粗的物料进行筛分,因为筛分效率随着细度的增加而迅速下降。

(4) 使用多层振动筛时,给料首先给入最上层的粗筛;筛下产品落入下层筛面上,因此,这就可生产一系列不同粒级的筛分产品。

3 GRINDING AND CLASSIFICATION (磨矿与分级)

3.1 Grinding Mills(磨矿机)

1. Introduction

Grinding is the last stage in the process of comminution; in this stage the particles are reduced in size by a combination of impact and abrasion, either dry or in suspension in water.

According to the ways by which motion is imparted to the charge, grinding mills are generally classified into two types: tumbling mills and stirred mills. In tumbling mills, the mill shell is rotated and motion is imparted to the charge via the mill shell. The grinding medium may be steel rods, balls, or rock itself. Tumbling mills are typically employed in the mineral industry for coarse-grinding processes, in which particles between 5 mm and 250 mm are reduced in size to between 40 μm and 300 μm. In stirred mills the mill shell with either a horizontal or a vertical orientation is stationary, and motion is imparted to the charge by the movement of an internal stirrer. Fine grinding media inside the mill are agitated or rotated by a stirrer, which typically comprises a central shaft to which are attached pins or discs of various designs. Stirred mills find application in fine (15～40 μm) and ultra-fine (<15 μm) grinding.

All ores have an economic optimum particle size, which will depend on many factors, including the extent to which the values are dispersed in the gangue, and the subsequent separation process to be used. It is the purpose of the grinding section to exercise close control on this product size and, for this reason, correct grinding is often said to be the key to good mineral processing. Undergrinding of the ore will, of course, result in a product which is too coarse, with a degree of liberation too low for economic separation; poor recovery and enrichment ratio will be achieved in the concentration stage. Over-grinding needlessly reduces the particle size of the subsequently liberated major constituent (usually the gangue) and may reduce the particle size of the minor constituent (usually the mineral value) below the size required for most efficient separation. Much expensive energy is wasted in the process. It is important to realize that grinding is the most energy-intensive operation in mineral processing. It has been estimated that 50% of the energy consumed in US mills is used in comminution.

Although tumbling mills have been developed to a high degree of mechanical efficiency and reliability, they are extremely wasteful in terms of energy expended, since the ore is mostly broken as a result of repeated, random impacts, which break liberated as well as unliberated particles. At present, there is no practical way that these impacts can be directed at the interfaces between the mineral grains, which would produce optimum liberation, although various ideas have been postulated.

Apart from laboratory testing, grinding in mineral processing is a continuous process, material being fed at a controlled rate from storage bins into one end of the mill and overflowing at the other end after a suitable dwell time. Control of product size is exercised by the type of medium used, the speed of rotation of the mill, the nature of the ore feed, and the type of circuit used.

2. Tumbling mills

Tumbling mills are of three basic types: rod, ball and autogenous. Structurally, each type of mill consists of a horizontal cylindrical shell, provided with renewable wearing liners and a charge of grinding medium. The drum is supported so as to rotate on its axis on hollow trunnions attached to the end walls (Figure 3. 1). The diameter of the mill determines the pressure that can be exerted by the medium on the ore particles and, in general, the larger the feed size the larger needs to be the mill diameter. The length of the mill, in conjunction with the diameter, determines the volume, and hence the capacity of the mill.

The feed material is usually fed to the mill continuously through one end trunnion, the ground product leaving via the other trunnion, although in certain applications the product may leave the mills through a number of ports spaced around the periphery of the shell. All types of mills can be used for wet or dry grinding by modification of feed and discharge equipment.

(1) Rod mills

These may be considered as either fine crushers or coarse grinding machines. They are capable of taking feed as large as 50 mm and making a product as fine as 300 μm, reduction ratios normally being in the range (15～20) : 1. They are often preferred to fine crushing machines when the ore is clayey or damp, thus tending to choke crushers.

The distinctive feature of a rod mill is that the length of the cylindrical shell is between 1. 5 and 2. 5 times its diameter (Figure 3. 2). This ratio is important because the rods, which are only a few centimeters shorter than the length of the shell, must be prevented from turning so that they become wedged across the diameter of the cylinder. Since rods longer than about 6 m will bend, this establishes the maximum length of mill. Thus, with a mill 6. 4 m long the diameter should not be over 4. 57 m. Rod mills of up to 4. 57 m in diameter by 6. 4 m in length are in use, run by 1,640 kW motors. Rod and other grinding mills are rated by power rather than capacity, since the capacity is determined by many factors, such as the grindability and the reduction in size required.

Figure 3.1 Tumbling mill

Figure 3.2 Rod mills

(2) Ball mills

The final stages of comminution are performed in tumbling mills using steel balls as the grinding medium and so designated ball mills.

Since balls have a greater surface area per unit weight than rods, they are better suited for fine finishing. The term ball mill is restricted to those having a length to diameter ratio of 1.5 to 1 and less. Ball mills in which the length to diameter ratio is between 3 and 5 are designated tube mills. These are sometimes divided into several longitudinal compartments, each having a different charge composition; the charges can be steel balls, rods, or pebbles, and they are often used dry to grind cement clinker, gypsum, and phosphate. Tube mills having only one compartment and a charge of hard, screened ore particles as the grinding medium, are known as pebble mills. They are widely used in the South African gold mines. Since the weight of pebbles per unit volume is 35% ~55% of that of steel balls, and as the power input is directly proportional to the volume weight of the grinding medium, the power input and capacity of pebble mills are correspondingly lower. Thus, in a given grinding circuit, for a certain feed rate, a pebble mill would be much larger than a ball mill, with correspondingly higher capital cost.

Ball mills are also classified by the nature of the discharge. They may be simple trunnion overflow mills, or grate discharge mills. The latter type is fitted with discharge grates between the cylindrical mill body and the discharge trunnion. The pulp can flow freely through the openings in the grate and is then lifted up to the level of the discharge trunnion (Figure 3.3). These mills have a lower pulp level than overflow mills, thus reducing the dwell time of particles in the mill, and very little overgrinding takes place in these mills. Grate discharge mills usually take a coarser feed than overflow mills and are not required to grind so finely, the main reason being that with many small balls forming the charge the grate open area plugs very quickly. The trunnion overflow mill is the simplest to operate and is used for most ball mill applications, especially for fine grinding and regrinding.

(3) Autogenous mills

Figure 3. 3 Grate discharge mill

There has been increasing emphasis during recent years towards the use of autogenous (AG) and semi-autogenous grinding (SAG) circuits. An AG mill is a tumbling mill that utilizes the ore itself as grinding media. The ore must contain sufficient competent pieces to act as grinding media. An SAG mill is an autogenous mill that utilizes steel balls in addition to the natural grinding media. Experience indicates that the ball charges used in SAG have generally been most effective in the range of 4%～15% of the mill volume.

On suitable ores, the advantages over conventional circuits include lower capital cost, the ability to handle wet and sticky material, relatively simple flowsheets, and the large size of available equipment, lower manpower requirements, and minimal grinding media expense. Autogenous grinding made a slow start in the 1950s , but by the 80s most large plants were using autogenous or SAG mills, and to reduce overall operating costs many older plants were at least considering replacing conventional mills with autogenous mills.

A schematic of various sections of the autogenous grinding mill is shown in Figure 3. 4. AG or SAG mills are defined by the aspect ratio of the mill shell design and the product discharge mechanisms. The aspect ratio is defined as the ratio of diameter to length. Aspect ratios generally fall into three main groups: high aspect ratio mills where the diameter is 1. 5～3 times of the length, "square mills" where the diameter is approximately equal to the length, and low aspect ratio mills where the length is 1. 5～3 times that of the diameter. Although Scandinavian and South African practice favours low aspect ratio AG or SAG mills, in North America and Australia AG or SAG mills are distinguished by high aspects. The largest SAG mill in present use is 12 m in diameter by 6. 1 m length driven by a motor power of more than 20 MW installed in an Australian gold-copper mine.

Inside the milling chamber the mill is lined with wearing plates held by lifter bars bolted to the shell. Lifter bars are essential to reduce slippage of the mill load, which causes rapid wear of the liners and also impairs the grinding action.

AG/SAG mills can handle feed ore as large as 200 mm, normally the product of the primary crusher or the run-of-mine ore, and achieve a product of 0. 1 mm in one piece of equipment. The particle size distribution of the product depends on the characteristics and

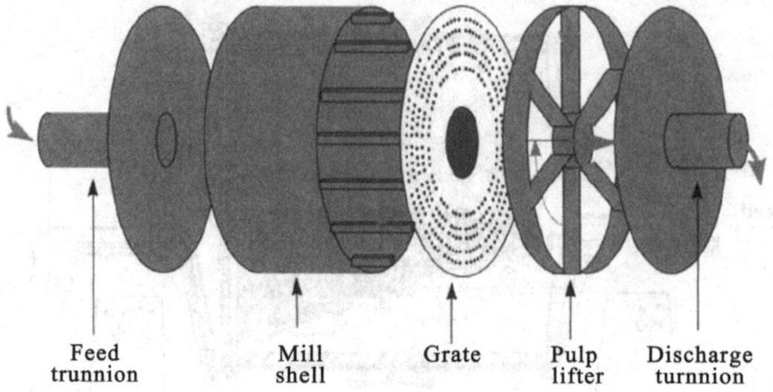

Figure 3.4 Schematic of various sections of an AG mill

structure of the ore being ground. The main mechanism of comminution in AG/SAG mills is considered to be abrasion and impact.

3. Stirred mills

(1) Tower mills

As tumbling mills such as ball and AG/SAG mills are not very effective in fine grinding due to their relatively low power intensity, alternatives for fine and ultra-fine grinding are tower mills and stirred mills. In contrast to tumbling mills where motion is imparted to the charge via the rotational mill shell, in tower and stirred mills motion is imparted to the charge by the movement of an internal stirrer while the shells are stationary. In a tower mill (Figure 3.5) such as that manufactured by Kubota in Japan and the Metso Vertimill, steel balls or pebbles are placed in a vertical grinding chamber in which an internal screw flight provides medium agitation. The feed enters at the top, with mill water, and is reduced in size by abrasion as it falls, the finely ground particles being carried upwards by pumped liquid and overflowing to a classifier. Oversize particles are returned to the bottom of the chamber, and final classification is by hydrocyclones, which return underflow to the mill sump for further grinding. According to the manufacturer's claims, advantages are a small installation area, low noise levels, efficient energy usage, minimal overgrinding and low capital and operating costs. Product sizes may be 1～100 μm at capacities of up to 100 t/h or more. The Japanese have been the primary users of these mills, grinding a variety of materials including limestone, silica, rock salt, coal and copper concentrates. The mill was introduced into the U. S. A. in 1979 for the purpose of grinding limestone, and has proved to provide a means of comminution in an area where the tumbling mill becomes inefficient.

(2) IsaMill

Despite the manufacturers claim that tower mills can achieve a product size of 1 μm, they are generally used at the coarse end of the fine grinding spectrum due to the use of coarse media (around 6 mm), and operation of their stirrer at a relatively slow velocity (less than 3 m/s tip speed). Finer grinding is normally achieved with stirred mills. These

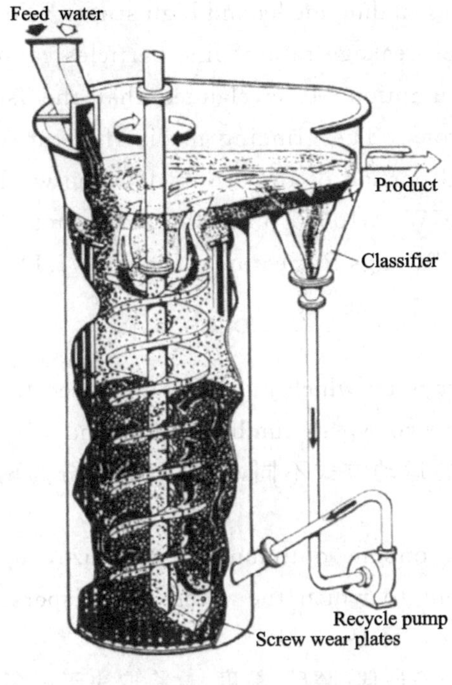

Figure 3.5 Tower mill

employ stirrers comprising a shaft with pins or disks. Stirred mills can be classified by their shell orientations: vertical such as the Sala Agitated Mill (SAM) or horizontal such as the IsaMill.

The IsaMill grinding chamber is a horizontally mounted shell, with a total volume of 10,000 litres in the currently largest IsaMill M10000. Inside the shell are rotating grinding discs mounted on a shaft which is coupled to a motor and gearbox (Figure 3.6). The grinding discs agitate the media and ore particles in a slurry that is continuously fed into the feed port. A patented product separator keeps the media inside the mill allowing only the product to exit.

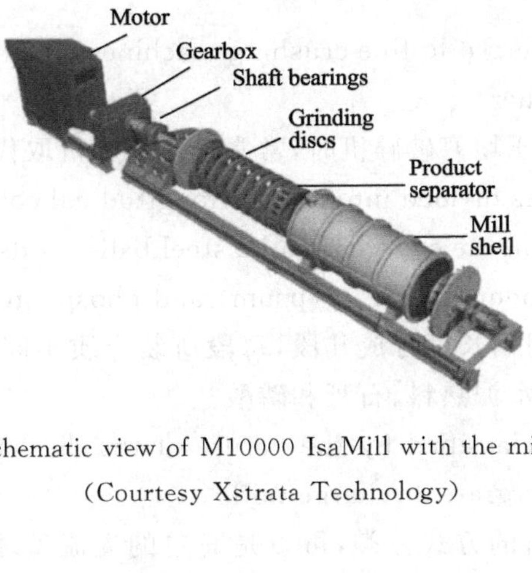

Figure 3.6 Schematic view of M10000 IsaMill with the mill shell removed
(Courtesy Xstrata Technology)

The IsaMill uses small grinding media and high stirrer velocity to impart energy to the media, which increases the breakage rate of fine particles caused by attrition/abrasion at relatively low power consumption. It is claimed that the IsaMill can efficiently grind minerals to below 10 microns. The grinding media that can be used include granulated slag, river sand, or a sized portion of the ore itself. The world's then largest ultra-fine grinding mill, the 2. 6 MW M10000 IsaMill, was commissioned in 2003 at Anglo Platinum's Western Limb Tailings Retreatment Plant (WLTRP) in South Africa.

Notes

1. According to the ways by which motion is imparted to the charge, grinding mills are generally classified into two types: tumbling mills and stirred mills.

根据对研磨物料施加的运动方式不同,磨机一般可分为两种类型:滚筒磨机和搅拌磨机。

2. All ores have an economic optimum particle size, which will depend on many factors, including the extent to which the values are dispersed in the gangue, and the subsequent separation process to be used.

所有矿石都具有一个经济最佳(磨矿)粒度,这将取决于许多因素,包括有用组分在脉石中的分散程度,以及随后要使用的分选方法。

3. It is the purpose of the grinding section to exercise close control on this product size and, for this reason, correct grinding is often said to be the key to good mineral processing.

磨矿阶段的目的就是严格控制产品的粒度,因此,准确的磨矿细度通常被认为是实现高效分选的关键。

4. The diameter of the mill determines the pressure that can be exerted by the medium on the ore particles and, in general, the larger the feed size the larger needs to be the mill diameter.

磨矿机的直径决定着介质可施加于矿石颗粒的压力。一般而言,给矿粒度愈大,所需磨矿机的直径也愈大。

5. They are often preferred to fine crushing machines when the ore is clayey or damp, thus tending to choke crushers.

当矿石发黏或潮湿,易于堵塞破碎机时,常常选用棒磨机取代细碎机。

6. These are sometimes divided into several longitudinal compartments, each having a different charge composition; the charges can be steel balls, rods, or pebbles, and they are often used dry to grind cement clinker, gypsum, and phosphate.

这些机器(管磨机)有时沿长度分成几段,每段所装介质不同,介质可以是钢球、钢棒或砾石,而且它们常用于干磨水泥熟料、石膏和磷酸盐。

7. Ball mills are also classified by the nature of the discharge. They may be simple trunnion overflow mills, or grate discharge mills.

球磨机也可以按照排料的方式分类,可以是简单的溢流型球磨机,也可以是格子型球磨机。

8. An AG mill is a tumbling mill that utilizes the ore itself as grinding media. The ore must contain sufficient competent pieces to act as grinding media.

自磨机是一种以矿石本身作为磨矿介质的滚筒磨机,矿石中必须含有足够多的能作为磨矿介质的矿石颗粒。

Exercises

1. Answer the following questions according to the text.

(1) In the process of comminution, what can be used as the grinding medium?

(2) There are three types of tumbling mills, what are they?

(3) What's the distinctive feature of a rod mill?

(4) Why are some tumbling mills called ball mills?

(5) How can comminution be achieved in autogenous mills and semi-autogenous mills.

2. Translate the following phrases into English.

(1) 球磨机	(2) 搅拌磨	(3) 自磨机	(4) 再磨
(5) 棒磨机	(6) 研磨介质	(7) 比表面积	(8) 给矿粒度
(9) 长径比	(10) 过磨		

3. Translate the following paragraphs into Chinese.

Paragraph 2 in section 1 "According to the ways by which motion is imparted to the charge ...".

4. Translate the following sentences into English.

(1) 磨矿是粉碎过程中的最后一个阶段,在这一阶段,颗粒受到冲击和磨损的联合作用而破碎,可以是干法也可以是悬浮在水中。

(2) 滚筒磨机有三种基本类型:棒磨机、球磨机和自磨机。

(3) 自磨/半自磨磨机的主要粉碎机理是研磨和冲击作用。

(4) 搅拌磨可以根据它们的外壳方向分类,包括:立式磨机,如萨拉搅拌磨(SAM);卧式磨机,如艾萨磨。

3.2 Grinding Circuits(磨矿流程)

1. Open and closed circuits

(1) Open circuits

The type of mill for a particular grind, and the circuit in which it is to be used, must be considered simultaneously. Circuits are divided into two broad classifications: open and closed. In open circuit the material is fed into the mill at a rate calculated to produce the correct product in one pass (Figure 3.7). This type of circuit is rarely used in mineral processing applications as there is no control on product size distribution. The feed rate must be low enough to ensure that every particle spends enough time in the mill to be broken down to product size. As a result, many particles in the product are overground, which consumes energy needlessly, and the product may subsequently be difficult to treat.

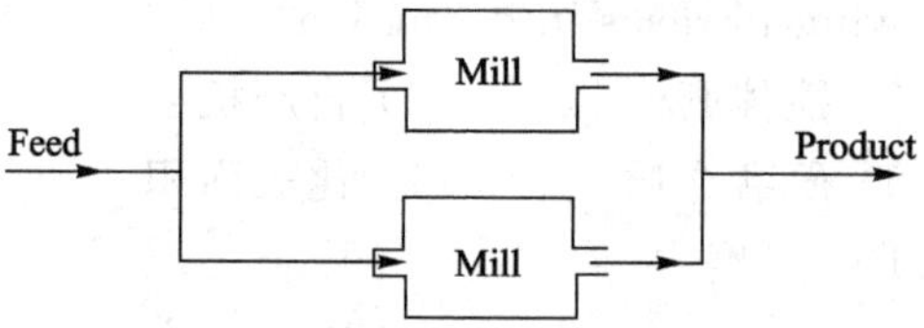

Figure 3.7 Single-stage open circuit

(2) Closed circuits

Grinding in the mining industry is almost always in closed circuit (Figure 3.8), in which material of the required size is removed by a classifier, which returns oversize to the mill. In closed-circuit operation, every effort is made to remove material from the circuit as soon as it reaches the required size. When grinding to a specified size, an increase in capacity of up to 35% can be obtained by closed-circuit operation.

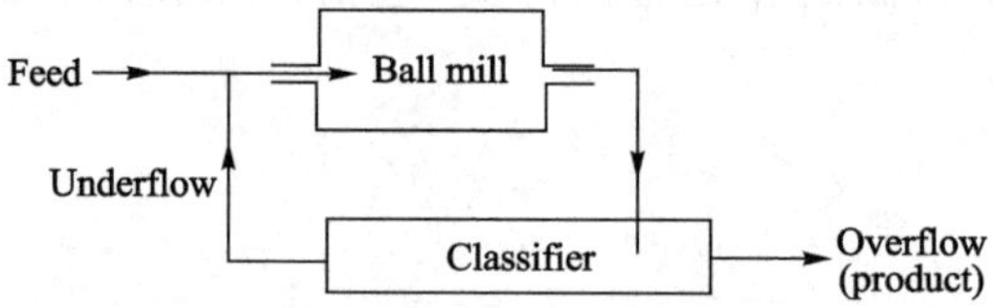

Figure 3.8 Simple closed-grinding circuit

Rod mills are generally operated in open circuit, especially when preparing feed for ball mills. The parallel grinding surfaces of rod mills simulate a slotted screen and tend to retard the larger particles until they are broken. Smaller particles slip through the spaces between the rods and are discharged without appreciable reduction. Ball mills, however, are virtually always operated in closed circuit with some form of classifier.

(3) Circulating load

The material returned to the mill by the classifier is known as the *circulating load*, and its weight is expressed as a percentage of the weight of new feed.

Closed-circuit grinding reduces the residence time of particles in each pass, and so the proportion of finished sizes in the mill, compared with open circuit grinding. This decreases overgrinding and increases the energy available for useful grinding. The optimum circulating load is usually in the range of 100%～350%, although it can be as high as 600%.

(4) Classifiers in closed circuits

Various types of classifying device can be used to close the circuit, mechanical classifiers being used in many of the older mills. They are robust, easy to control, and smooth out, and tolerate surges well. They can handle very coarse sands products and are still used on many coarse-grinding circuits. They suffer from the severe disadvantage of classifying by gravitational force, which restricts their capacity when dealing with extremely fine material.

Hydrocyclones classify by centrifugal action, which speeds up the classification of fine particles, giving much sharper separations and increasing the optimum circulating load. They occupy much less floor space than mechanical classifiers of the same capacity and have lower capital and installation costs. Due to their much quicker action, the grinding circuit can rapidly be brought into balance if changes are made, and the reduced dwell time of particles in the return load gives less time for oxidation to occur, which is important with sulphide minerals which are to be subsequently floated. They have therefore become widely used in fine grinding circuits preceding flotation operations.

The action of all classifiers in grinding circuits is dependent on the differential settling rates of particles in a fluid, which means that particles are classified not only by size but also by specific gravity. A small dense particle may therefore behave in a similar way to a large, low-density particle. Thus, when an ore containing a heavy valuable mineral is ground, overgrinding of this material is likely to occur, due to it being returned in the circulating load, even though it is below the required product size.

Selective grinding of heavy sulphides in this manner prior to flotation can allow a coarser overall grind, the light gangue reporting to the classifier overflow, while the heavy particles containing valuable mineral are selectively reground. Such circuits are often closed by screens rather than classifiers. Fine screens have the disadvantages, however, of being relatively inefficient and delicate, and often a combination of classification and screening is used to reduce the load on the screens.

Such a circuit was used at Wheal Jane Ltd, UK, for the primary grinding of the complex copper-zinc-tin ore. The secondary crusher product was fed on to a wedge-wire screen, which routed the oversize into the primary ball mill; the mill discharge was pumped to a cyclone, and the underflow passed on to the screen to remove any fine heavy minerals before being returned to the ball mill (Figure 3.9).

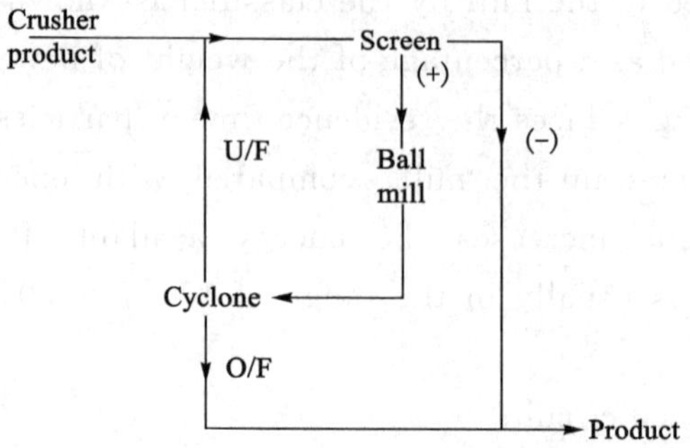

Figure 3.9 Cyclone and screen in the closed circuit

(5) Multi-stage grinding circuits

Grinding circuits are fed at a uniform rate from bins holding the crusher plant product. There may be a number of ball mills in parallel, each circuit being closed by its own classifier, and taking a definite fraction of the feed.

Parallel mill circuits increase circuit flexibility, since individual units can be shut down or the feed rate can be changed, with little effect on the flowsheet. Fewer mills are, however, easier to control and capital and installation costs are lower, so the optimum number of mills must be decided at the design stage.

Multi-stage grinding in which mills are arranged in series can be used to produce successively finer products but the present trend is towards large single-stage primary ball mills, which greatly reduces capital and operating costs, and facilitates automatic control. The disadvantage of single-stage milling is that if too high a reduction ratio is produced, then relatively large balls are required for coarse feed, which may not efficiently grind the finer particles.

Two-stage grinding is used where rod milling is substituted for tertiary crushing (Figure 3.10). The crusher product is fed to the rod mills, which operate in open circuit, producing ball-mill feed. This should, if capacity permits, be fed to the classifier rather than directly to the ball mill, as any finished material is immediately removed.

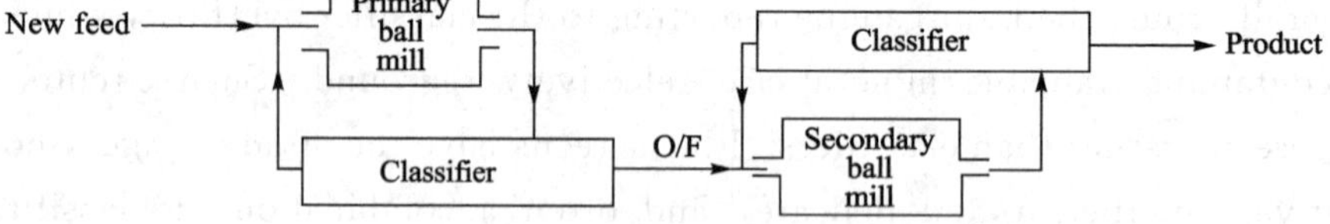

Figure 3.10 Two-stage grinding circuit

2. SAG milling circuits

(1) Closed circuit configurations

In recent years SAG milling has revolutionized the mining industry. Its high unit capacity has contributed towards substantial savings in capital and operating costs, which has in turn made many low-grade, high-tonnage operations such as copper and gold ores grinding economically feasible. SAG milling is a mature technology, and there is now a generation of metallurgists who believe that "conventional" grinding implies a SAG ball-mill circuit instead of three stages of crushing followed by rod and ball milling.

The AG/SAG mills can be operated in open or closed circuit. However, even in open circuit a coarse classifier such as a trommel attached to the mill, or a vibrating screen can be used. The oversize material is recycled either externally or internally. The external recycling can be achieved by conveyor belt in a continuous manner or in a batch mode where the material is stockpiled and periodically fed back into the mill by front-end loader. In the internal recycling the coarse material is conveyed by a reverse spiral or water jet back down the centre of the trommel into the mill.

There are two closed circuit configurations that are common in AG/SAG operations. A single stage AG or SAG mill in closed circuit with hydrocyclone is favoured in the South African gold industry to process the run-of-mine ore and to produce a final grind. The mill discharge is pumped into a cluster of hydrocyclones, and the hydrocyclone underflow, together with the fresh ore, feeds the mill. The South African RoM mills are often designed with a low aspect ratio (defined as a ratio of shell diameter to length), and operated at a high filling level and a high mill rotational speed. As a general rule an AG/SAG mill in closed circuit with a fine classifier will produce a finer product, but will have a lower throughput capacity than an open circuit unit of the same size treating the same ore.

Another circuit configuration is the AG/SAG mill closed with a recycle or pebble crusher. In

AG/SAG mill operation, the "critical size" of 25～50 mm is too small to act as grinding media to provide sufficient kinetic energy for other rock breakage, but is too large to be broken. If the critical size material is accumulated inside an AG/SAG mill, the mill energy efficiency will deteriorate, and the mill feed rate decreases. As a solution, additional large holes or pebble ports (40～100 mm) are cut into the mill grate allowing coarse material to exit the mill. The crusher in closed circuit is then used to reduce the size of the critical size material and return it to the mill. As the pebble ports also allow steel balls to exit the mill, a steel removal system (such as an electromagnet) must be installed to prevent them from entering the crusher. This circuit configuration has become increasingly popular in recent years as it usually produces a significant increase in throughput due to the removal of the critical size material. Compared with the closed circuit, however, it gives a coarser product.

(2) The SABC circuit

In many recently designed plants, the traditional three stages crushing followed by rod and ball milling circuit has been replaced by the popular SABC circuit, denoting a circuit comprising SAG mill followed by Ball mill closed with Cyclone. Figure 3.11 shows the comminution circuit at Newcrest Mining's Cadia Hill Gold Mine in New South Wales. The Cadia comminution circuit comprises a primary crusher, a single SAG mill, two pebble crushers and two parallel ball mills in closed circuit with cyclones.

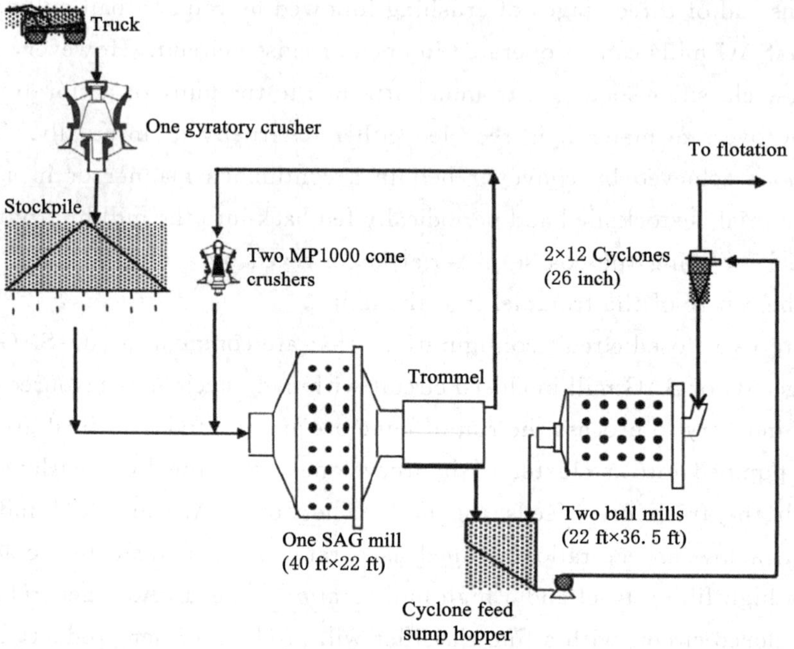

Figure 3.11 Diagram of the SABC circuit at Newcrest Mining's Cadia Hill Gold Mine

Ore from the open cut mine is directly dump-trucked into a Fuller 1.5 m×2.8 m (60 inch×110 inch) gyratory crusher with a gap setting of 130～200 mm. Maximum design capacity of the primary crusher is 5,800 t/h with a product size of 80% passing 200 mm. The primary crusher product is conveyed to a coarse ore stockpile with a live capacity of 41,000 t. A concrete tunnel under the stockpile houses three hydraulically driven belt feeders with variable speed drives. Each feeder draws ore from a mass flow hopper and has a capacity of 1,800 t/h. A minimum of two feeders operating at 82% capacity is required to feed the SAG mill. The belt feeders discharge onto the main SAG mill feed conveyor. This conveyor has a capacity of 3,700 t/h.

The SAG mill is of Metso design and 12 m diameter by 6.1 m in length (belly inside liners) operating in semi-autogenous mode. It is fitted with a 20 MW Siemens gearless drive motor with bi-directional rotational capacity. The SAG mill was designed to treat 2,065 t/h of monzonite ore at a ball charge of 8% volume, total filling of 25% volume, and an operating mill speed of 4% of critical. A 4.5 m diameter by 5.2 m long trommel screens the discharge product. Discharge material less than 12 mm falls directly into a cyclone feed sump hopper, where it is combined with discharge from the ball mills. Oversize pebbles from the trommel are conveyed to a surge bin of 735 t capacity, adjacent to the recycle crushers. Two Nordberg MP1000 cone crushers with a closed side setting gap of 12～16 mm are used to crush the pebbles with a designed product P_{80} being 12 mm and an expected total recycle pebble rate of 725 t/h. The crushed pebbles fall directly onto the SAG mill feed belt and return to the SAG mill.

SAG mill product feeds two parallel Metso ball mills of 6. 6 m×11. 1 m (internal diameter×length), each with a 9. 7 MW twin pinion drive. The ball mills are operated at a ball charge volume of 30%～32% and 78. 5% mill critical speed. The SAG mill trommel undersize combined with the ball mills discharge are pumped by two parallel cyclone feed pumps, each delivering to a pack of 12 Krebs 660 mm diameter cyclones. The cyclone underflow from each line reports to a ball mill while the cyclone overflow is directed to the flotation circuit. The designed ball milling circuit product is 80% passing 150 microns.

3. Control of the grinding circuit

The purpose of milling is to reduce the size of the ore particles to the point where economic liberation of the valuable minerals takes place. Thus, it is essential that not only should a mill accept a certain tonnage of ore per day, but should also yield a product that is of known and controllable particle size. The principal variables which can affect control are: changes in the new feed rate and the circulating load, size distribution and hardness of the ore, and the rate of water addition to the circuit. Also important are interruptions in the operation of the circuit, such as stoppages for such reasons as to feed new grinding medium or to clear a choked cyclone. For the purposes of stabilizing control, feed rate, densities, and circulating loads can be maintained at values which experience has shown will produce the required product, but this method fails when disturbances cause deviations from normal operation. Fluctuation in feed size and hardness are probably the most significant factors disturbing the balance of a grinding circuit. Such fluctuations may arise from differences in composition, mineralization, grain size, and crystallisation of the ore from different parts of the mine, from changes in the crusher settings, often due to wear, and from damage to screens in the crushing circuit. Minor fluctuations can be smoothed out by blending material excavated at different points in space and time. Ore storage tends to smooth out variations providing that segregation does not take place in the bins, and the amount of storage capacity available depends on the nature of the ore (such as its tendency to oxidize) and on the mill economics.

The most important variables associated with a conventional grinding circuit are shown in Figure 3. 12. Of the variables shown, only the ore feed rate and the water addition rates can be varied independently, as the other variables depend upon and respond to changes in these items. It is therefore ore feed rate and water addition to the mill and classifier which are the major variables used to control grinding circuits. In recent years, it has been reported that slurry viscosity, rather than or in addition to the density, should be used as a key operating variable for milling circuit control. For a copper slurry, 1%

increase from 55% to 56% density results in a change in viscosity from 329 mPa · s to 390 mPa · s. The plant control system may not be able to distinguish the increment of 1% solids, but the viscometer can easily identify the associated change in viscosity. In a closed grinding circuit, since hydrocyclone performance exerts a significant influence on solids concentration and particle size of grinding slurry, and the solids concentration and particle size affect the rheological characteristics of the slurry, the hydrocyclone classifier could be used as a rheological controller for the grinding mill. The use of a variable speed pump feeding the classifiers introduces another independent variable, and has a significant effect on the stability of the circuit. Pump speed should, however, be viewed as a variable that provides the conditions under which particular control objectives can be achieved, rather than as a variable that actually achieves them.

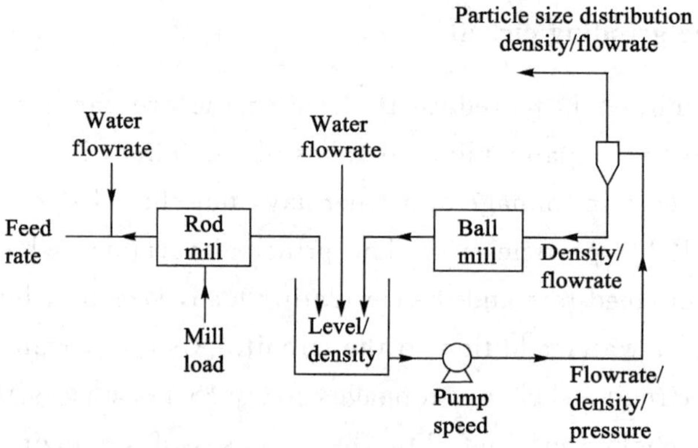

Figure 3.12 Grinding circuit control variables

Notes

1. The type of mill for a particular grind, and the circuit in which it is to be used, must be considered simultaneously. Circuits are divided into two broad classifications: open and closed.

对于特定的磨矿作业而言,必须同时考虑磨机的类型和磨矿的流程。磨矿流程可分为两大类:开路流程和闭路流程。

2. Grinding in the mining industry is almost always in closed circuit (Figure 3.8), in which material of the required size is removed by a classifier, which returns oversize to the mill.

在矿业生产中,磨矿几乎总是在闭路流程中进行的(图 3.8),该流程将达到所需粒度的物料由分级机脱除,而将粗颗粒返回磨矿机中。

3. They suffer from the severe disadvantage of classifying by gravitational force, which restricts their capacity when dealing with extremely fine material.

该设备(机械分级机)的主要缺点在于,依靠重力进行分级,在处理极细物料时其处理能力较低。

4. Hydrocyclones classify by centrifugal action, which speeds up the classification of fine particles, giving much sharper separations and increasing the optimum circulating load.

水力旋流器通过离心作用进行分级,这加快了细颗粒的分级,使分离更加准确,同时增加了最佳循环负荷。

5. The action of all classifiers in grinding circuits is dependent on the differential settling rates of particles in a fluid, which means that particles are classified not only by size but also by specific gravity.

在磨矿回路中,所有分级机的作用都取决于颗粒在流体中沉降速度的差异,这意味着颗粒分级不仅与粒度有关,还与密度有关。

6. Parallel mill circuits increase circuit flexibility, since individual units can be shut down or the feed rate can be changed, with little effect on the flowsheet.

并联磨矿回路增加了流程的灵活性,因为关闭个别磨矿单元或改变给矿速度,对流程几乎没有影响。

7. Its high unit capacity has contributed towards substantial savings in capital and operating costs, which has in turn made many low-grade, high-tonnage operations such as copper and gold ores grinding economically feasible.

由于半自磨技术的单位生产能力高,大大节省了投资和生产成本,从而使许多像铜矿石和金矿石这样低品位、大处理量的矿石的磨矿变得经济可行。

8. This circuit configuration has become increasingly popular in recent years as it usually produces a significant increase in throughput due to the removal of the critical size material.

这种流程结构(自磨/半自磨磨机采用循环破碎机构成闭路)近年来越来越受欢迎,因为它通常能去除临界粒度的颗粒而显著提高处理能力。

Exercises

1. Answer the following questions according to the text.

(1) What are the advantages of hydrocyclone as a classifier in a closed circuit?

(2) What are the disadvantages of fine screens?

(3) What are the advantages and disadvantages of single-stage milling?

(4) What are the common closed-circuit configurations in AG/SAG operations?

(5) What is SABC circuit?

2. Translate the following phrases into English.

(1) 循环负荷	(2) 闭路磨矿流程	(3) 分级机	(4) 半自磨磨矿流程
(5) 离心作用	(6) 并联磨矿回路	(7) 临界粒度	(8) 沉降速度

3. Translate the following paragraphs into Chinese.

Paragraphs 1 and 2 in section 1 (2) "Grinding in the mining industry is almost always ...".

4. Translate the following sentences into English.

（1）棒磨机通常在开路中运行，特别是在为球磨机准备给料时。

（2）分级机返回磨机的物料称为循环负荷，其质量表示为新进料质量的百分比。

（3）闭路磨矿流程可以使用各种类型的分级机，机械分级机在许多老选矿厂应用。

（4）磨矿的目的是将矿石颗粒的粒度减小到有用矿物经济解离的程度。

3.3 Principles of Classification(分级原理)

Classification is a method of separating mixtures of minerals into two or more products on the basis of the velocity with which the grains fall through a fluid medium. In mineral processing, this is usually water, and wet classification is generally applied to mineral particles which are considered too fine to be sorted efficiently by screening. Since the velocity of particles in a fluid medium is dependent not only on the size, but also on the specific gravity and shape of the particles, the principles of classification are important in mineral separations utilizing gravityconcentrators. Classifiers also strongly influence the performance of grinding circuits.

1. Terminal velocity

When a solid particle falls freely in a vacuum, it is subject to constant acceleration and its velocity increases indefinitely, being independent of size and density. Thus, a lump of lead and a feather fall at exactly the same rate.

In a viscous medium, such as air or water, there is resistance to this movement and the value increases with velocity. When equilibrium is attained between the gravitational and fluid resistance forces, the body reaches its terminal velocity and thereafter falls at a uniform rate.

The nature of the resistance depends on the velocity of the descent. At low velocities motion is smooth because the layer of fluid in contact with the body moves with it, while the fluid a short distance away is motionless. Between these two positions is a zone of intense shear in the fluid all around the descending particle. Effectively all resistance to motion is due to the shear forces or viscosity of the fluid and is hence called viscous resistance [Figure 3.13(a)]. At high velocities the main resistance is due to the displacement of fluid by the body, and viscous resistance is relatively small; this is known as turbulent resistance [Figure 3.13(b)].

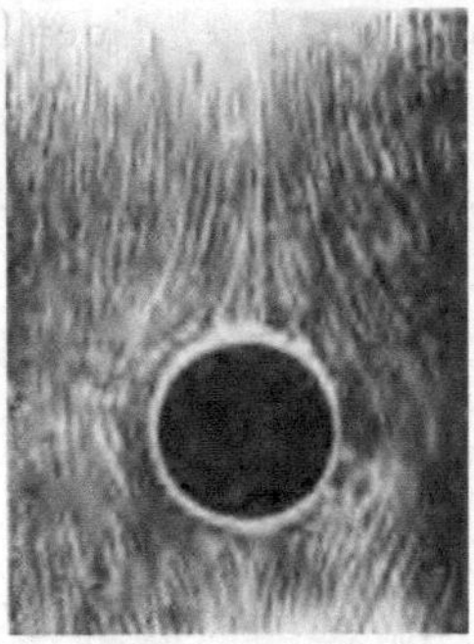

(a) Viscous resistance

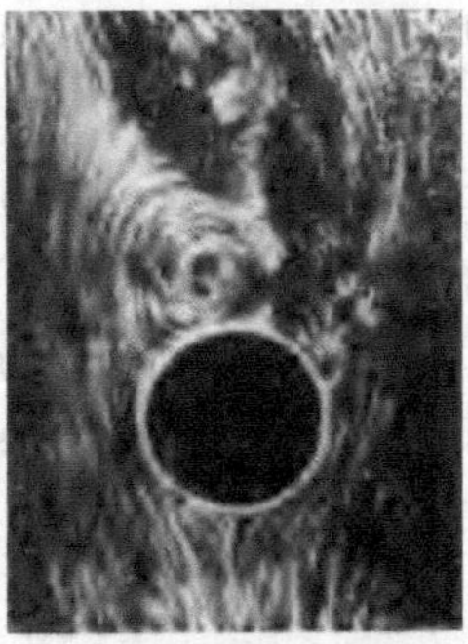

(b) Turbulent resistance

Figure 3.13

Whether viscous or turbulent resistance predominates, the acceleration of particles in a fluid rapidly decreases and the terminal velocity is quickly reached.

Classifiers consist essentially of a sorting column in which a fluid is rising at a uniform rate (Figure 3.14). Particles introduced into the sorting column either sink or rise according to whether their terminal velocities are greater or lesser than the upward velocity of the fluid. The sorting column therefore separates the feed into two products—an overflow consisting of particles with terminal velocities lesser than the velocity of the fluid and an underflow or spigot product of particles with terminal velocities greater than the rising velocity.

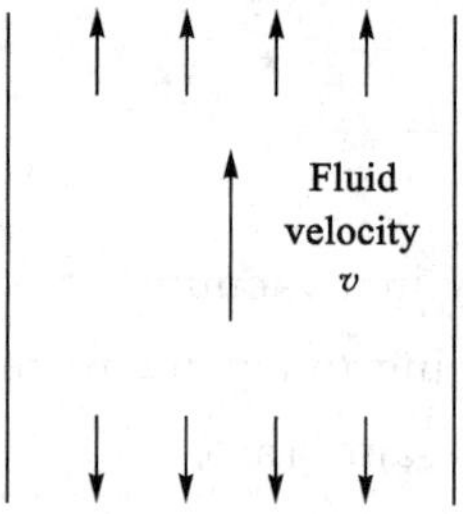

Figure 3.14 Classifier sorting column

2. Free settling

(1) Stokes' law

Free settling refers to the sinking of particles in a volume of fluid which is large with respect to the total volume of particle, hence particle crowding is negligible. For well-dispersed ore pulps, free settling predominates when the percentage of solids by weight is less than about 15.

Consider a spherical particle of diameter d and density D_s falling under gravity in a viscous fluid of density D_f under free-settling conditions, i.e., ideally in a fluid of infinite extent. The particle is acted upon by three forces: a gravitational force acting downwards, an upward buoyant force due to the displaced fluid, and a drag force D acting upwards. The equation of motion of the particle is

$$mg - m'g - D = m\mathrm{d}x/\mathrm{d}t \tag{3.1}$$

where m is the mass of the particle, m' is the mass of the displaced fluid, x is the particle velocity, and g is the acceleration due to gravity.

When the terminal velocity is reached, $\mathrm{d}x/\mathrm{d}t=0$, and hence $D=(m-m')g$.

Therefore

$$D = (\pi/6)gd^3(D_s - D_f) \tag{3.2}$$

Stokes assumed the drag force on a spherical particle to be entirely due to viscous resistance and deduced the expression

$$D = 3\pi d\eta v \tag{3.3}$$

where η is the fluid viscosity and v is the terminal velocity.

Hence, substituting in Equation (3.2) gives

$$3\pi d\eta v = (\pi/6) g d^3 (D_s - D_f)$$

and

$$v = g d^2 (D_s - D_f)/18\eta \tag{3.4}$$

This expression is known as Stokes' law.

(2) Newton's law

Newton assumed that the drag force was entirely due to turbulent resistance, and deduced:

$$D = 0.055\pi d^2 v^2 D_f \tag{3.5}$$

Substituting in Equation (3.2) gives

$$v = [3gd(D_s - D_f)/D_f]^{1/2} \tag{3.6}$$

This is Newton's law for turbulent resistance.

Stokes' law is valid for particles below about 50 μm in diameter. The upper size limit is determined by the dimensionless Reynolds number. Newton's law holds for particles larger than about 0.5 cm in diameter. There is, therefore, an intermediate range of particle size, which corresponds to the range in which most wet classification is performed, in which neither law fits experimental data.

Stokes' law (3.4) for a particular fluid can be simplified to

$$v = k_1 d^2 (D_s - D_f) \tag{3.7}$$

and Newton's law (3.6) can be simplified to

$$v = k_2 [d(D_s - D_f)]^{1/2} \tag{3.8}$$

where k_1 and k_2 are constants, and $(D_s - D_f)$ is known as the effective density of a particle of density D_s in a fluid of density D_f.

Both laws show that the terminal velocity of a particle in a particular fluid is a function only of the particle size and density. It can be seen that:

① If two particles have the same density, then the particle with the larger diameter has the higher terminal velocity.

② If two particles have the same diameter, then the heavier particle has the higher terminal velocity.

(3) Free-settling ratio

Consider two mineral particles of densities D_a and D_b and diameters d_a and d_b respectively, falling in a fluid of density D_f, at exactly the same settling rate. Their terminal velocities must be the same, and hence from Stokes' law (3.7)

$$d_a^2 (D_a - D_f) = d_b^2 (D_b - D_f)$$

or

$$d_a/d_b = [(D_b - D_f)/(D_a - D_f)]^{1/2} \tag{3.9}$$

This expression is known as the free-settling ratio of the two minerals, i.e., the ratio of particle size required for the two minerals to fall at equal rates.

Similarly, from Newton's law (3.8), the free settling ratio of large particle is

$$d_a/d_b = [(D_b - D_f)/(D_a - D_f)] \quad (3.10)$$

Consider a mixture of galena (density 7.5) and quartz (density 2.65) particles classifying in water. For small particles, obeying Stokes' law, the free settling ratio [Equation (3.9)] is

$$[(7.5-1)/(2.65-1)]^{1/2} = 1.99$$

i.e., a small particle of galena will settle at the same rate as a small particle of quartz which has a diameter 1.99 times as large.

For large particles obeying Newton's law, the free-settling ratio [Equation (3.10)] is

$$(7.5-1)/(2.65-1) = 3.94$$

The free-settling ratio is therefore larger for coarse particles obeying Newton's law than for fine particles obeying Stokes's law. This means that the density difference between the mineral particles has a more pronounced effect on classification at coarser size ranges. This is important where gravity concentration is being utilized. Overgrinding of the ore must be avoided, such that particles are fed to the separator in as coarse a state as possible, so that a rapid separation can be made, exploiting the enhanced effect of specific gravity difference. The enhanced gravity effect does, however, mean that fine heavy minerals are more likely to be overground in conventional ball mill-classifier circuits, so it is preferable where possible to use rod mills for primary coarse grind.

The general expression for free-settling ratio can be deduced from Equations (3.9) and (3.10) as

$$d_a/d_b = [(D_b - D_f)/(D_a - D_f)]^n \quad (3.11)$$

where $n=0.5$ for small particles obeying Stokes's law and $n=1$ for large particles obeying Newton's law.

The value of n lies in the range 0.5～1 for particles in the intermediate size range of 50 μm～0.5 cm.

3. Hindered settling

(1) Terminal velocity

As the proportion of solids in the pulp increases, the effect of particle crowding becomes more apparent and the falling rate of particles begins to decrease. The system begins to behave as a heavy liquid whose density is that of the pulp rather than that of the carrier liquid; hindered-settling conditions now prevail. Because of the high density and viscosity of the slurry through which a particle must fall in a separation by hindered settling, the resistance to fall is mainly due to the turbulence created, and a modified form of Newton's law can be used to determine the approximate falling rate of the particles

$$v = k[d(D_s - D_p)]^{1/2} \quad (3.12)$$

where D_p is the pulp density.

The lower the density of the particle the more marked is the effect of reduction of the effective density, $D_s - D_p$, and the greater is the reduction of falling rate. Similarly, the larger the particle, the greater is the reduction in falling rate as the pulp density increases.

This is important in classifier design; in effect, hindered-settling reduces the effect of size, while increasing the effect of density on classification.

(2) Hindered-settling ratio

This is illustrated by considering a mixture of quartz and galena particles settling in a pulp of density 1.5. The hindered-settling ratio can be derived from Equation (3.12) as

$$d_a/d_b = (D_b - D_p)/(D_a - D_p) \quad (3.13)$$

Therefore, in this system

$$d_a/d_b = (7.5 - 1.5)/(2.65 - 1.5) = 5.22$$

A particle of galena will thus fall in the pulp at the same rate as a particle of quartz which has a diameter 5.22 times as large. This compares with the free-settling ratio, calculated as 3.94 for turbulent resistance.

The hindered-settling ratio is always greater than the free-settling ratio, and the denser the pulp, the greater is the ratio of the diameter of equal settling particles. For quartz and galena, the greatest hindered-settling ratio that we can attain practically is about 7.5. Hindered-settling classifiers are used to increase the effect of density on the separation, whereas free-settling classifiers use relatively dilute suspensions to increase the effect of size on the separation (Figure 3.15). Relatively dense slurries are fed to certain gravity concentrators, particularly those treating heavy alluvial sands. This allows high tonnages to be treated, and enhances the effect of specific gravity difference on the separation. The efficiency of separation, however, may be reduced since the viscosity of a slurry increases with density. For separations involving feeds with a high proportion of particles close to the required density of separation, lower slurry densities may be necessary, even though the density difference effect is reduced.

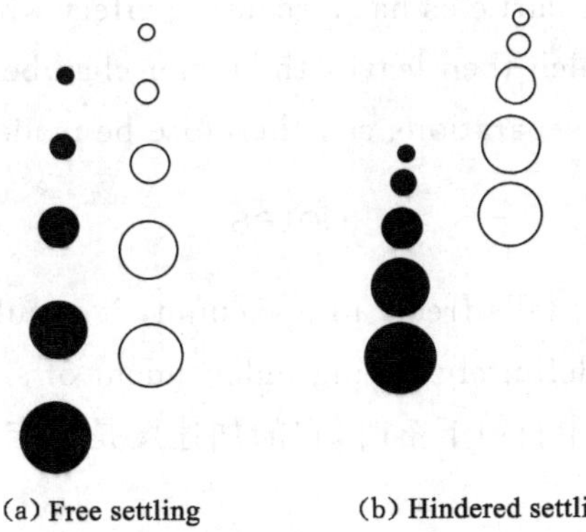

Figure 3.15

As the pulp density increases a point is reached where each mineral particle is covered only with a thin film of water. This condition is known as a quicksand, and because of surface tension, the mixture is a perfect suspension and does not tend to separate. The solids are in a condition of full teeter, which means that each grain is free to move, but is unable to do so without colliding with other grains and as a result stay in place. The mass acts as a viscous liquid and can be penetrated by solids with higher specific gravity than that of the mass, which will then move at a velocity impeded by the viscosity of the mass.

A condition of teeter can be produced in a classifier sorting column by putting a constriction in the column, either by tapering the column or by inserting a grid into the base (Figure 3.16).

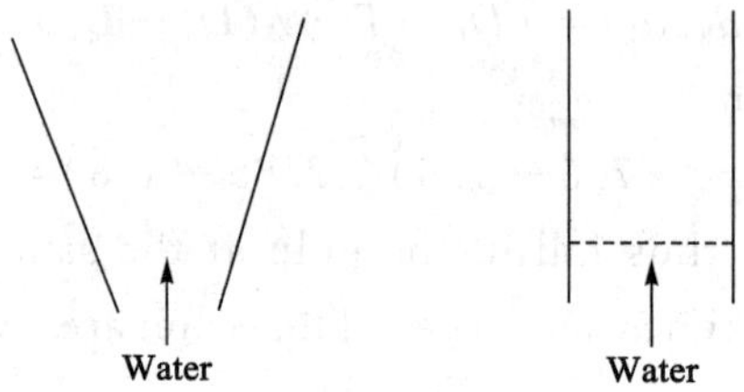

Figure 3.16 Teeter chambers

Such hindered-settling sorting columns are known as teeter chambers. Due to the constriction, the velocity of the introduced water current is greatest at the bottom of the column. A particle falls until it reaches a point where its falling velocity equals that of the rising current. The particle can now fall no further. Many particles reach this condition, and as a result, a mass of particles becomes trapped above the constriction and pressure builds up in the mass. Particles move upward along the path of least resistance, which is usually the centre of the column, until they reach a region of lower pressure at or near the top of the settled mass; here, under conditions in which they previously fell, they fall again. As particles from the bottom rise at the centre, those from the sides fall into the resulting void. A general circulation is built up, the particles being said to teeter. The constant jostling of teetering particles has a scouring effect which removes any entrained or adhering slime particles, which then leaves the teeter chamber and passes out through the classifier overflow. Cleaner separations can therefore be made in such classifiers.

Notes

1. When a solid particle falls freely in a vacuum, it is subject to constant acceleration and its velocity increases indefinitely, being independent of size and density.

当一个固体颗粒在真空中自由下落时，它的加速度是恒定的，它的速度会无限地增加，与大小和密度无关。

2. When equilibrium is attained between the gravitational and fluid resistance forces, the body reaches its terminal velocity and thereafter falls at a uniform rate.

当重力和流体阻力之间达到平衡时，物体就会达到它的末速度，然后匀速下落。

3. The sorting column therefore separates the feed into two products — an overflow consisting of particles with terminal velocities lesser than the velocity of the fluid and an

underflow or spigot product of particles with terminal velocities greater than the rising velocity.

分级柱把给矿分离成两种产品：一种是溢流产品，其由末速比液体上升速度小的颗粒组成；另一种是底流或沉砂产品，其由末速大于上升水流速度的颗粒组成。

4. Free settling refers to the sinking of particles in a volume of fluid which is large with respect to the total volume of particle, hence particle crowding is negligible.

自由沉降是指颗粒在相对于颗粒总体积较大的流体中沉降，颗粒之间拥挤可以忽略不计。

5. The system begins to behave as a heavy liquid whose density is that of the pulp rather than that of the carrier liquid; hindered-settling conditions now prevail.

矿浆体系开始变得如同重液，其密度主要是矿浆的密度而不是载流液体的密度，此时干涉沉降条件占优势。

6. A general circulation is built up, the particles being said to teeter. The constant jostling of teetering particles has a scouring effect which removes any entrained or adhering slime particles, which then leaves the teeter chamber and passes out through the classifier overflow.

这样就形成了一个整体环流，矿粒呈现出摇摆悬浮状态。摇摆颗粒的不断碰撞具有冲刷作用，能去除任何夹带或黏附的矿泥颗粒，使之离开悬浮室，通过分级机溢流流出。

Exercises

1. Answer the following questions

(1) What is classification?

(2) What will sorting column do when particles are introduced?

(3) What is free settling?

(4) What is the difference between free-settling and hindered-settling?

(5) How to produce teeter condition in a classifier sorting column?

2. Translate the following phrases into English.

(1) 分级	(2) 末速度	(3) 干涉沉降	(4) 自由沉降
(5) 分级柱	(6) 矿浆密度	(7) 等沉颗粒	(8) 黏滞阻力

3. Translate the following paragraphs into Chinese.

Paragraph 1 of the text“Classification is a method of separating mixtures of ...”.

4. Translate the following sentences into English.

(1) 对于特定流体，斯托克斯定律可简化。

(2) 事实上，干涉沉降在分级中减小了粒度的影响，但增强了密度的影响。

(3) 干涉沉降分级机用于增加密度对分离的影响，而自由沉降分级机使用相对稀释的悬浮液来增加粒度对分离的影响。

(4) 当矿浆中固体的比例增加时，颗粒拥挤的影响便会变得更加明显，颗粒下降的速度开始减小。

(5) 干涉沉降比总是大于自由沉降比，而且浆体密度越大，等降颗粒的直径之比越大。

3.4 Classifiers(分级机)

Many different types of classifier have been designed and built. They may be grouped, however, into two broad classes depending on the direction of flow of the carrying current. Horizontal current classifiers such as mechanical classifiers are essentially of the free-settling type and accentuate the sizing function; vertical current or hydraulic classifiers are usually hindered settling types and so increase the effect of density on the separation.

The hydrocyclone is also one of the most important classifiers in the minerals industry, which utilizes centrifugal force to accelerate the settling rate of particles.

1. Vertical current classifiers-Hydraulic classifiers

These are characterized by the use of water additional to that of the feed pulp, introduced so that its direction of flow opposes that of the settling particles. They normally consist of series of sorting columns through each of which a vertical current of water is rising and particles are settling out (Figure 3.17).

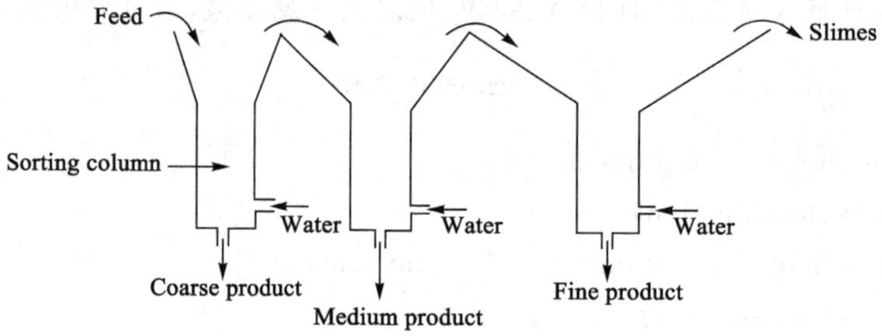

Figure 3.17 Principles of hydraulic classifier

The rising currents are graded from a relatively high velocity in the first sorting column, to a relatively low velocity in the last, so that a series of spigot products can be obtained, with the coarser, denser particles in the first spigot and the fines in the latter spigots. Very fine slimes overflow the final sorting column of the classifier. The size of each successive vessel is increased, partly because the amount of liquid to be handled includes all the water used for classifying in the previous vessels and partly because it is desired to reduce, in stages, the surface velocity of the fluid flowing from one vessel to the next.

The greatest use for hydraulic classifiers in the mineral industry is for sorting the feed to certain gravity concentration processes. Such classifiers are of the hindered-settling type. These differ from the free-settling classifiers in that the sorting column is constricted at the bottom in order to produce a teeter chamber. The hindered-settling classifier uses much less water than the free-settling type, and is more selective in its action, due to the scouring action in the teeter chamber, and the buoyancy effect of the pulp. Since the ratio of sizes of equally falling particles is high, the classifier is capable of performing a concentrating effect, and the first spigot product is normally of higher grade than the other products (Figure 3. 18).

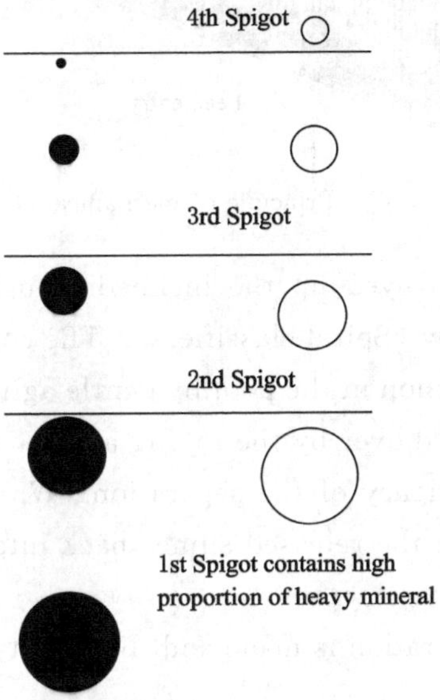

Figure 3. 18 Spigot products of hindered-settling classifier

2. Horizontal current classifiers-Mechanical classifiers

(1) The principle

Several forms of mechanical classifier exist in which the material of lower settling velocity is carried away in a liquid overflow, and the material of higher settling velocity is deposited on the bottom of the equipment and is dragged upwards against the flow of liquid by some mechanical means.

Mechanical classifiers have widespread use in closed-circuit grinding operations and in the classification of products from ore-washing plants. In washing plants, they act more or less as sizing devices, as the particles are essentially unliberated, so are of similar density. In closed-circuit grinding they have a tendency to return small dense particles to the mill, causing overgrinding. The principle of the mechanical classifier is shown in Figure 3. 19.

The pulp feed is introduced into the inclined trough and forms a settling pool in which particles of high falling velocity quickly fall to the bottom of the trough. Above this coarse sand is a quicksand zone where essentially hindered settling takes place. The depth and shape of this zone depends on the classifier action, and on the feed pulp density. Above the quicksand is a zone of essentially free settling material, comprising a stream of pulp flowing horizontally across the top of the quicksand zone from the feed inlet to the overflow weir, where the fines are removed.

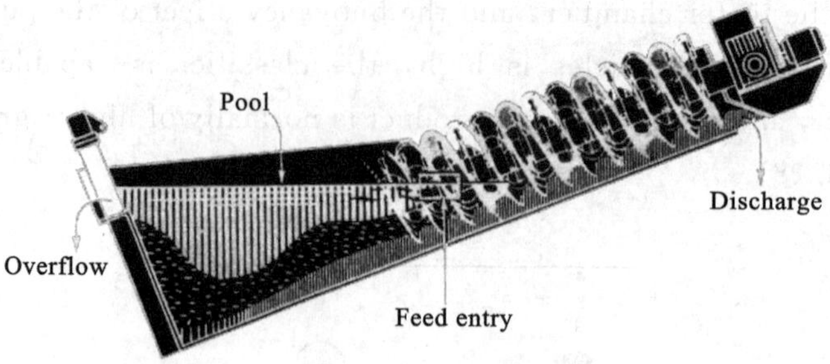

Figure 3. 19 Principle of mechanical classifier

The settled sands are conveyed up the inclined trough by mechanical rakes (Rake classifiers) or by a helical screw (Spiral classifiers). The conveying mechanism also serves to keep fine particles in suspension in the pool by gentle agitation and when the sands leave the pool, they are slowly turned over by the raking action, thus releasing entrained slimes and water, increasing the efficiency of the separation. Washing sprays are often directed on the emergent sands to wash the released slimes back into the pool.

(2) Effect of operational parameters

The size at which the separation is made and the quality of the separation depend on a number of factors.

Increasing the feed rate increases the horizontal carrying velocity and thus increases the size of particle leaving in the overflow. The speed of the rakes or spiral determines the degree of agitation of the pulp and the tonnage rate of sand removal. For coarse separations, a high degree of agitation may be necessary to keep the coarse particles in suspension in the pool, whereas for finer separations, less agitation and thus lower raking speeds are required.

The height of the overflow weir is an operating variable in some mechanical classifiers. Increasing the weir height increases the pool volume, and hence allows more settling time and decreases the surface agitation, thus reducing the pulp density at overflow level, where the final separation is made. High weirs are thus used for fine separations.

Dilution of the pulp is the most important variable in the operation of mechanical classifiers. In closed-circuit grinding operations, ball mills rarely discharge at less than 65% solids by weight, whereas mechanical classifiers never operate at more than about 50% solids. Water to control dilution is added in the feed launder. Water addition determines the rate of settling of the particles; increased dilution reduces the density of the weir overflow, and increases free settling, allowing finer particles to settle out of the influence of the horizontal current. Finer separations are thus produced, providing that the overflow pulp density is above a value known as the critical dilution, which is normally about 10% solids. Below this density, the effect of increasing rising velocity with dilution becomes more important than the increase in particle settling rates produced by decrease of pulp density. The overflow therefore becomes coarser with increasing dilution (Figure 3.20). In mineral processing applications, however, very rarely is the overflow density less than the critical dilution.

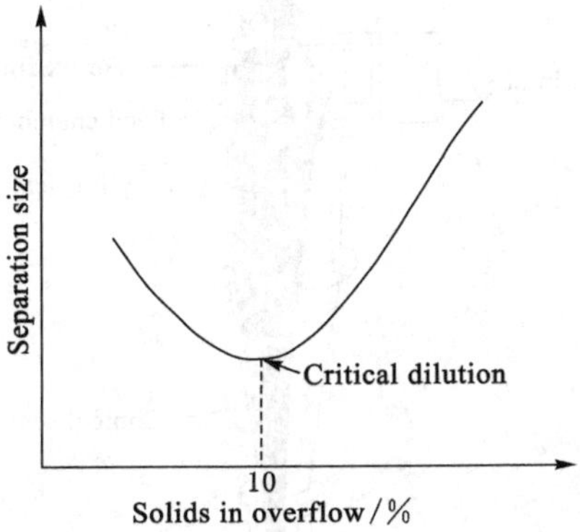

Figure 3.20 Effect of dilution of overflow on classifier separation

One of the major disadvantages of the mechanical classifier is its inability to produce overflows of very fine particle size at reasonable pulp densities. To produce such separations, the pulp may have to be diluted to such an extent to increase particle settling rates that it becomes too thin for subsequent operations.

3. The hydrocyclone

(1) The structure

The hydrocyclone is a continuously operating classifying device that utilizes centrifugal force to accelerate the settling rate of particles. It is one of the most important devices in the minerals industry, its main use in mineral processing being as a classifier, which has proved extremely efficient at fine separation sizes. It is widely used in closed-circuit grinding operations but has found many other uses, such as de-sliming, de-gritting, and thickening. It has also found wide acceptance recently for the washing of fine coal.

A typical hydrocyclone (Figure 3.21) consists of a conically shaped vessel, open at its apex, joined to a cylindrical section, which has a tangential feed inlet. The top of the cylindrical section is closed with a plate through which passes an axially mounted overflow pipe. The pipe is extended into the body of the cyclone by a short, removable section known as the vortex finder, which prevents short-circuiting of feed directly into the overflow. The feed is introduced under pressure through the tangential entry which imparts a swirling motion to the pulp. This generates a vortex in the cyclone, with a low-pressure zone along the vertical axis. An air core develops along the axis, normally connected to the atmosphere through the apex opening, but in part created by dissolved air coming out of solution in the zone of low pressure.

Hydroclones have replaced mechanical classifiers in many grinding plants, as they are more efficient, especially in the finer size ranges. They also require floor space and

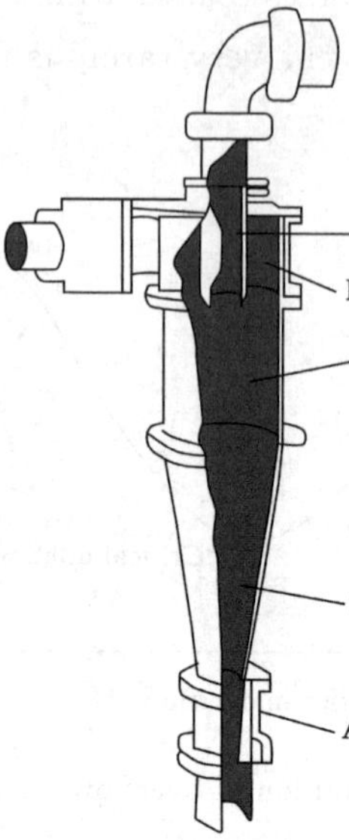

Figure 3. 21　Hydrocyclone (Courtesy of JKMRC)

consume less power. Due to the relatively short residence time of particles within the cyclone, the mill circuit can rapidly be brought into balance if any changes are made and oxidation of particles within the circuit is reduced. They are almost universally used for classification between 150 μm and 5 μm, although coarser separations are possible.

(2) The principle of classification

The classical theory of hydrocyclone action is that particles within the flow pattern are subjected to two opposing forces—an outward centrifugal force and an inwardly acting drag. The centrifugal force developed accelerates the settling rate of the particles thereby separating particles according to size, shape, and specific gravity. Faster settling particles move to the wall of the cyclone, where the velocity is lowest, and migrate to the apex opening. Due to the action of the drag force, the slower-settling particles move towards the zone of low pressure along the axis and are carried upward through the vortex-finder to the overflow.

The existence of an outer region of downward flow and an inner region of upward flow implies a position at which there is no vertical velocity. This applies throughout the greater part of the cyclone body, and an envelope of zero vertical velocity should exist throughout the body of the cyclone (Figure 3. 22). Particles thrown outside the envelope of zero vertical velocity by the greater centrifugal force exit via the underflow, while particles swept to the centre by the greater drag force leave in the overflow. Particles lying on the envelope of zero velocity are acted upon by equal centrifugal and drag forces and have an equal chance of reporting either to the underflow or overflow.

(3) Cyclone efficiency

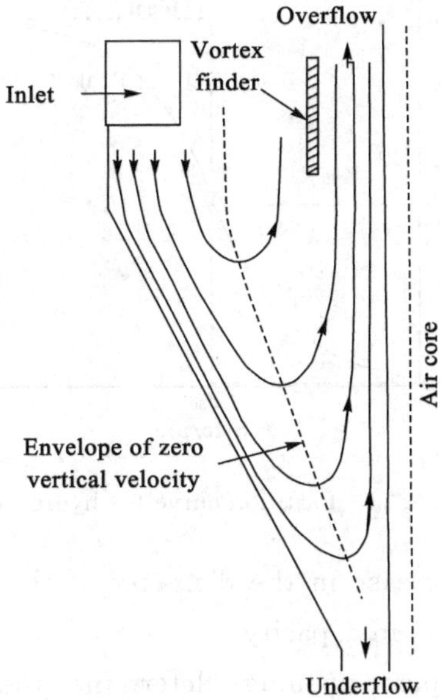

Figure 3. 22 Distribution of the vertical and radial components of velocity in a hydrocyclone

The commonest method of representing cyclone efficiency is by a performance or partition curve (Figure 3. 23), which relates the weight fraction, or percentage, of each particle size in the feed which reports to the apex, or underflow, to the particle size. The cut point, or separation size, of the cyclone is defined as the size for which 50% of the particles in the feed report to the underflow, i. e. , particles of this size have an equal chance of going either with the overflow or underflow (Svarovsky, 1984). This point is usually referred to as the d_{50} size. The sharpness of the cut depends on the slope of the central section of the partition curve; the closer to vertical is the slope, the higher is the efficiency. The slope of the curve can be expressed by taking the points at which 75% and 25% of the feed particles report to the underflow. These are the d_{75} and d_{25} sizes, respectively. The efficiency of separation, or the so-called imperfection I, is then given by $I=(d_{75}-d_{25})/(2d_{50})$.

In practice, the cut-point is mainly controlled by the cyclone design variables, such as inlet, vortex-finder, and apex openings, and most cyclones are designed such that these are easily changed.

The area of the inlet determines the entrance velocity and an increase in area increases the flow rate. Also important is the geometry of the feed inlet. In most cyclones the shape of the entry is developed from circular cross-section to rectangular cross-section at the entrance to the cylindrical section of the cyclone. This helps to "spread" the flow along the wall of the chamber. The inlet is normally tangential, but involuted feed entries are also common.

The diameter of the vortex finder is a very important variable. At a given pressure

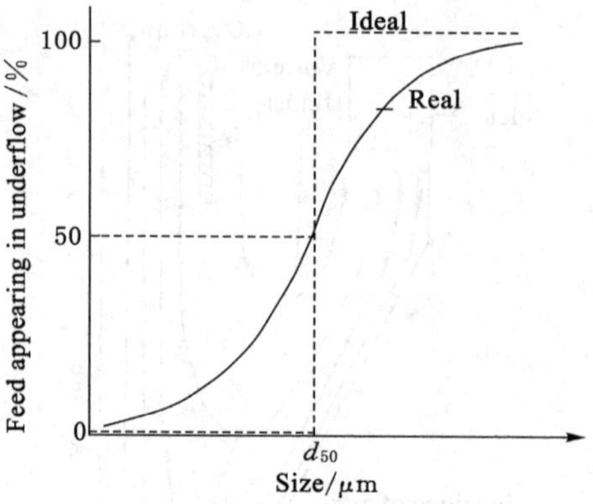

Figure 3.23 Partition curve for hydrocyclone

drop across the cyclone, an increase in the diameter of the vortex finder will result in a coarser cut-point and an increase in capacity.

The size of the apex, or spigot opening, determines the underflow density, and must be large enough to discharge the coarse solids that are being separated by the cyclone. The orifice must also permit the entry of air along the axis of the cyclone in order to establish the air vortex. Cyclones should be operated at the highest possible underflow density, since unclassified material leaves the underflow in proportion to the fraction of feed water leaving via the underflow. Under correct operating conditions, the discharge should form a hollow cone spray with a 20°～30° included angle (Figure 3.24). Air can then enter the cyclone, the classified coarse particles will discharge freely, and solids concentrations greater than 50% by weight can be achieved. Too small an apex opening can lead to the condition known as "roping", where an extremely thick pulp stream of the same diameter as the apex is formed, and the air vortex may be lost, the separation efficiency will fall, and oversize material will discharge through the vortex finder (This condition is sometimes encouraged where a very high underflow solids concentration is required, but is otherwise deleterious). Too large an apex orifice results in the larger hollow cone pattern seen in

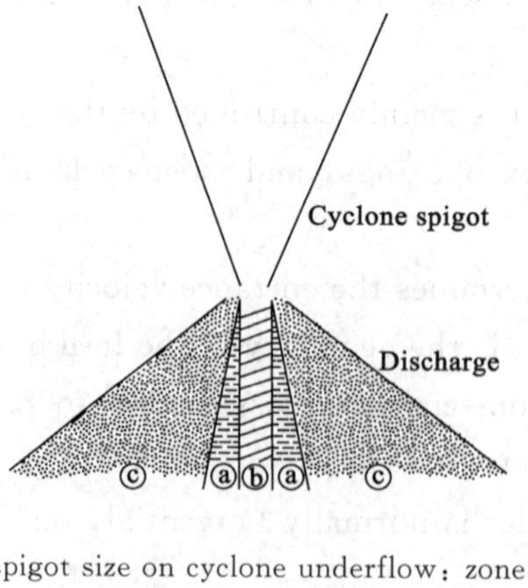

Figure 3.24 Effect of spigot size on cyclone underflow: zone ⓐ, correct operation; zone ⓑ, "roping"-spigot too small; zone ⓒ, excessively dilute-spigot too large

Figure 3.24. The underflow will be excessively dilute and the additional water will carry unclassified fine solids that would otherwise report to the overflow.

Notes

1. Horizontal current classifiers such as mechanical classifiers are essentially of the free-settling type and accentuate the sizing function; vertical current or hydraulic classifiers are usually hindered settling types and so increase the effect of density on the separation.

水平流分级机，如机械分级机，本质上是自由沉降类型并强调粒度对分离的影响；垂直水流或水力分级机通常是干涉沉降类型，因此增加了密度对分离的影响。

2. These are characterized by the use of water additional to that of the feed pulp, introduced so that its direction of flow opposes that of the settling particle.

水力分级机的特征是除使用矿浆外还要外加水，水的流动方向与矿粒沉降方向相反。

3. The rising currents are graded from a relatively high velocity in the first sorting column, to a relatively low velocity in the last, so that a series of spigot products can be obtained, with the coarser, denser particles in the first spigot and the fines in the latter spigots.

上升水流速度依次减小，第一分级柱内的流速最高，最末一个分级柱内的流速最低，这样便可获得一系列的分级产品：粒度粗、密度高的颗粒在第一分级柱内排出，细颗粒则在后续的分级柱内排出。

4. Increasing the feed rate increases the horizontal carrying velocity and thus increases the size of particle leaving in the overflow.

增加给矿速度增加了水平流的携带速度，从而增加了溢流中颗粒的粒度。

5. In mineral processing applications, however, very rarely is the overflow density less than the critical dilution.

然而，在矿物加工应用中，溢流浓度很少低于临界稀释浓度。

6. It is widely used in closed-circuit grinding operations but has found many other uses, such as de-sliming, de-gritting, and thickening. It has also found wide acceptance recently for the washing of fine coal.

它(水力旋流器)广泛应用于闭路磨矿作业，但也有许多其他用途，如脱泥、脱砂和浓缩。最近，在细粒煤的分选中也得到了广泛的应用。

7. The classical theory of hydrocyclone action is that particles within the flow pattern are subjected to two opposing forces—an outward centrifugal force and an inwardly acting drag.

水力旋流器作用的经典理论是，流体中的颗粒受到两种相反的力——向外的离心力和向内的阻力。

8. Particles thrown outside the envelope of zero vertical velocity by the greater centrifugal force exit via the underflow, while particles swept to the centre by the greater drag force leave in the overflow. Particles lying on the envelope of zero velocity are acted

upon by equal centrifugal and drag forces and have an equal chance of reporting either to the underflow or overflow.

由较大的离心力抛到轴向零速度包络面外的颗粒通过底流口排出，而被较大阻力掠至中心的颗粒则从溢流口排出。位于轴向零速度包络线上的颗粒受到相等的离心力和阻力的作用，并有均等的机会进入底流或溢流。

9. The commonest method of representing cyclone efficiency is by a performance or partition curve (Figure 3. 23), which relates the weight fraction, or percentage, of each particle size in the feed which reports to the apex, or underflow, to the particle size.

描述旋流器作业效率最常用的方法是性能曲线或分配曲线（图 3. 23），该曲线是底流中不同粒度颗粒占给料的质量百分比与粒度的关系曲线。

10. The cut point, or separation size, of the cyclone is defined as the size for which 50% of the particles in the feed report to the underflow, i. e. , particles of this size have an equal chance of going either with the overflow or underflow.

旋流器的分割点或分离粒度的定义是，给料中 50%的颗粒进入底流的粒度值，即该粒度的颗粒有相同的机会进入溢流或底流。

Exercise

1. Answer the following questions according to the text.

(1) What are the two main classes of classifiers?

(2) What are the characteristics of the hydraulic classifiers?

(3) What is the principle of mechanical classifiers?

(4) What is the principle of hydrocyclone?

(5) What is the classical theory of hydrocyclone action?

2. Translate the following phrases into English.

(1) 零速包络面　(2) 水力旋流器　(3) 水力分级机　(4) 沉砂产品

(5) 离心力　(6) 水平流　(7) 垂直流　(8) 机械分级机

3. Translate the following paragraphs into Chinese.

Paragraph 1 in section 3 (1)"The hydrocyclone is a continuously operating classifying ...".

4. Translate the following sentences into English.

(1) 干涉沉降分级机比自由沉降分级机耗水量少得多，由于悬浮室中的擦洗作用和浆料的浮力作用，其更具选择性。

(2) 在机械分级中，具有较高沉降速度的物料沉降到设备的底部，然后通过某些机械手段逆液体流动方向向上搬运。

(3) 机械分级机广泛应用于闭路磨矿作业和选矿厂产品的分级。

(4) 水力旋流器是一种连续作业的分级设备，它利用离心力来加速颗粒的沉降速度。

(5) 给料在压力条件下通过切向入口进入水力旋流器，引起矿浆产生旋涡运动。

(6) 机械分级机的主要缺点之一是它不能在合理的矿浆密度下产生非常细的溢流。

4 GRAVITY CONCENTRATION(重力分选)

4.1 Principle and Application of Gravity Concentration (重选原理及应用)

1. Introduction

Gravity concentration is the separation of minerals based upon the difference in density. Techniques of gravity concentration have been around for millennia.

Gravity methods of separation are used to treat a great variety of materials, ranging from heavy metal sulphides such as galena (sp. gr. 7.5) to coal (sp. gr. 1.3), at particle sizes in some cases below 50 μm.

These methods declined in importance in the first half of the twentieth century due to the development of the froth-flotation process, which allows the selective treatment of low-grade complex ores. They remained, however, the main concentrating methods for iron and tungsten ores and are used extensively for treating tin ores, coal and many industrial minerals.

In recent years, many companies have re-evaluated gravity systems due to increasing costs of flotation reagents, the relative simplicity of gravity processes, and the fact that they produce comparatively little environmental pollution. Modern gravity techniques have proved efficient for concentration of minerals having particle sizes in the 50 μm range and, when coupled with improved pumping technology and instrumentation, have been incorporated in high-capacity plants. In many cases a high proportion of the mineral in an orebody can at least be pre-concentrated effectively by cheap and ecologically acceptable gravity systems; the amount of reagents and fuel used can be cut significantly when the more expensive methods are restricted to the processing of gravity concentrate. Gravity separation of minerals at coarser sizes as soon as liberation is achieved can also have significant advantages for later treatment stages due to decreased surface area, more efficient dewatering, and the absence of adhering chemicals which could interfere with further processing.

Gravity techniques to recover residual valuable heavy minerals in flotation tailings are being increasingly used. Apart from current production, there are many large tailings dumps which could be excavated cheaply and processed to give high value concentrates using recently developed technology.

2. Principles of gravity concentration

Gravity concentration methods separate minerals of different specific gravity by their relative movement in response to gravity and one or more other forces, the latter often being the resistance to motion offered by a viscous fluid, such as water or air.

It is essential for effective separation that a marked density difference exists between the mineral and the gangue. Some idea of the type of separation possible can be gained from the concentration criterion, $\Delta\rho$.

$$\Delta\rho = (\rho_h - \rho_f)/(\rho_l - \rho_f) \tag{4.1}$$

where ρ_h is the density of the heavy mineral, ρ_l is the density of the light mineral, and ρ_f is the density of the fluid medium.

In general terms, if the quotient has a magnitude greater than 2.5, then gravity separation is relatively easy, with the efficiency of separation decreasing as the value of the quotient decreases. To give an example, if gold is being separated from quartz using water as the carrier fluid, the concentration criterion is 11.1 (density of gold being 19,300 kg/m^3; quartz being 2,650 kg/m^3), which is why panning for gold has been so successful.

The motion of a particle in a fluid is dependent not only on its specific gravity, but also on its size—large particles will be affected more than smaller ones. The efficiency of gravity processes therefore increases with particle size, and the particles should be sufficiently coarse to move in accordance with Newton's law. Particles small enough that their movement is dominated mainly by surface friction respond relatively poorly to commercial high-capacity gravity methods. In practice, close size control of feeds to gravity processes is required to reduce the size effect and make the relative motion of the particles specific gravity dependent. The incorporation of enhanced gravity concentrators, which impart additional centrifugal acceleration to the particles, have been utilized in order to overcome some of the drawbacks of fine particle processing.

3. Gravity concentrators

It is essential for the efficient operation of all gravity separators that the feed is carefully prepared. Grinding is particularly important to provide particles of adequate liberation; successive regrinding of middlings is required in most operations. Primary grinding should be performed where possible in open-circuit rod mills, but if fine grinding is required, closed-circuit ball milling should be used, preferably with screens closing the circuits rather than hydrocyclones, in order to reduce selective overgrinding of heavy friable valuable minerals. Other methods of comminution, such as semi-autogeneous grinding mills and high-pressure grinding rolls, may find application in preparing gravity feeds.

Gravity separators are sensitive to the presence of slimes (ultrafine particles), which increase the viscosity of the slurry and hence reduce the sharpness of separation. It has been common practice to remove particles less than about 10 μm from the feed (i. e., de-slime), and divert this fraction to the tailings, which can incur considerable loss of values. De-sliming is often achieved by the use of hydrocyclones, although hydraulic classifiers may be preferable in some cases since the high shear forces produced in hydrocyclones tend to cause degradation of friable minerals (and create more loss of slimed values).

The feed to jigs and spirals should, if possible, be screened before separation takes place, each fraction being treated separately. In most cases, however, removal of the oversize by screening, in conjunction with de-sliming, is adequate. Processes which employ flowing film separation, such as shaking tables, should always be preceded by good hydraulic classification in multi-spigot hydrosizers.

Although most slurry transportation is achieved by centrifugal pumps and pipelines, as much as possible use should be made of natural gravity flow; many old gravity concentrators were built on hillsides to achieve this. Reduction of slurry pumping to a minimum not only lowers energy consumption, but also reduces slimes production in the circuit. To minimize degradation of friable minerals, pumping velocities should be as low as possible, consistent with maintaining the solids in suspension.

One of the most important aspects of gravity circuit operations is correct water balance within the plant. Almost all gravity concentrators have an optimum feed pulp density, and relatively little deviation from this density causes a rapid decline in efficiency. Accurate pulp density control is therefore essential, and this is most important on the raw feed. Automatic density control should be used where possible, and the best way of achieving this is by the use of nucleonic density gauges controlling the water addition to the new feed. Although such instrumentation is expensive, it is usually economic in the long term. Control of pulp density within the circuit can be made by the use of settling cones preceding the gravity device. These thicken the pulp, but the overflow often contains solids, and should be directed to a central large sump or thickener. For substantial increase in pulp density, hydrocyclones or thickeners may be used. The latter are the more expensive, but produce less particle degradation and also provide substantial surge capacity. It is usually necessary to recycle water in most plants, so adequate thickener or cyclone capacity should be provided, and slimes build-up in the recycled water must be minimized.

If the ore contains an appreciable amount of sulfide minerals then, because they are relatively dense and tend to report with the "heavy" product, they need to be removed. If the primary grind is finer than 300 μm, the sulfides should be removed by froth flotation prior to gravity concentration. If the primary grind is too coarse for effective sulfide flotation, then the gravity concentrate must be reground prior to removal of the sulfides. The sulfide flotation tailing is then usually cleaned by further gravity concentration.

The final gravity concentrate often needs cleaning by magnetic separation, leaching, or some other method, in order to remove certain mineral contaminants. For instance, at the South Crofty tin mine in Cornwall, the gravity concentrate was subjected to cleaning by magnetic separators, which removed wolframite from the cassiterite product.

4. Application of gravity concentration

There are some metallic minerals where gravity concentration remains as the only, or at least primary, means of concentration. These include for example cassiterite, chromite, wolframite and tantalum, where modem plants remain essentially "gravity plants". Even where flotation is employed, generally on the ultrafine fractions, it is common to combine gravity concentration for concentrate clean-up, as at the old Wheal Jane plant in Cornwall.

The gravity recovery of free milling gold prior to cyanidation circuits has a long history: jigs were used in North America, and Johnson barrels in South Africa. However, today, the Knelson concentrator, and to a lesser extent the Falcon, have become the unit of choice. Knelson concentrators are being installed even in the largest of plants, such as the Bajo de la Alumbrera plant in Argentina.

Dense medium separation (DMS), and to a lesser extent other forms of gravity concentration have a role in the industrial minerals field, where throughput and product quality often override recovery in importance. One novel application is the use of a Floatex density separator to successfully recover zircon from a Florida beach sand tailings. An "industrial" mineral where recovery is paramount is the diamond, and Canada's first diamond mine, along with diamond mines in Africa and Australia, use dense medium cyclones for primary recovery. The use of high density DMS as a first cleaner of primary DMS concentrates, to remove minerals heavier than diamonds, with equal attention to recovery, has recently been proposed.

More of a challenge are base metal ores where conventional wisdom remains that, considering the efficiency of flotation, gravity has little part to play. Nevertheless, it is considered that gravity preconcentration, probably with DMS, could, and should have a much greater role than it does. As stated earlier, rejection of a reasonable proportion of the barren (and harder) feed prior to grinding of the softer sulphides would have a major impact on costs, with some minor impact on recovery. In addition, there is a role for gravity concentration in the recovery of non-floatable by-products from sulphide mills, for example in the recovery of heavy minerals.

Undoubtedly the largest tonnages treated by gravity concentration are in the iron ore and coal industries, where the process remains, and will remain, of primary importance. In both these areas, very high capacity equipment is required, and DMS, jigs and spirals are regarded as standard units of choice for coal, with these plus the Reichert cone generally being employed in iron ore processing. The new generation of centrifugal separators are being seriously considered at least by the coal industry, with both Knelson and Falcon units being tested.

Enhanced gravity separation (EGS) is a physical separation technique that employs centrifugation to separate fine valuable minerals from their gangue phases based on their specific gravities. In terms of performance, separation efficiency, and processing of finer feed particle size ranges, this method outperforms standard gravity-based separation. Falcon concentrator (FC), as an EGS, is a fast-spinning bowl that uses centrifugal force to drain the slurry in a thin film along its wall. Due to the rapid rotation rate in the flowing film, centrifugal force can be many orders of magnitude larger than earth's gravity. In general, two separation mechanisms can be identified inside the FCs, including differential particle settling within the liquid film flowing on the surface of the rotating bowl, and particle rearrangement inside the granular bed (retention zone). The second separation mechanism works for dense particles, such as fine gold, while differential settling controls the separation process for fine light particles.

Notes

1. Gravity methods of separation are used to treat a great variety of materials, ranging from heavy metal sulphides such as galena (sp. gr. 7.5) to coal (sp. gr. 1.3), at particle sizes in some cases below 50 μm.

重选方法用于处理多种物料，这些物料的范围从重金属硫化物如方铅矿(相对密度 7.5)到煤(相对密度 1.3)，在某些情况下颗粒粒度小于 50 μm。

2. Modern gravity techniques have proved efficient for concentration of minerals having particle sizes in the 50 μm range and, when coupled with improved pumping technology and instrumentation, have been incorporated in high-capacity plants.

现代重选技术已被证明对粒径在 50 μm 范围内的矿物的富集是有效的，当与改进的泵送技术和仪器相结合时，已被用于高处理能力的选矿厂。

3. Gravity concentration methods separate minerals of different specific gravity by their relative movement in response to gravity and one or more other forces, the latter often being the resistance to motion offered by a viscous fluid, such as water or air.

重选方法通过因重力和一种或多种其他力的作用而产生的相对运动来分选不同相对密度的矿物，其他力往往指像空气或水这样的黏性流体对运动产生的阻力。

4. In practice, close size control of feeds to gravity processes is required to reduce the size effect and make the relative motion of the particles specific gravity dependent.

在实践中，对重选过程的给矿必须进行严格的粒度控制，以缩小粒度影响，使颗粒的相对运动同其相对密度有关.

5. Grinding is particularly important to provide particles of adequate liberation; successive regrinding of middlings is required in most operations.

磨矿对于提供充分解离的颗粒特别重要；在大多数生产中都需要对中矿进行连续再磨。

6. It is usually necessary to recycle water in most plants, so adequate thickener or cyclone capacity should be provided, and slimes build-up in the recycled water must be minimized.

在大多数选矿厂中，通常需要对水进行循环利用，因此应提供足够处理能力的浓密机和旋流器，并且必须尽量减少循环水中的矿泥积聚。

7. Undoubtedly the largest tonnages treated by gravity concentration are in the iron ore and coal industries, where the process remains, and will remain, of primary importance.

毫无疑问，通过重选方法处理的最大量是在铁矿石和煤炭工业，在这些领域中，重选工艺过程仍然是而且将继续是最重要的。

Exercises

1. Answer the following questions according to the text.

(1) What is gravity concentration?

(2) Why the importance of gravity concentration has declined in the twentieth century?

(3) What are the principles of gravity concentration?

(4) How to prepare the feed to jigs, spirals, and shaking tables?

(5) What kinds of gravity separators can be used to remove slimes from the feed?

2. Translate the following phrases into English.

(1) 重选 (2) 密度差异 (3) 预选 (4) 脱泥

(5) 矿物相对密度 (6) 中矿再磨 (7) 重选设备 (8) 分选效率

3. Translate the following paragraphs into Chinese.

Paragraph 4 in section 2 "The motion of a particle in a fluid is dependent not only ...".

4. Translate the following sentences into English.

(1) 重力分选是一种基于密度差异的矿物分选方法。

(2) 近年来，由于浮选药剂成本增加、重力分选过程相对简单以及它们产生的环境污染相对较少，许多公司已重新评估了重力分选系统。

(3) 重选仍然是一些金属矿石唯一或者主要的富集方法。

(4) 流体中颗粒的运动不仅取决于它的密度，还取决于它的粒度；大颗粒比小颗粒受到的影响更大。

(5) 重选设备对矿泥(超细颗粒)的存在极为敏感，矿泥增加了矿浆的黏度，从而降低了分离精度。

4.2 Jigs (跳汰机)

1. Introduction

In the jig, the separation of minerals of different specific gravity is accomplished in a particle bed, which is fluidized by a pulsating current of water, producing stratification based upon density. The bed may be a specific mineral added to and retained in the jig, called ragging, composed of a certain density and shape through which the dense particles penetrate and the light particles pass over the top. The aim is to dilate the bed and to control the dilation so that the heavier, smaller particles penetrate the interstices of the bed and the larger high-specific gravity particles fall under a condition similar to hindered settling.

On the pulsion stroke the bed is normally lifted as a mass, then as the velocity decreases it tends to dilate, the bottom particles falling first until the whole bed is loosened. On the suction stroke it then closes slowly again, and this is repeated at every stroke. Fine particles tend to pass through the interstices after the large ones have become immobile. The motion can be obtained either by using a fixed sieve jig, and pulsating the water, or by employing a moving sieve, as in the simple hand-jig shown in Figure 4. 1.

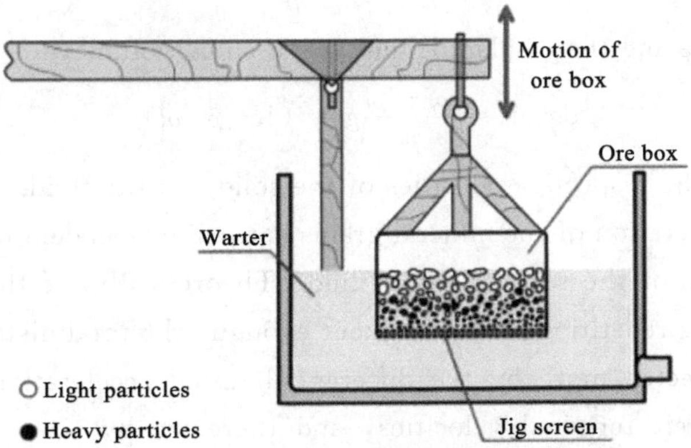

Figure 4. 1 Jigging action: by moving box up and down the particles segregate by density (Adapted from Priester et al. (1993))

The jig is normally used to concentrate relatively coarse material and, if the feed is fairly close-sized (e. g. , 3～10 mm), it is not difficult to achieve good separation of a fairly narrow specific gravity range in minerals in the feed (e. g. , fluorite, s. g. 3. 2, from quartz, s. g. 2. 7). When the specific gravity difference is large, good concentration is possible with a wider size range. Many large jig circuits are still operated in coal, cassiterite, tungsten, gold, barytes, and iron ore concentrators. They have a relatively high unit capacity on classified feed and can achieve good recovery of values down to

150 μm and acceptable recoveries often down to 75 μm. High proportions of fine sand and slime interfere with performance and the fines content should be controlled to provide optimum bed conditions.

Jigs are widely used in coal cleaning (also referred to as "coal washing") and are preferred to the more expensive DMS when the coal has relatively little middlings, or "near-gravity" material. No feed preparation is required, as is necessary with DMS, and for coals that are easily washed, that is, those consisting predominantly of liberated coal and denser "rock" particles, the lack of close density control is not a disadvantage.

2. Jigging action

It was shown that the equation of motion of a particle settling in a viscous fluid is:

$$m\mathrm{d}x/\mathrm{d}t = mg - m'g - D \tag{4.2}$$

where m is the mass of the mineral particle, $\mathrm{d}x/\mathrm{d}t$ is the acceleration, g is the acceleration due to gravity, m' is the mass of displaced fluid, and D is the fluid resistance due to the particle movement.

At the beginning of the particle movement, since the velocity x is very small, D can be ignored, as it is a function of velocity.

Therefore

$$\mathrm{d}x/\mathrm{d}t = \frac{m - m'}{m}g \tag{4.3}$$

and since the particle and the displaced fluid are of equal volume,

$$\mathrm{d}x/\mathrm{d}t = \frac{\rho_s - \rho_f}{\rho_s}g = (1 - \rho_f/\rho_s)g \tag{4.4}$$

where ρ_s and ρ_f are the respective densities of the solid and the fluid.

The initial acceleration of the mineral grains is thus independent of size and dependent only on the densities of the solid and the fluid. Theoretically, if the duration of fall is short enough and the repetition of fall frequent enough, the total distance travelled by the particles will be affected more by the differential initial acceleration, and therefore by density, than by their terminal velocities, and therefore by size. In other words, to separate small heavy mineral particles from large light particles, a short jigging cycle is necessary. Although relatively short, fast strokes are used to separate fine minerals, more control and better stratification can be achieved by using longer, slower strokes, especially with the coarser particle sizes.

It is therefore good practice to screen the feed to jigs into different size ranges and treat these separately. The effect of differential initial acceleration is shown in Figure 4.2.

If the mineral particles are examined after a longer time, they will have attained their terminal velocities and will be moving at a rate dependent on their specific gravity and size. Since the bed is really a loosely packed mass with interstitial water providing a very thick suspension of high density, hindered-settling conditions prevail, and the settling ratio of heavy to light minerals is higher than that for free settling. Figure 4.3 shows the effect of hindered settling on the separation.

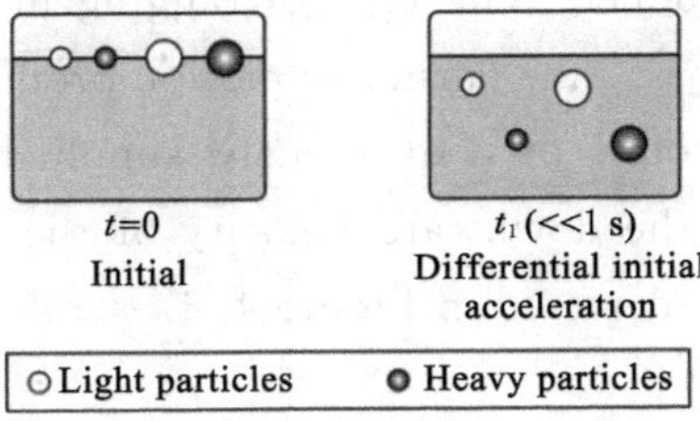

Figure 4.2 Differential initial acceleration

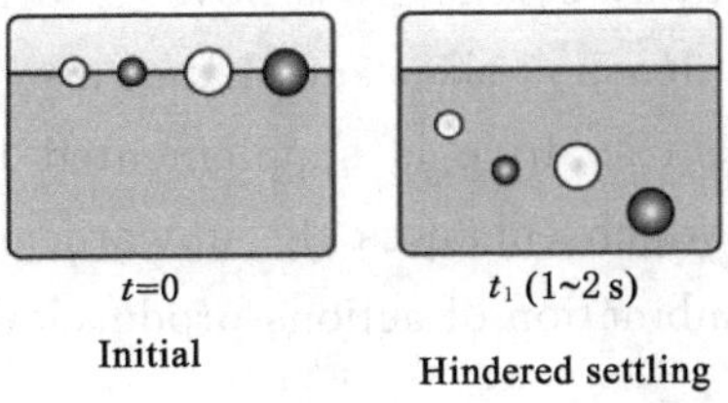

Figure 4.3 Hindered settling

The upward flow can be adjusted so that it overcomes the downward velocity of the fine light particles and carries them away, thus achieving separation. It can be increased further so that only large heavy particles settle, but it is apparent that it will not be possible to separate the small heavy and large light particles of similar terminal velocity.

Hindered settling has a marked effect on the separation of coarse minerals, for which longer, slower strokes should be used, although in practice, with coarser feeds, it is unlikely that the larger particles have sufficient time to reach their terminal velocities.

At the end of a pulsion stroke, as the bed begins to compact, the larger particles interlock, with the smaller particles still moving downward through the interstices under the influence of gravity. The fine particles may not settle as rapidly during this consolidation trickling phase as during the initial acceleration or suspension, but if consolidation trickling can be made to last long enough, the effect, especially in the recovery of the fine heavy minerals, can be considerable.

Figure 4.4 shows an idealized jigging process by the described phenomena.

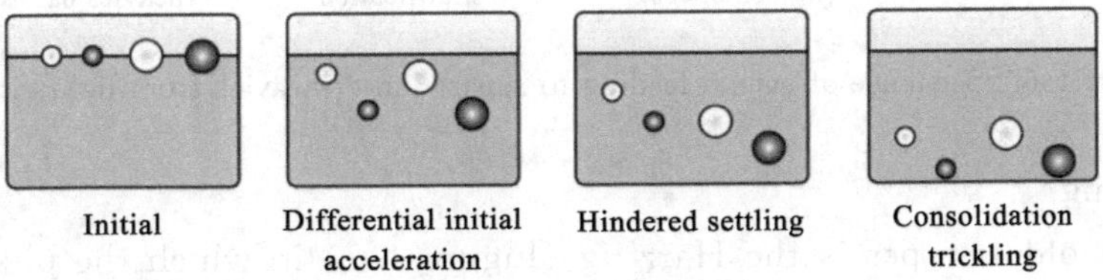

Figure 4.4 Ideal jigging process

3. Structure and types of jig

Essentially, the jig is an open tank filled with water, with a horizontal jig screen at the top supporting the jig bed, and provided with a spigot in the bottom, or hutch compartment, for "heavies" removal (Figure 4.5). The jig bed consists of a layer of coarse, heavy particles, or ragging, placed on the jig screen on to which the slurry is fed. The feed flows across the ragging and the separation takes place in the jig bed (i. e., in the ragging), so that particles with a high specific gravity penetrate through the ragging and then pass the screen to be drawn off as the heavy product, while the light particles are carried away by the cross-flow to form the light product. The type of ragging material, particle density, size, and shape, are important factors. The harmonic motion produced by the eccentric drive is supplemented by a large amount of continuously supplied hutch water, which enhances the upward and diminishes the downward velocity of the water. The combination of actions produces the segregation depicted in Figure 4.6.

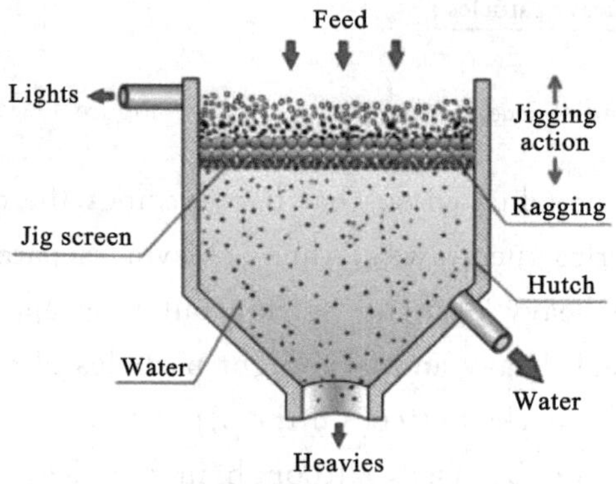

Figure 4.5 Basic jig construction

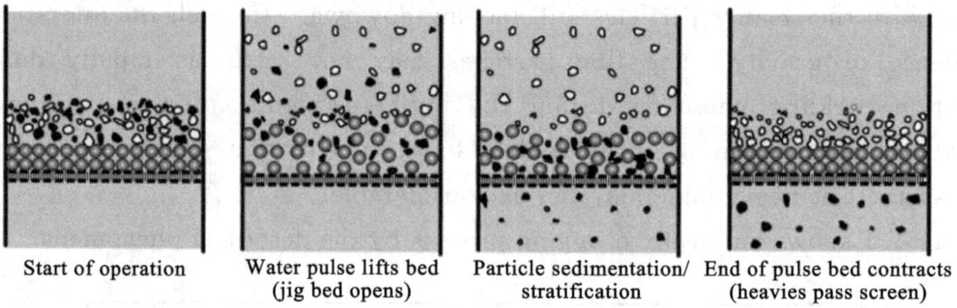

Figure 4.6 Sequence of events leading to separation of heavies from lights in a jig

(1) Harz jig

One of the oldest types is the Harz jig (Figure 4.7) in which the plunger moves up and down vertically in a separate compartment. Up to four successive compartments are placed in series in the hutch. A high-grade heavy product is produced in the first compartment, successively lower grades being produced in the other compartments, and the light product overflowing the final compartment. If the feed particles are larger than the apertures of the screen, jigging "over the screen" is used, and the concentrate grade is partly governed by the thickness of the bottom layer, determined by the rate of withdrawal through the concentrate discharge port.

(2) Denver mineral jig

The Denver mineral jig (Figure 4.8) is widely used, especially for removing heavy

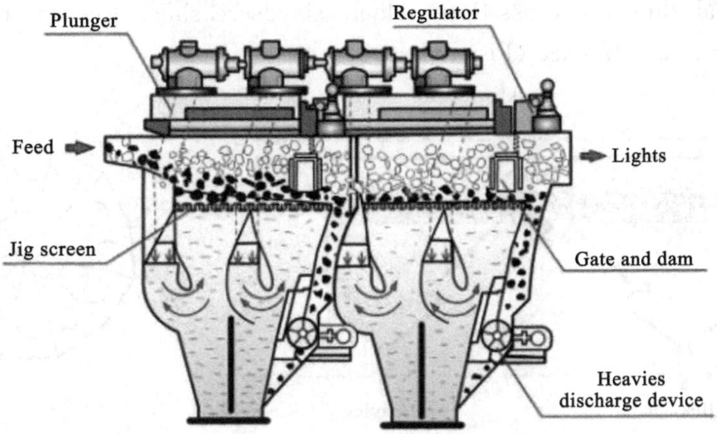

Figure 4.7 Harz jig (Adapted from *The Great Soviet Encyclopedia*, 3rd ed. (1970—1979))

minerals from closed grinding circuits, thus preventing over-grinding. The rotary water valve can be adjusted so as to open at any desired part of the jig cycle, synchronization between the valve and the plungers being achieved by a rubber timing belt. By suitable adjustment of the valve, any desired variation can be achieved, from complete neutralization of the suction stroke with hydraulic water to a full balance between suction and pulsion.

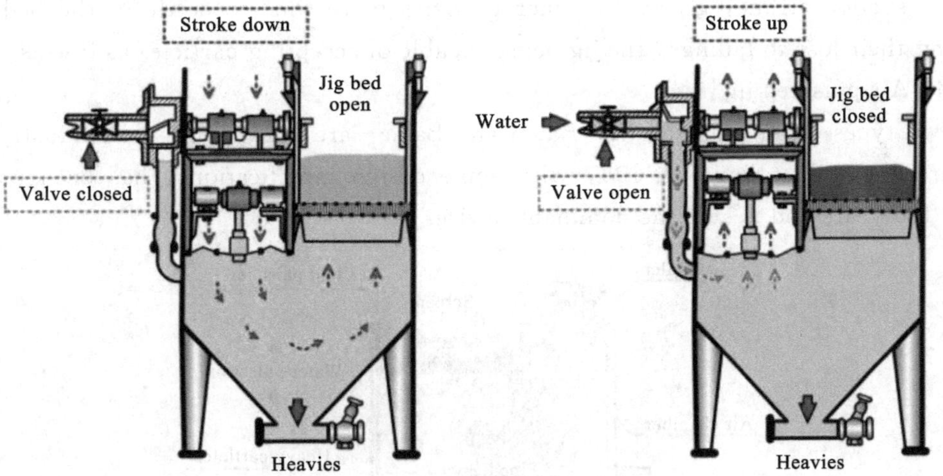

Figure 4.8 Denver mineral jig

(3) Circular jig

Conventional mineral jigs consist of square or rectangular tanks, sometimes combined to form two, three, or four cells in series. In order to compensate for the increase in cross-flow velocity over the jig bed, caused by the addition of hutch water, trapezoidal-shaped jigs were developed. By arranging these as sectors of a circle, the modular circular, or radial, jig was introduced, in which the feed enters in the center and flows radially over the jig bed (and thus the cross-flow velocity is decreasing) toward the light product discharge at the circumference (Figure 4. 9).

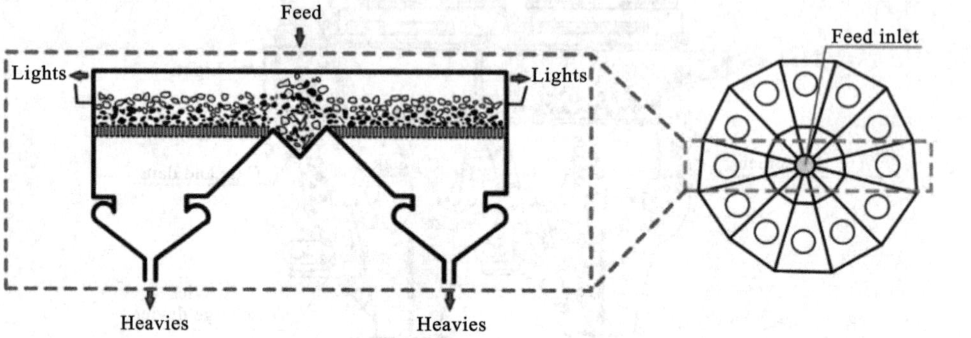

(a) Cross section through a circular jig

(b) Radial jig up to 12 modules

Figure 4. 9

The advantage of the circular jig is its large capacity. Since their development in 1970, IHC Radial Jigs were installed on most newly built tin dredges in Malaysia and Thailand. In the IHC jig, the harmonic motion of the conventional eccentric-driven jig is replaced by an asymmetrical "saw-tooth" movement of the diaphragm, with a rapid upward, followed by a slow downward, stroke. This produces a much larger and more constant suction stroke, giving the finer particles more time to settle in the bed, thus reducing their loss to tailings, the jig being capable of accepting particles as fine as 60 μm.

(4) Air-pulsated jig

Two types of air-pulsated jig—Baum and Batac—are used in the coal industry. The standard Baum jig (Figure 4. 10), with some design modifications, has been used for nearly 100 years and is still the dominant device.

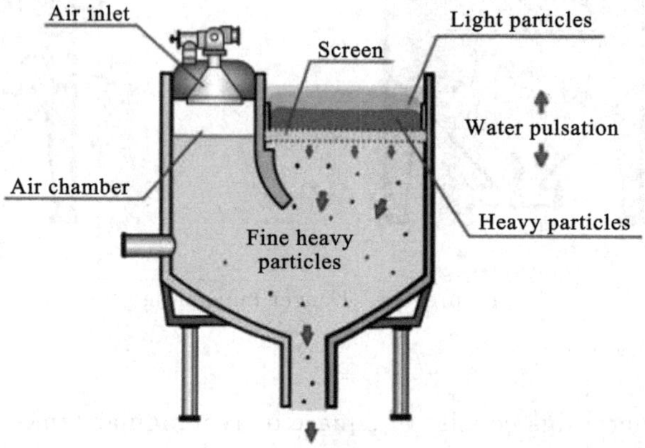

Figure 4. 10 Baum jig

Air under pressure is forced into a large air chamber on one side of the jig vessel, causing pulsation and suction to the jig water, which in turn causes pulsation and suction through the screen plates upon which the raw coal is fed, thus causing stratification. Various methods are used to continuously separate the refuse (heavy non-coal matter) from the lighter coal product, and all the modern Baum jigs are fitted with some form of automatic refuse extraction. One form of control incorporates a float immersed in the bed of material. The float is suitably weighted to settle on the dense layer of refuse moving across the screen plates. An increase in the depth of refuse raises the float, which automatically controls the refuse discharge, either by adjusting the height of a moving gate, or by controlling the pulsating water, which lifts the rejects over a fixed weir plate. This system is reported to respond quickly and accurately.

Notes

1. On the pulsion stroke the bed is normally lifted as a mass, and then as the velocity decreases it tends to dilate, the bottom particles falling first until the whole bed is loosened. On the suction stroke it then closes slowly again and this is repeated at every stroke.

在推进冲程阶段,床层通常作为整体抬升,然后随着速度下降床层趋于扩张,直到整个床层松散后底部颗粒首先开始下落。在吸入冲程阶段,床层(从松散状态)再次慢慢闭合,每个冲程重复进行这种过程。

2. Jigs are widely used in coal cleaning (also referred to as "coal washing") and are preferred to the more expensive DMS when the coal has relatively little middlings, or "near-gravity" material.

跳汰机广泛用于选煤(也被称为"洗煤"),当煤中的中煤或"邻近密度"物料含量相对少时,该方法优于成本较高的重介工艺。

3. Hindered settling has a marked effect on the separation of coarse minerals, for which longer, slower strokes should be used, although in practice, with coarser feeds, it is unlikely that the larger particles have sufficient time to reach their terminal velocities.

干扰沉降对粗矿物的分离有显著影响,为此应该使用更长、更慢的冲程,但是实际情况是使用较粗的给料时,较大的颗粒不太可能有足够的时间达到其最终速度。

4. The jig bed consists of a layer of coarse, heavy particles, or ragging, placed on the jig screen on to which the slurry is fed.

跳汰机床层由一层粗的、重的颗粒或人工床层组成,它们放置在给入矿浆的跳汰机筛板上。

5. The feed flows across the ragging and the separation takes place in the jig bed (i. e. , in the ragging), so that particles with a high specific gravity penetrate through the ragging and then pass the screen to be drawn off as the heavy product, while the light particles are carried away by the cross-flow to form the light product.

进料流过床层,分离在跳汰机床层(即人工床层)中进行,因此具有高相对密度的颗粒穿透人工床层,然后透过筛子,作为重产品被排除,而轻颗粒则被横流带走,形成轻产品。

6. The type of ragging material, particle density, size, and shape, are important factors.

人工床层材料的类型、颗粒密度、粒度和形状是重要因素。

Exercises

1. Answer the following questions according to the text.

(1) What is ragging, and what is the aim to use it in jigs?

(2) How does the particle bed become loose and closed during every stroke?

(3) What are the applications of jig?

(4) How does the hindered settling affect the separation of coarse minerals?

(5) What is the basic construction of a jig?

2. Translate the following phrases into English.

(1) 吸气冲程　(2) 跳汰周期　(3) 动筛跳汰机　(4) 人工床层

(5) 跳汰筛　(6) 空气脉动跳汰机　(7) 脉动水流　(8) 邻近密度物料

3. Translate the following paragraphs into Chinese.

Paragraphs 1 and 2 in section 1 "In the jig, the separation of minerals of different specific gravity ...".

4. Translate the following sentences into English.

(1) 在跳汰机中,不同密度矿物的分离是在颗粒床层中完成的,颗粒床层被脉动水流流化,根据密度产生分层。

(2) 在大颗粒不移动后,细颗粒倾向于穿过间隙。

(3) 矿物颗粒的初始加速度与粒度无关,只与固体和流体的密度有关。

(4) 从本质上讲,跳汰机是一个装满水的开放式水槽,顶部有一个支撑跳汰机床层的水平跳汰筛,底部有一个沉砂口,或称为筛下室,用于"重物"的移除。

4.3 Spirals and Shaking Tables(螺旋溜槽与摇床)

1. Spirals

(1) Structure

Spiral concentrators have found many varied applications in mineral processing, but perhaps their most extensive application has been in the treatment of heavy mineral sand deposits, such as those carrying ilmenite, rutile, zircon, and monazite, and in recent years in the recovery of fine coal.

The Humphreys spiral was introduced in 1943, its first commercial use being on chrome-bearing sands. It is composed of a helical conduit of modified semicircular cross section. Feed pulp of between 15% and 45% solids by weight and in the size range from 3 mm to 75 μm is introduced at the top of the spiral. As it flows spirally downward, the particles stratify due to the combined effect of centrifugal force, the differential settling rates of the particles, and the effect of interstitial trickling through the flowing particle bed. The result of this action is depicted in Figure 4. 11. Figure 4. 12 shows the stratification across a spiral trough, with the darker heavy mineral toward the center, with the band becoming increasingly lighter radially where the less dense material flows.

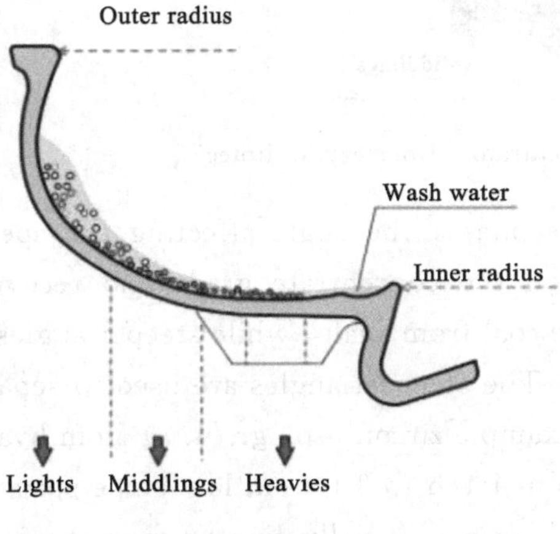

Figure 4. 11 Cross section of spiral stream

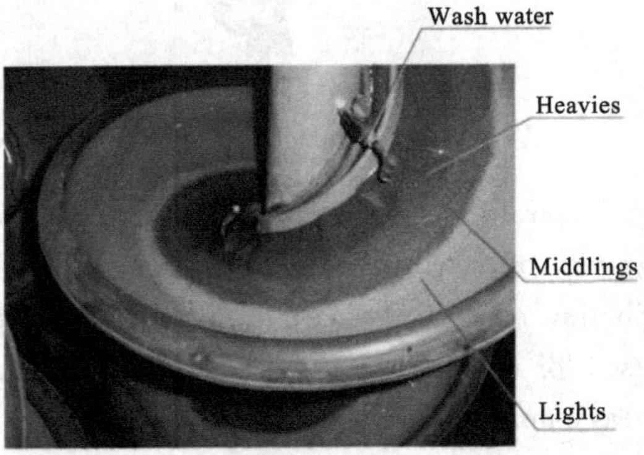

Figure 4. 12 Example of stratification across a spiral trough, with dense, dark material in the centra and lighter, less dense extending out radially (Courtesy Multotec)

Ports for the removal of the higher specific gravity particles are located at the lowest points in the cross section. Wash water is added at the inner edge of the stream and flows outwardly across the concentrate band to aid in flushing out entrapped light particles. Adjustable splitters control the width of the heavy product band removed at the ports. The heavy product taken via descending ports is of a progressively decreasing grade, with the light product discharged from the lower end of the spiral. Splitters at the end of the spiral are often used to give three products: heavies, lights, and middlings, giving possibilities for recycle and retreatment in other spirals (e. g. , rougher-cleaner spiral combination) or to feed other separation units.

(2) Application

One of the most important developments in fine coal washing was the introduction in the 1980s of spiral separators specifically designed for coal. It is common practice to separate down to 0. 5 mm coal using dense medium cyclones, and below this by froth flotation. Spiral circuits have been installed to process the size range which is least effectively treated by these two methods, typically 0. 1～2 mm.

Double-spiral concentrators, with two starts integrated into the one space around a common column, have been used in Australia for many years and have also been accepted elsewhere (Figure 4. 13). At Mount Wright in Canada 4, 300 double-start spirals are upgrading specular hematite ore at 6,900 t/h at 86% recovery.

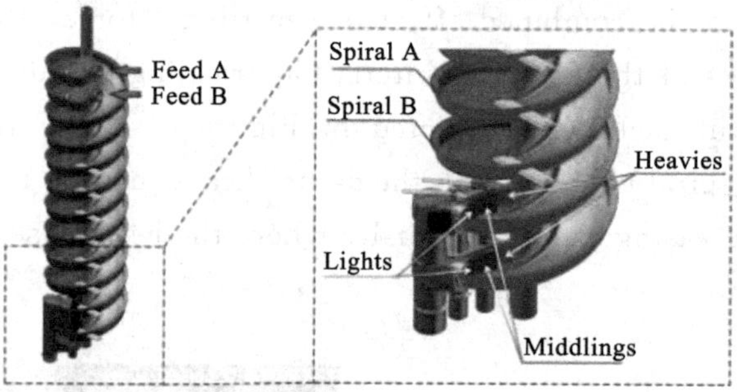

Figure 4. 13 Double-start spiral concentrator (Courtesy Multotec)

Spirals are made with slopes of varying steepness, the angle affecting the specific gravity of separation, but having little effect on the concentrate grade and recovery. Shallow angles are used, for example, to separate coal from shale, while steeper angles are used for normal heavy mineral-silica separations. The steepest angles are used to separate heavy minerals from heavy waste minerals, for example zircon (sp. gr. 4. 7) from kyanite and staurolite (sp. gr. 3. 6). Capacity ranges from 1 t/h to 3 t/h on low slope spirals to about double this for the steeper units. Spiral length is usually five or more turns for roughing duty and three turns in some cleaning units. Because treatment by spiral separators involves a multiplicity of units, the separation efficiency is very sensitive to the pulp distribution system employed. Lack of uniformity in feeding results in substantial falls in operating efficiency and can lead to severe losses in recovery, this is especially true with coal spirals.

2. Shaking tables

(1) Structure

When a film of water flows over a flat, inclined surface, the water closest to the surface is retarded by the friction of the water absorbed on the surface and the velocity increases toward the water surface. If mineral particles are introduced into the film, small particles will not move as rapidly as large particles, since they will be submerged in the slower-moving portion of the film. Particles of high specific gravity will move more slowly than lighter particles, and so a lateral displacement of the material will be produced (Figure 4.14).

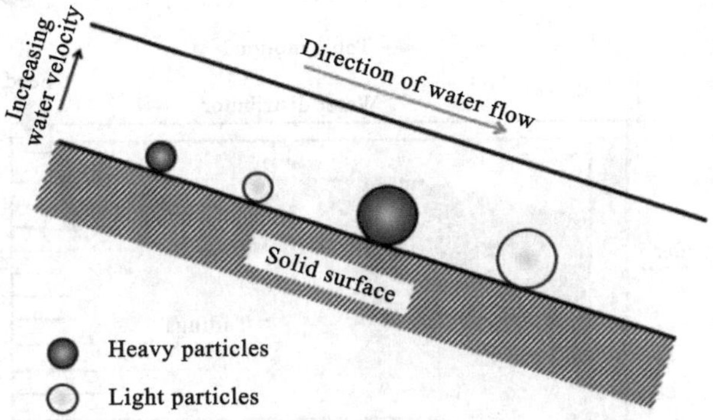

Figure 4.14 Action in a flowing film

The flowing film effectively separates coarse light particles from small dense particles, and this mechanism is exploited to some extent in the shaking-table concentrator (Figure 4.15), which is perhaps the most metallurgically efficient form of gravity concentrator, being used to treat small, more difficult flow-streams, and to produce finished concentrates from the products of other forms of gravity system.

The shaking table consists of a slightly inclined deck, on to which feed, at about 25% solids by weight, is introduced at the feed box and is distributed across the table by the combination of table motion and flow of water (wash water). Wash water is distributed along the length of the feed side, and the table is vibrated longitudinally, using a slow forward stroke and a rapid return, which causes the mineral particles to "crawl" along the deck parallel to the direction of motion. The minerals are thus subjected to two forces, that due to the table motion and that, at right angles to it, due to the flowing film of water. The net effect is that the particles move diagonally across the deck from the feed end and, since the effect of the flowing film depends on the size and density of the particles, they will fan out on the table, the smaller, denser particles riding highest toward the concentrate launder at the far end, while the larger lighter particles are washed into the tailings launder, which runs along the length of the table. An adjustable splitter at the concentrate end is often used to separate this product into two fractions—a high-grade concentrate (heavy product) and a middlings fraction.

Although true flowing film concentration requires a single layer of feed, in practice, a multilayered feed is introduced onto the table, enabling much larger tonnages to be

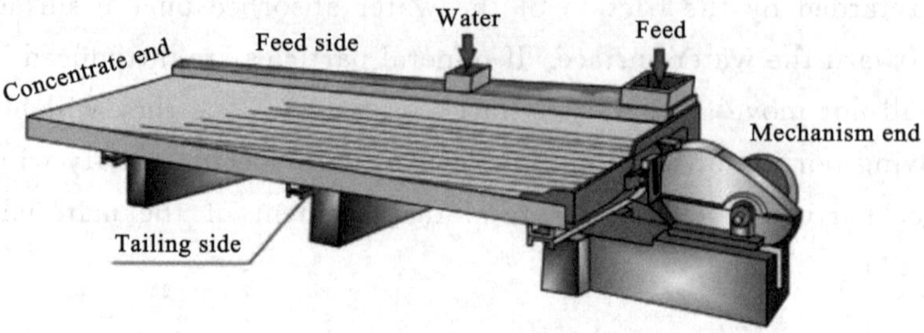

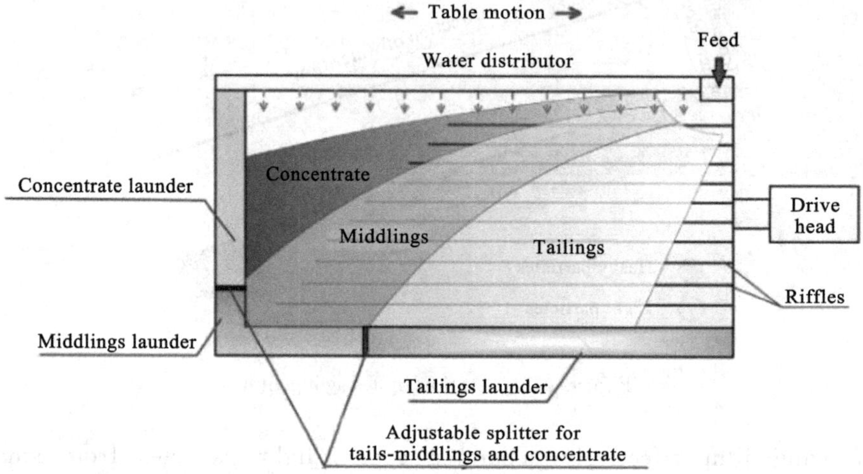

Figure 4.15 Shaking table (concentrate refers to heavy product)

processed. Vertical stratification due to shaking action takes place behind the riffles, which generally run parallel with the long axis of the table and are tapered from a maximum height on the feed side, till they die out near the opposite side, part of which is left smooth. In the protected pockets behind the riffles, the particles stratify so that the finest and heaviest particles are at the bottom and the coarsest and lightest particles are at the top (Figure 4.16). Layers of particles are moved across the riffles by the crowding action of new feed and by the flowing film of wash water. Due to the taper of the riffles, progressively finer sized and higher density particles are continuously being brought into contact with the flowing film of water that tops the riffles. Final concentration takes place at the unriffled area at the end of the deck, where the layer of material is at this stage usually only one or two particles deep.

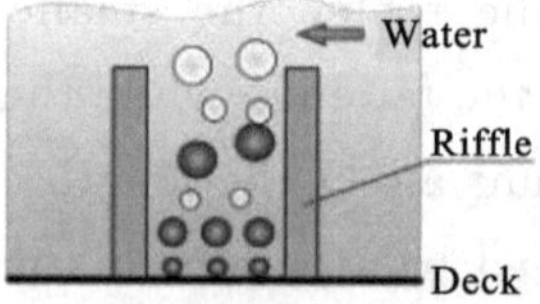

Figure 4.16 Vertical stratification between riffles

(2) Application

The separation on a shaking table is controlled by a number of operating variables, such as wash water, feed pulp density, deck slope, amplitude, and feed rate. Other factors, including particle shape and the type of deck, play an important part in table separations. Flat particles, such as mica, although light, do not roll easily across the deck in the water film; such particles cling to the deck and are carried down to the concentrate discharge. Likewise, spherical dense particles may move easily in the film toward the tailings launder.

The table decks are usually constructed of wood, lined with materials with a high coefficient of friction, such as linoleum, rubber, and plastics. Decks made from fibre-glass are also used which, although more expensive, are extremely hard wearing. The riffles on such decks are incorporated as part of the mold.

Particle size plays an important role in table separations; as the range of sizes in a table feed increases, the efficiency of separation decreases. If a table feed is made up of a wide range of particle sizes, some of these sizes will be cleaned inefficiently. The middlings produced may not be "true middlings", that is, locked particles of associated mineral and gangue, but relatively coarse dense particles and fine light particles. If these particles are returned to the grinding circuit, together with the true middlings, then they will be needlessly reground.

Since the shaking table effectively separates coarse light from fine dense particles, it is common practice to classify the feed, since classifiers put such particles into the same product, on the basis of their equal settling rates. In order to feed as narrow a size range as possible onto the table, classification is usually performed in multi-spigot hydrosizers, each spigot product, comprising a narrow range of equally settling particles, being fed to a separate set of shaking tables. A typical gravity concentrator employing shaking tables may have an initial grind in rod mills to liberate as much mineral at as coarse a size as possible to aid separation, with middlings being reground before returning to the hydrosizer. Tables operating on feed sizes in the range 3 mm to 100 μm are sometimes referred to as sand tables, and the hydrosizer overflow, consisting primarily of particles finer than 100 μm, is usually thickened and then distributed to tables whose decks have a series of planes, rather than riffles, and are designated slime tables.

The capacity of a table varies according to size of feed particles and the concentration criteria. Tables can handle up to 2 t/h of 1.5 mm sand and perhaps 1 t/h of fine sand. On 100～150 μm feed materials, table capacities may be as low as 0.5 t/h. On coal feeds, however, which are often tabled at sizes of up to 15 mm, much higher capacities are common. A normal 5 mm raw coal feed can be tabled with high efficiency at 12.5 t/h per deck, while tonnages as high as 15 t/h per deck are not uncommon when the feed top size is 15 mm. The introduction of double and triple-deck units has improved the area/capacity ratio at the expense of some flexibility and control.

Separation can be influenced by the length of stroke, which can be altered by means of a hand-wheel on the vibrator, or head motion, and by the reciprocating speed. The length of stroke usually varies within the range of 10～25 mm or more, the speed being in the range 240～325 strokes per minute. Generally, a fine feed requires a higher speed and shorter stroke.

The quantity of water used in the feed pulp varies, but for ore-tables normal feed dilution is 20%～25% solids by weight, while for coal tables pulps of 33%～40% solids are used. In addition to the water in the feed pulp, clear water flows over the table for final concentrate cleaning. This varies from a few liters to almost 100 L/min according to the nature of the feed material.

Ore-concentrating tables are used primarily for the concentration of minerals of tin, iron, tungsten, tantalum, mica, barium, titanium, zirconium, and, to a lesser extent, gold, silver, thorium, and uranium. Tables are now being used in the recycling of electronic scrap to recover precious metals.

Notes

1. Spiral concentrators have found many varied applications in mineral processing, but perhaps their most extensive application has been in the treatment of heavy mineral sand deposits, such as those carrying ilmenite, rutile, zircon, and monazite, and in recent years in the recovery of fine coal.

螺旋选矿机在选矿中有许多不同的应用,但也许它们最广泛的应用是处理那些含有钛铁矿、金红石、锆石和独居石的重砂矿床,以及近年来用来回收细粒煤。

2. Wash water is added at the inner edge of the stream and flows outwardly across the concentrate band to aid in flushing out entrapped light particles.

从矿浆流内缘加入的洗涤水横穿精矿带而向外流,这样有助于冲洗掉夹带的轻颗粒。

3. Double-spiral concentrators, with two starts integrated into the one space around a common column, have been used in Australia for many years and have also been accepted elsewhere.

双头螺旋选矿机,即围绕同一圆柱组装两圈螺旋的选矿装置,在澳大利亚已使用多年,在其他地方也已被应用。

4. Spirals are made with slopes of varying steepness, the angle affecting the specific gravity of separation, but having little effect on the concentrate grade and recovery.

螺旋选矿机的坡度不一,倾斜角度影响分选密度,但对精矿品位和回收率影响不大。

5. Shallow angles are used, for example, to separate coal from shale, while steeper angles are used for normal heavy mineral-silica separations.

例如,缓倾角螺旋选矿机用于从页岩中分离煤,而较陡的倾角用于常见的重矿物与二氧化硅的分离。

6. Particles of high specific gravity will move more slowly than lighter particles, and so a lateral displacement of the material will be produced.

重颗粒比轻颗粒移动得更慢,因此会产生物料的横向位移。

7. The flowing film effectively separates coarse light particles from small dense particles, and this mechanism is exploited to some extent in the shaking-table concentrator.

流膜有效地将粗而轻的颗粒与细而重的颗粒分离,这一机制在摇床选矿机中得到了一定程度的应用。

8. The shaking table consists of a slightly inclined deck, on to which feed, at about 25% solids by weight, is introduced at the feed box and is distributed across the table by the combination of table motion and flow of water (wash water).

摇床由一个略微倾斜的床板组成,在床板上的进料箱中引入质量浓度约 25%的固体给料,并通过摇床运动和水流(洗涤水)的组合作用分布在整个床面上。

9. Tables operating on feed sizes in the range 3 mm to 100 μm are sometimes referred to as sand tables, and the hydrosizer overflow, consisting primarily of particles finer than 100 μm, is usually thickened and then distributed to tables whose decks have a series of planes, rather than riffles, and are designated slime tables.

入料粒度在 3 mm 至 100 μm 范围内运行的摇床有时称为砂矿摇床;主要由小于 100 μm 的颗粒组成的水力分级机溢流一般经过浓缩后分配到摇床分选,这种摇床称为矿泥摇床,它由一系列平面而不是床条所构成。

Exercises

1. Answer the following questions according to the text.

(1) What is the major use of spirals?

(2) What causes the particles stratification as the feed pulp flows spirally downward?

(3) What are the factors that control the separation on a shaking table?

(4) What is the net effect of the table motion and the flowing film of water?

(5) What is the application of ore-concentrating tables?

2. Translate the following phrases into English.

(1) 螺旋选矿机　(2) 矿砂摇床　(3) 垂直分层　(4) 流膜分选

(5) 给矿矿浆浓度　(6) 小而重的颗粒　(7) 冲洗水　(8) 细粒煤

3. Translate the following paragraphs into Chinese.

Paragraph 4 in section 2 (1) "Although true flowing film concentration requires ...".

4. Translate the following sentences into English.

(1) 螺旋选矿机在选矿中有许多不同的应用,但其最广泛的应用可能是在处理重砂矿床。

(2) 选煤最重要的发展之一是在 20 世纪 80 年代引入了专门为煤炭设计的螺旋分选机。

(3) 摇床分选由许多操作变量控制,例如洗涤水、给料浆密度、床面坡度、振幅和给料速率。

4.4 Other Gravity Concentrators (其他重选设备)

1. Centrifugal concentrators

In an attempt to recover fine particles using gravity concentration methods, devices have been developed to make use of centrifugal force.

(1) Knelson concentrator

This is a compact batch centrifugal separator with an active fluidized bed to capture heavy minerals (Figure 4.17). A centrifugal force up to 60 times that of gravity acts on the particles, trapping denser particles in a series of rings (riffles) located in the machine, while the low-density particles are flushed out. Unit capacities range from laboratory scale to 150 t/h for particles ranging in size from 10 μm to a maximum of 6 mm. It is generally used for feeds in which the dense component to be recovered is a very small fraction of the total material, less than 500 g/t (0.05% by weight).

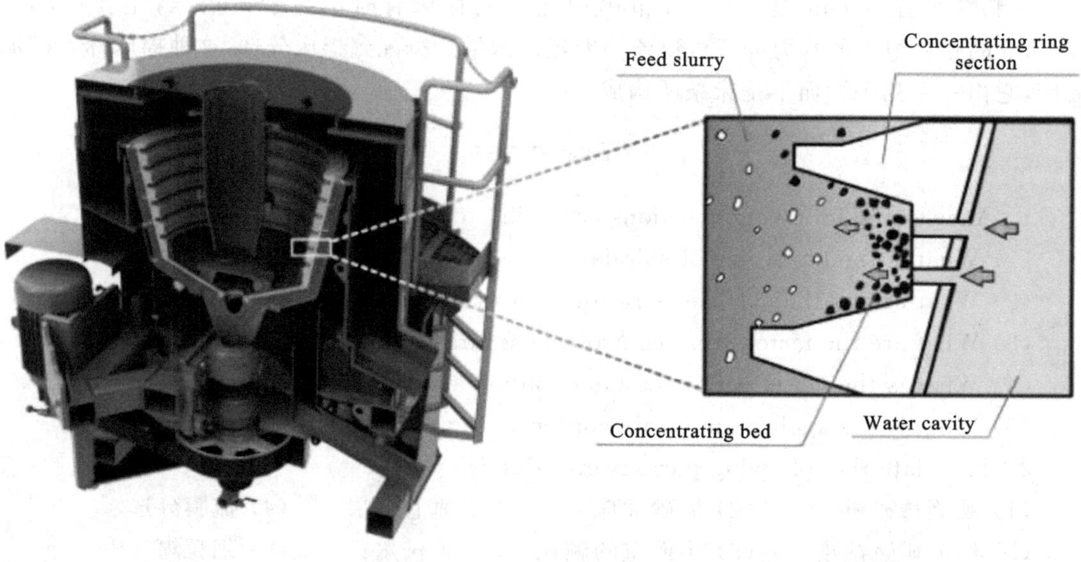

Figure 4.17 Knelson concentrator cutaway; and action inside the riffle (Courtesy FLSmidth)

Feed slurry is introduced through a stationary feed tube and into the concentrate cone. When the slurry reaches the bottom of the cone it is forced outward and up the cone wall under the influence of centrifugal force. Fluidization water is introduced into the concentrate cone through a series of fluidization holes (see inset in Figure 4.17). The slurry fills each ring to capacity to create a concentrating bed, with compaction of the bed prevented by the fluidization water. The flow of water that is injected into the rings is controlled to achieve optimum bed fluidization. High specific gravity particles are captured and retained in the concentrating cone; the high-density material may also substitute for the low-density material that was previously in the riffles—made possible by the fluidization of the bed. When the concentrate cycle is complete, concentrates are flushed from the cone into the concentrate launder. Under normal operating conditions, this automated procedure is achieved in less than 2 min in a secure environment. The first installation was in the grinding circuit at Camchib Mine, Chibougamau, Quebec, Canada, in 1987.

(2) Falcon concentrator

It is another spinning batch concentrator (Figure 4.18) designed principally for the recovery of free gold from classifier underflows in grinding circuit where, again, a very small (<1%) mass pull to concentrate is required. The feed first flows up the sides of a cone-shaped bowl, where it stratifies according to particle density before passing over a concentrate bed fluidized from behind by back-pressure (process) water. The bed retains dense particles such as gold, and lighter gangue particles are washed over the top. Periodically the feed is stopped, the bed rinsed to remove any remaining lights and is then flushed out as the heavy product. Rinsing/flushing frequency, which is under automatic control, is determined from grade and recovery requirements. The first installation was at the Blackdome Gold Mine, British Columbia, Canada, in 1986.

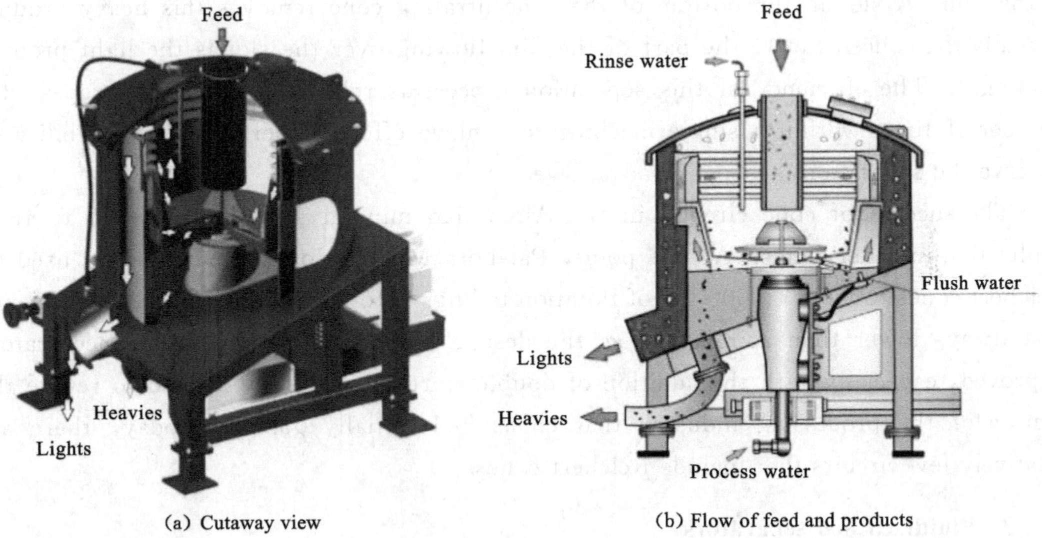

Figure 4.18 Falcon SB concentrator

These two batch centrifugal concentrators have been widely applied in the recovery of gold, platinum, silver, mercury, and native copper; continuous versions, the Knelson Continuous Variable Discharge (CVD) and the Falcon Continuous (C), are also operational.

2. Sluices and cones

(1) Pinched sluices

Pinched sluices of various forms have been used for heavy mineral separations for centuries and are familiar in many western movies. In its simplest form (Figure 4.19), it is an inclined launder about 1 m long, narrowing from about 200 mm in width at the feed end to about 25 mm at the discharge. Pulp of between 50% and 65% solids by weight is fed with minimal turbulence and stratifies as it descends; at the discharge end these strata are separated by various means, such as by splitters, or by some type of tray. The simple sluice box can be a relatively efficient gravity concentrator, provided it is correctly operated.

(2) Reichert cone

The Reichert cone is a wet gravity concentrating device that was designed for high-capacity applications in the early 1960s, primarily to treat titanium-bearing beach sands. Its principle of operation is similar to the pinched sluice. Figure 4.20 shows a schematic of a Reichert cone, with the feed pulp being distributed evenly around the periphery of the cone. As it flows toward the center of the cone the heavy particles separate to the bottom of the film. A slot in the bottom of the concentrating cone removes this heavy product (usually the concentrate); the part of the film flowing over the slot is the light product (tailings). The efficiency of this separation process is relatively low and is repeated a number of times within a single machine to achieve effective performance. Inability to observe the separation is also a disadvantage.

The success of cone circuits in the Australian mineral sand industry led to their application in other fields. At one point, Palabora Mining Co. in South Africa used 68 Reichert cones to treat 34,000 t/d of flotation tailings. However, the separation efficiency was always lower than spirals and as the design and efficiency of spiral concentrators improved, especially with the addition of double-start spirals, they started to retake the section of the processing industry that cones had initially gained. Today, there are relatively few circuits that include Reichert cones.

3. Fluidized bed separators

(1) Hindered-bed separators

Fluidized bed separators (FBS), also known as teetered-bed or hindered-bed separators, have been in mineral processing plants for over a century. Initially used for size separation, FBS units can be operated to provide efficient density-based separation in the particle size range 1～0.15 mm by exploiting hindered settling and the autogenous dense medium naturally provided by fine high-density particles in the feed.

A typical arrangement [Figure 4.21(a)] is to feed slurry into the vessel and let the particles descend against an upward current of water. The upward velocity is set to match

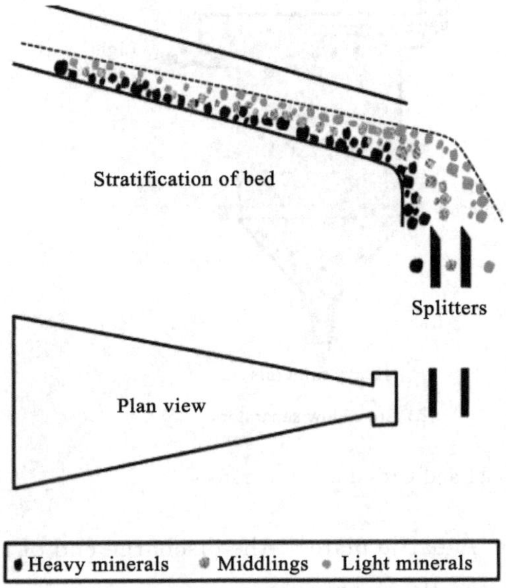

Figure 4.19 Pinched sluice

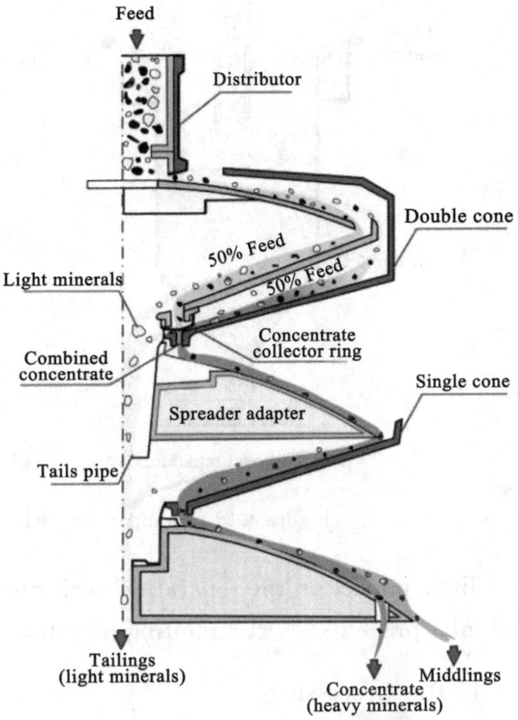

Figure 4.20 Schematic of the Reichert cone concentrator

the settling velocity of the finest fraction of the dense particles, resulting in accumulation to form the fluidized bed. The bed level is monitored, and underflow discharge controlled to remove the heavy product at a rate dependent on the mass of heavies in the incoming feed. The lights cannot penetrate the bed and report as overflow.

Some positive features of FBS units include efficient separation at flow rates up to 20 t/(h · m^2), the capability to adjust to variations in feed characteristics, and general simplicity of operation. There is a need for close control of the top size and for clean fluidization water to avoid plugging the injection system.

There have been some significant advances in design. Noting that entering feed slurry could cause some disruption to the teeter bed, Mankosa and Luttrell(1999) developed a feed system to gently introduce the feed across the top of an FBS unit, now known as the CrossFlow Separator.

(2) CrossFlow separator

In this device, rather than the feed entering the teeter bed, a tangential feed inlet, which increases in area to the full width of the separator to reduce input turbulence, directs the feed slurry across the top of the chamber, leaving chamber contents largely undisturbed [Figure 4.21(b)]. The upward velocity in the separator is thus constant and because the feed does not directly enter the teeter bed, variations in feed characteristics

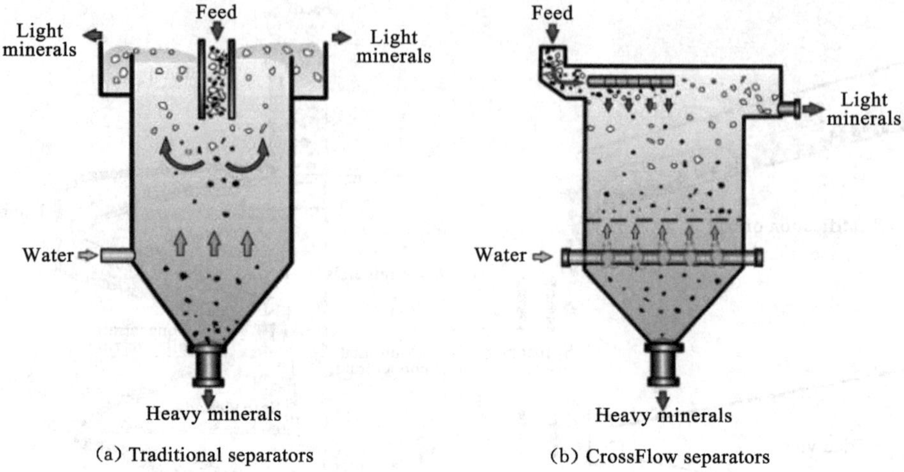

Figure 4.21 Comparison of traditional and CrossFlow separators

have little impact on the separation performance. A baffle plate at the discharge end of the feed inlet prevents short-circuiting of solids into the floats product.

4. Dry processing

The pneumatic technologies follow the same basic mechanisms described for wet processing. With demand to reduce water usage growing, these technologies are worth reexamining.

(1) Pneumatic tables

Originally developed for seed separation, pneumatic or air tables have an important use in the treatment of heavy mineral sand deposits, and in the upgrading of asbestos, and in applications where water is at a premium. Pneumatic tables use a throwing motion to move the feed along a flat riffled deck, and blow air continuously up through a porous bed. The stratification produced is somewhat different from that of wet tables. Whereas in wet tabling the particle size increases and the density decreases from the top of the concentrate band to the tailings, on an air table both particle size and density decrease from the top down, the coarsest particles in the middlings band having the lowest density. Pneumatic tabling is therefore similar in effect to hydraulic classification. They are commonly used in combination with wet tables to clean zircon concentrates, one of the products obtained from heavy mineral sand deposits. Such concentrates are often contaminated with small amounts of fine quartz, which can effectively be separated from the coarse zircon particles by air tabling. Some fine zircon may be lost in the tailings and can be recovered by treatment on wet shaking tables.

(2) Air jigs

These units employ a constant air flow through a jig bed supported on a fixed screen in order to open the bed and allow stratification. Bed level is sensed and used to control the discharge rate in proportion to the amount of material to be rejected. There are several installations around the world, mostly for coal cleaning. Commercial units, for example, can treat up to 60 t/h of 75～12 mm coal.

(3) Other pneumatic-based devices

Various units have been modified to operate dry including: the Reflux Classifier, Knelson concentrator, Reichert cone, and fluidized bed devices. The latter devices overlap with DMS. For example, applications in coal employ a dense medium of air and magnetite—the air dense medium fluidized bed (ADMFB) process. The first commercial ADMFB installation was a 50 t/h unit in China in 1994.

Notes

1. In an attempt to recover fine particles using gravity concentration methods, devices have been developed to make use of centrifugal force.

为了使用重力选矿方法回收细颗粒,已经开发了利用离心力的设备。

2. These two batch centrifugal concentrators have been widely applied in the recovery of gold, platinum, silver, mercury, and native copper.

这两款间歇式离心选矿机已广泛应用于金、铂、银、汞、天然铜的回收。

3. The Reichert cone is a wet gravity concentrating device that was designed for high-capacity applications in the early 1960s, primarily to treat titanium-bearing beach sands.

赖克特圆锥选矿机是一种湿式重力选矿设备,是为 20 世纪 60 年代初期的高处理量应用而设计的,主要用于处理含钛的海滩砂矿。

4. Its principle of operation is similar to the pinched sluice. Figure 4.20 shows a schematic of a Reichert cone, with the feed pulp being distributed evenly around the periphery of the cone.

其(赖克特圆锥选矿机)操作原理类似于尖缩溜槽。图 4.20 为赖克特圆锥选矿机示意图,入料矿均匀地分布在锥体的外围。

5. Initially used for size separation, FBS units can be operated to provide efficient density-based separation in the particle size range 1～0.15 mm by exploiting hindered settling and the autogenous dense medium naturally provided by fine high-density particles in the feed.

FBS 设备最初用于分级,可通过利用干扰沉降和入料中细而密度大的颗粒作为自生重介质,在 1～0.15 mm 粒径范围内提供有效的基于密度的分离。

6. Originally developed for seed separation, pneumatic or air tables have an important use in the treatment of heavy mineral sand deposits, and in the upgrading of asbestos, and in applications where water is at a premium.

风力摇床或空气摇床最初是为种子分选而开发的,它们在处理重砂矿床、石棉提质以及水资源非常宝贵的地区的应用中具有重要用途。

7. They are commonly used in combination with wet tables to clean zircon concentrates, one of the products obtained from heavy mineral sand deposits.

风力摇床通常与湿式摇床结合使用，对锆石精矿进行精选，这些精矿是从重砂矿床中获得的产品之一。

Exercises

1. Answer the following questions according to the text.

(1) What are the applications of these two batch centrifugal concentrators (Knelson concentrator and Falcon concentrator)?

(2) What is the simplest form of pinched sluices?

(3) How does the Reichert cone realize the separation of minerals?

(4) What is the typical arrangement of the fluidized bed separators?

(5) What is the difference between pneumatic tables and wet tables in stratification?

2. Translate the following phrases into English.

(1) 离心选矿机　(2) 尖缩溜槽　(3) 空气摇床　(4) 流化床分选机

(5) 圆锥选矿机　(6) 细颗粒　(7) 重砂矿床　(8) 重矿物

3. Translate the following paragraphs into Chinese.

Paragraph 2 in section 1 (1) "Feed slurry is introduced through a stationary feed tube ...".

4. Translate the following sentences into English.

(1) 为了使用重力分选方法回收细颗粒矿物，开发了利用离心力的设备。

(2) 这是一种紧凑的间歇式离心分选机，有活动的流化床以捕获重矿物。

(3) 进料矿浆通过固定进料管进入分选锥。当矿浆到达锥底时，在离心力的作用下，矿浆被强制沿锥壁向上向外流动。

(4) 流化床分选机(FBS)，也称为干扰床分选机，已经在选矿厂使用了一个多世纪。

5 DENSE MEDIUM SEPARATION(重介质分选)

5.1 Principle and Application of Dense Medium Separation (重介质分选原理及应用)

1. Basic principles

Dense medium separation (or the sink-and-float process) is applied to the pre-concentration of minerals, i. e., the rejection of gangue prior to grinding for final liberation. It is also used in coal preparation to produce a commercially graded end-product, clean coal being separated from the heavier shale or high-ash coal.

In principle, it is the simplest of all gravity processes and has long been a standard laboratory method for separating minerals of different specific gravity. Heavy liquids of suitable density are used, so that those minerals lighter than the liquid float, while those denser than it sink (Figure 5.1).

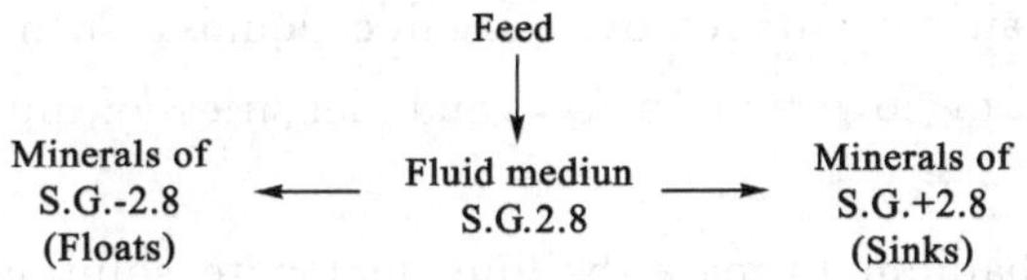

Figure 5.1 Principle of heavy medium separation

Since most of the liquids used in the laboratory are toxic, or give off harmful vapours, the heavy medium used in industrial separations is a thick suspension, or pulp, of some heavy mineral in water, which behaves as a heavy liquid.

The process offers some advantages over other gravity processes. It has the ability to make sharp separations at any required density, with a high degree of efficiency even in the presence of high percentages of near-density material. The density of separation can be closely controlled, within a relative density of ±0.005 kg/L and can be maintained, under normal conditions, for indefinite periods. The separating density can, however, be changed at will and fairly quickly, to meet varying requirements. The process is, however, rather expensive, mainly due to the ancillary equipment needed to clean the medium and the cost of the medium itself.

The process is most widely applied when the density difference occurs at a coarse particle size, as separation efficiency decreases with size due to the slower rate of settling of the particles. Particles should preferably be larger than about 4 mm in diameter, in which case separation can be effective on a difference in specific gravity of 0.1 or less. Separation down to 500 μm, and less, in size can, however, be made by the use of centrifugal separators. Providing a density difference exists, there is no upper size limit except that determined by the ability of the plant to handle the material.

2. The dense medium

(1) Liquids

Heavy liquids have wide use in the laboratory for the appraisal of gravity-separation techniques on ores. Heavy liquid testing may be performed to determine the feasibility of dense medium separation on a particular ore, and to determine the economic separating density, or it may be used to assess the efficiency of an existing dense medium circuit by carrying out tests on the sink and float products. The aim is to separate the ore samples into a series of fractions according to density, establishing the relationship between the high and the low specific gravity minerals.

Tetrabromoethane (TBE), having a specific gravity of 2.96, is commonly used and may be diluted with white spirit or carbon tetrachloride (sp. gr. 1.58) to give a range of densities below 2.96. Bromoform (sp. gr. 2.89) may be mixed with carbon tetrachloride to give densities in the range 1.58～2.89. For densities of up to 3.3, diiodomethane is useful, diluted as required with triethyl orthophosphate. Aqueous solutions of sodium polytungstate have certain advantages over organic liquids, such as being virtually non-volatile, non-toxic and of lower viscosity, and densities of up to 3.1 can easily be achieved.

Clerici solution (thallium formate-thallium malonate solution) allows separation at densities up to 4.2 at 20 ℃ or 5.0 at 90 ℃. Separations of up to 18 kg/L can be achieved by the use of magnetohydrostatics, i.e., the utilization of the supplementary weighting force produced in a solution of a paramagnetic salt or ferrofluid when situated in a magnetic field gradient. This type of separation is applicable primarily to non-magnetic minerals with a lower limiting particle size of about 50 μm.

Many heavy liquids give off toxic fumes and must be used with adequate ventilation: the Clerici liquids are extremely poisonous and must be handled with extreme care. The use of pure liquids on a commercial scale has therefore not been found practicable, and industrial processes employ finely ground solids suspended in water.

(2) Suspensions

Below a concentration of about 15% by volume, finely ground suspensions in water behave essentially as simple Newtonian fluids. Above this concentration, however, the suspension becomes non-Newtonian and a certain minimum stress, or yield stress, has to be applied before shear will occur and the movement of a particle can commence. Thus, small particles, or those close to the medium density, are unable to overcome the rigidity offered by the medium before movement can be achieved. This can be overcome to some extent either by increasing the shearing forces on the particles, or by decreasing the apparent viscosity of the suspension. The shearing force may be increased by substituting centrifugal force for gravity. The viscous effect may be decreased by agitating the medium, which causes elements of liquid to be sheared relative to each other. In practice the medium is never static, as motion is imparted to it by paddles, air, etc. , and also by the sinking material itself. All these factors, by reducing the yield stress, tend to bring the parting or separating density as close as possible to the density of the medium in the bath.

In order to produce a stable suspension of sufficiently high density, with a reasonably low viscosity, it is necessary to use fine, high specific gravity solid particles, agitation being necessary to maintain the suspension and to lower the apparent viscosity. The solids comprising the medium must be hard, with no tendency to slime, as degradation increases the apparent viscosity by increasing the surface area of the medium. The medium must be easily removed from the mineral surfaces by washing, and must be easily recoverable from the fine-ore particles washed from the surfaces. It must not be affected by the constituents of the ore and must resist chemical attack, such as corrosion.

The most widely used medium for metalliferous ores is now ferrosilicon, whilst magnetite is used in coal preparation. Recovery of medium in both cases is by magnetic separation.

Magnetite (sp. gr. 5.1) is relatively cheap, and is used to maintain bath densities of up to 2.5 kg/L. Ferrosilicon (sp. gr. 6.7～6.9) is an alloy of iron and silicon which should contain not less than 82% Fe and 15%～16% Si. Milled ferrosilicon is produced in a range of size distributions, from 30 to 95% −45 μm, the finer grades being used for finer ores and centrifugal separators. The coarser, lower viscosity grades can achieve medium densities up to about 3.3. Atomised ferrosilicon consists of rounded particles which produce media of lower viscosity and can be used to achieve densities as high as 4.

3. Laboratory heavy liquid tests

Laboratory testing may be performed on ores in order to assess the suitability of dense medium separation and other gravity methods, and to determine the economic separating density.

Liquids covering a range of densities in incremental steps are prepared, and the representative sample of crushed ore is introduced into the liquid of highest density. The floats product is removed and washed and placed in the liquid of next lower density, whose float product is then transferred to the next lower density and so on. The sinks product is finally drained, washed, and dried, and then weighed, together with the final floats product, to give the density distribution of the sample by weight (Figure 5.2).

Care should be taken when evaluating ores of fine particle size that sufficient time is given for the particles to settle into the appropriate fraction. Centrifuging is often carried

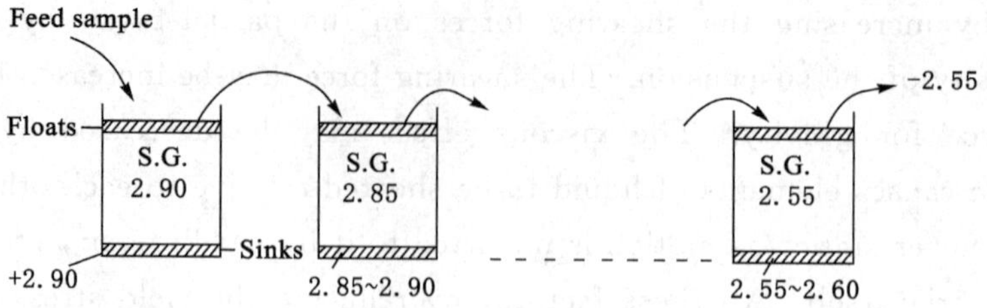

Figure 5.2 Heavy liquid testing

out on fine materials to reduce the settling time, but this should be done with care, as there is a tendency for the floats to become entrained in the sinks fraction. Unsatisfactory results are often obtained with porous materials, such as magnesite ores, due to the entrainment of liquid in the pores, which changes the apparent density of the particles.

After assaying the fractions for metal content, the distribution of material and metal in the density fractions of the sample can be tabulated. Table 5.1 shows such a distribution from tests performed on a tin ore. The computations are easily accomplished in a spreadsheet. It can be seen from columns 3 and 6 of the table that if a separation density of 2.75 was chosen, then 68.48% of the material, being lighter than 2.75, would be discarded as a float product, and only 3.78% of the tin would be lost in this fraction. Similarly, 96.22% of the tin would be recovered into the sink product, which accounts for 31.52% of the original total feed weight.

Table 5.1 Heavy liquid test results

1	2	3	4	5	6
Specific Gravity fraction	Weight /%	Cumulative Weight/%	Assay /(% Sn)	Distribution /(% Sn)	Cumulative distribution/(% Sn)
−2.55	1.57	1.57	0.003	0.00	0.00
2.55～2.60	9.22	10.79	0.04	0.33	0.33
2.60～2.65	26.11	36.90	0.04	0.93	1.27
2.65～2.70	19.67	56.57	0.04	0.70	1.97
2.70～2.75	11.91	68.48	0.17	1.81	3.78
2.75～2.80	10.92	79.40	0.34	3.32	7.10
2.80～2.85	7.87	87.27	0.37	2.60	9.70
2.85～2.90	2.55	89.82	1.30	2.96	12.66
+2.90	10.18	100.00	9.60	87.34	100.00

Heavy liquid tests are important in coal preparation in order to determine the required density of separation and the expected yield of coal of the required ash content. The "ash" content refers to the amount of incombustible material in the coal. Since coal is lighter than the contained minerals, the higher the density of separation the higher is the yield, but the higher is the ash content. The ash content of each density fraction from heavy liquid testing is determined by taking about 1 g of the fraction, placing it in a cold well-ventilated furnace, and slowly raising the temperature to 815 ℃, maintaining the sample at this temperature until constant weight is obtained. The residue is cooled and then weighed. The ash content is the mass of ash expressed as a percentage of the initial sample weight taken.

Table 5. 2 shows the results of heavy liquid tests performed on a coal sample. The coal was separated into the density fractions shown in column 1, and the weight fractions and ash contents are tabulated in columns 2 and 3 respectively. The weight per cent of each product is multiplied by the ash content to give the ash product (column 4).

Table 5. 2 Results of heavy liquid tests on a coal sample

1	2	3	4	5	6	7	8	9	10	11
Sp. gr fraction	Wt/%	Ash/%	Ash product	Separation density	Cumulative float (Clean coal)			Cumulative sink (Discard)		
					Wt/%	Ash product	Ash/%	Wt/%	Ash product	Ash/%
−1. 30	0. 77	4. 4	3. 39	1. 30	0. 77	3. 39	4. 4	99. 23	2,213. 16	22. 3
1. 30～1. 32	0. 73	5. 6	4. 09	1. 32	1. 50	7. 48	5. 0	98. 50	2,209. 67	22. 4
1. 32～1. 34	1. 26	6. 5	8. 19	1. 34	2. 76	15. 67	5. 7	97. 24	2,201. 48	22. 6
1. 34～1. 36	4. 01	7. 2	28. 87	1. 36	6. 77	44. 54	6. 6	93. 23	2,172. 61	23. 3
1. 36～1. 38	8. 92	9. 2	82. 06	1. 38	15. 69	126. 60	8. 1	84. 31	2,090. 55	24. 8
1. 38～1. 40	10. 33	11. 0	113. 63	1. 40	26. 02	240. 23	9. 2	73. 98	1,976. 92	26. 7
1. 40～1. 42	9. 28	12. 1	112. 29	1. 42	35. 30	352. 52	10. 0	64. 70	1,864. 63	28. 8
1. 42～1. 44	9. 00	14. 1	126. 90	1. 44	44. 30	479. 42	10. 8	55. 70	1,737. 73	31. 2
1. 44～1. 46	8. 58	16. 0	137. 28	1. 46	52. 88	616. 70	11. 7	47. 12	1,600. 45	34. 0
1. 46～1. 48	7. 79	17. 9	139. 44	1. 48	60. 67	756. 14	12. 5	39. 33	1,461. 01	37. 1
1. 48～1. 50	6. 42	21. 5	138. 03	1. 50	67. 09	894. 17	13. 3	32. 91	1,322. 98	40. 2
+1. 50	32. 91	40. 2	1,322. 98	—	100. 00	2,217. 15	22. 2	0. 00	0. 00	0. 0
Total	100. 0	22. 2	2,217. 15							

The total floats and sinks products at the various separating densities shown in column 5 are tabulated in columns 6～11. To obtain the cumulative per cent for each gravity fraction, columns 2 and 4 are cumulated from top to bottom to give columns 6 and 7 respectively. Column 7 is then divided by column 6 to obtain the cumulative per cent ash (column 8). Cumulative sink ash is obtained in essentially the same manner, except that columns 2 and 4 are cumulated from bottom to top to give columns 9 and 10 respectively.

The results of Table 5. 2 are plotted in Figure 5. 3 as typical washability curves. Suppose an ash content of 12% is required in the coal product. It can be seen from the washability curves that such a coal would be produced at a yield of 55% (cumulative percent floats), and the required density of separation is 1.465.

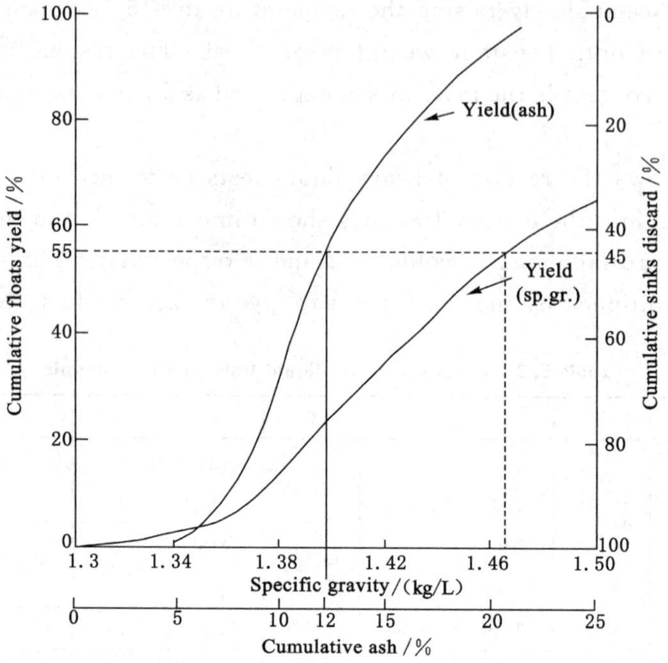

Figure 5.3 Typical coal washability curves

The amount of near-gravity material present is sometimes regarded as being the weight of material in the range of ±0.1 kg/L or ±0.05 kg/L of the separating density, and separations involving feeds with less than about 7% of ±0.1 near-gravity material are regarded by coal preparation engineers as being fairly easy to control. Such separations are often performed in Baum jigs, as these are cheaper than dense medium plants, which require expensive media-cleaning facilities, and no feed preparation, i. e., removal of the fine particles by screening, is required. However, the density of separation in jigs is not as easy to control to fine limits, as it is in dense medium baths, and for near-gravity material much above 7%, dense medium separation is preferred.

Notes

1. Dense medium separation (or the sink-and-float process) is applied to the pre-concentration of minerals, i. e., the rejection of gangue prior to grinding for final liberation.

重介质分选(或浮沉选矿法),用于矿物预选,也就是在最终解离磨矿前丢弃脉石。

2. In principle, it is the simplest of all gravity processes and has long been a standard laboratory method for separating minerals of different specific gravity. Heavy liquids of suitable density are used, so that those minerals lighter than the liquid float, while those denser than it sink.

从原理来讲,该方法是所有重选工艺中最简单的方法,长期以来一直是分离不同密度矿物的标准实验室方法。使用密度适当的重液,使那些比重液轻的矿物上浮,而密度比重液大的矿物下沉。

3. It has the ability to make sharp separations at any required density, with a high degree of efficiency even in the presence of high percentages of near-density material.

该方法(重介质选矿)能够在任何要求的密度下进行精确的分离,即使在邻近密度物料百分含量较高时也具有很高的效率。

4. Particles should preferably be larger than about 4 mm in diameter, in which case separation can be effective on a difference in specific gravity of 0.1 or less.

颗粒直径最好是大于 4 mm,在这种情况下,对于相对密度差只有 0.1 或更小时也可有效分离。

5. Laboratory testing may be performed on ores in order to assess the suitability of dense medium separation and other gravity methods, and to determine the economic separating density.

为了评估重介质分选和其他重选方法的适用性,并确定经济分选密度,可以对矿石进行实验室试验。

6. The floats product is removed and washed and placed in the liquid of next lower density, whose float product is then transferred to the next lower density and so on.

浮物被取出,经洗涤后放入相邻的低密度的重介质中分选,然后再将浮物转移到下一个较低的密度,以此类推。

7. Heavy liquid tests are important in coal preparation in order to determine the require density of separation and the expected yield of coal of the required ash content.

重液试验在选煤中非常重要,以确定所需的分选密度和所要求灰分下的预期产率。

Exercises

1. Answer the following questions according to the text.

(1) What are the advantages of DMS over other gravity processes?

(2) What is the principle of DMS?

(3) What can be done if small particles are unable to overcome the rigidity offered by the medium?

(4) What is the purpose of heavy liquid testing?

(5) What are the experimental steps of the laboratory heavy liquid tests?

2. Translate the following phrases into English.

(1) 重介质分选　(2) 密度　(3) 可选性曲线　(4) 经济分选密度

(5) 重液　(6) 粒度上限　(7) 浮沉实验　(8) 沉降时间

(9) 沉物产品　(10) 悬浮液

3. Translate the following paragraphs into Chinese.

Paragraphs 4 and 5 in section 1 “The process offers some advantages over other ...”.

4. Translate the following sentences into English.

(1) 重介质分选(DMS)还用于煤炭分选,从较重的页岩或高灰分煤中分选出商业品级的最终产品。

(2) 重介质分选工艺比其他选煤工艺具有一定的优势。它能够在任何要求的密度下进行精确的分离,即使在邻近密度物料含量很高的情况下也具有很高的效率。

(3) 通过进行重液实验,以确定在特定矿石上进行重介质分离的可行性,并确定经济分选密度,也可以通过对沉物和浮物产品进行测试来评估现有重介质回路的效率。

(4) 跳汰机中的分离密度不像在重介质槽中那样容易控制到精细极限,对于邻近密度组分远高于7%的物料,首选重介质分离。

5.2 Separating Vessels(分选设备)

1. Gravitational vessels

Several types of separating vessel are in use, and these may be classified into gravitational ("static-baths") and centrifugal (dynamic) vessels.

Gravitational units comprise some form of vessel into which the feed and medium are introduced and the floats are removed by paddles, or merely by overflow. Removal of the sinks is the most difficult part of separator design. The aim is to discharge the sinks particles without removing sufficient of the medium to cause disturbing downward currents in the vessel.

(1) Wemco cone separator

Wemco cone separator (Figure 5.4) is widely used for ore treatment, since it has a relatively high sinks capacity. The cone, which has a diameter of up to 6 m, accommodates feed particles of up to 10 cm in diameter, with capacities of up to 500 t/h.

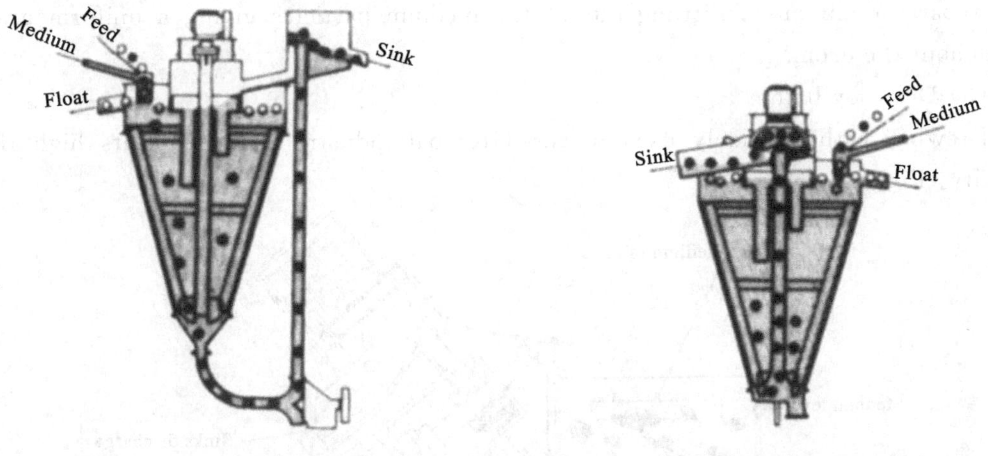

(a) Single-gravity, two-product system with torque-flow-pump sink removal

(b) Single-gravity, two-product system with compressed-air sink removal

Figure 5.4 Wemco cone separator

The feed is introduced on to the surface of the medium by free-fall, which allows it to plunge several centimetres into the medium. Gentle agitation by rakes mounted on the central shaft helps keep the medium in suspension. The float fraction simply overflows a weir, whilst the sinks are removed by pump or by external or internal air lift.

(2) Drum separators

Drum separators (Figure 5.5) are built in several sizes, up to 4.3 m diameter by 6 m long, with maximum capacities of 450 t/h, and can treat feeds of up to 30 cm in diameter. Separation is accomplished by the continuous removal of the sink product through the action of lifters fixed to the inside of the rotating drum. The lifters empty into the sink launder when they pass the horizontal position. The float product overflows a weir at the opposite end of the drum from the feed chute.

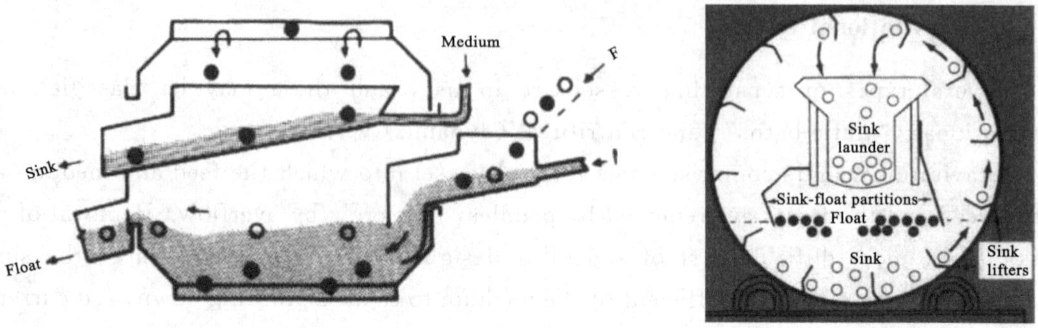

(a) Side view (b) End view

Figure 5.5 Drum separator

Longitudinal partitions separate the float surface from the sink-discharge action of the revolving lifters. The comparatively shallow pool depth in the drum compared with the cone separator minimises settling out of the medium particles giving a uniform gravity throughout the drum.

(3) Drewboy bath

Drewboy bath is widely used in the UK coal industry because of its high floats capacity (Figure 5.6).

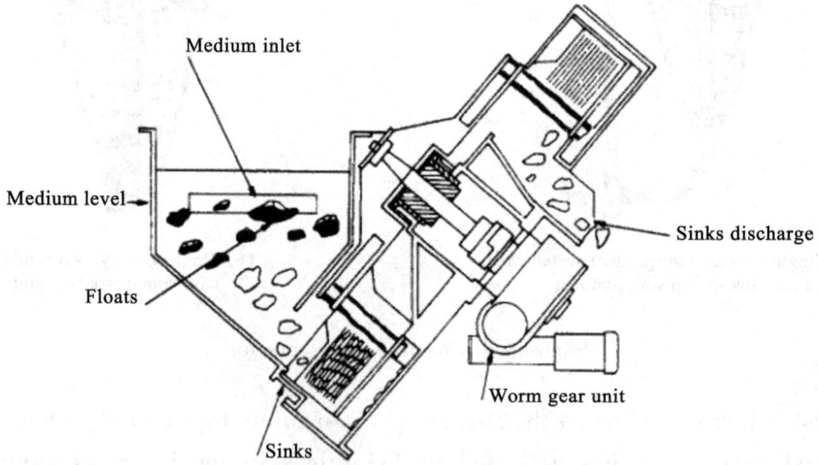

Figure 5.6 Drewboy bath

The raw coal is fed into the separator at one end, and the floats are discharged from the opposite end by a star-wheel with suspended rubber, or chain straps, while the sinks are lifted out from the bottom of the bath by a radial-vaned wheel mounted on an inclined shaft. The medium is fed into the bath at two points—at the bottom of the vessel and with the raw coal—the proportion being controlled by valves.

(4) Norwalt washer

Norwalt washer was developed in South Africa, and most installations are to be found in that country. Raw coal is introduced into the centre of the annular separating vessel, which is provided with stirring arms (Figure 5.7).

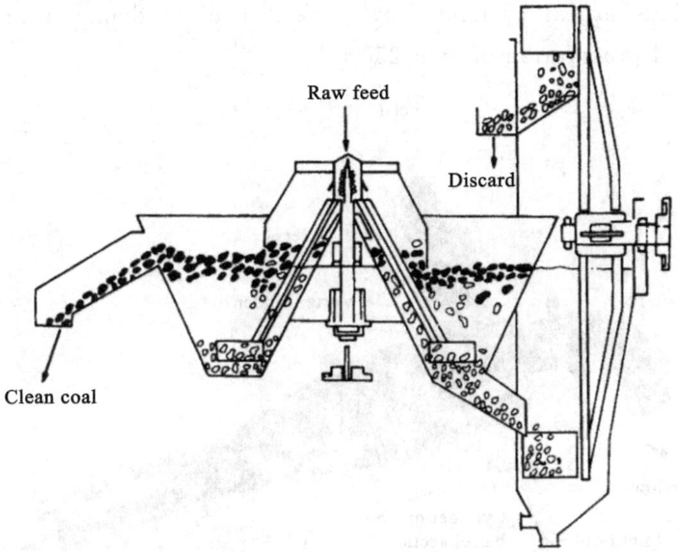

Figure 5.7 Norwalt bath

The floats are carried round by the stirrers, and are discharged over a weir on the other side of the vessel, being carried out of the vessel by the medium flow. The discard sinks to the bottom of the vessel and is dragged along by scrapers attached to the bottom of the stirring arms, and is discharged via a hole in the bottom of the bath into a sealed elevator, either of the wheel or bucket type, which continuously removes the sinks product.

2. Centrifugal separators

Cyclone dense medium separators have now become widely used in the treatment of ores and coal. They provide a high centrifugal force and a low viscosity in the medium, enabling much finer separations to be achieved than in gravitational separators. Feed to these devices is typically de-slimed at about 0.5 mm, to avoid contamination of the medium with slimes, and to minimise medium consumption. A finer medium is required than with gravitational vessels, to avoid medium instability. In recent years work has been carried out in many parts of the world to extend the range of particle size treated by centrifugal separators, particularly those operating in coal preparation plants, where advantages to be gained are elimination of de-sliming screens and reduced froth flotation of the screen undersize, as well as more accurate separation of fine coal.

(1) The dense medium cyclone

By far the most widely used centrifugal DM separator is the cyclone (Figure 5. 8) whose principle of operation is similar to that of the conventional hydrocyclone. The commonest form of DM cyclone is that developed by the Dutch State Mines in the 1940s, which has an included cone angle of 20°. Cyclones typically treat ores and coal in the range 0. 5 ~ 40 mm. The largest cyclones now exceed 1 m in diameter and are capable of throughputs in coal preparation of over 250 t/h.

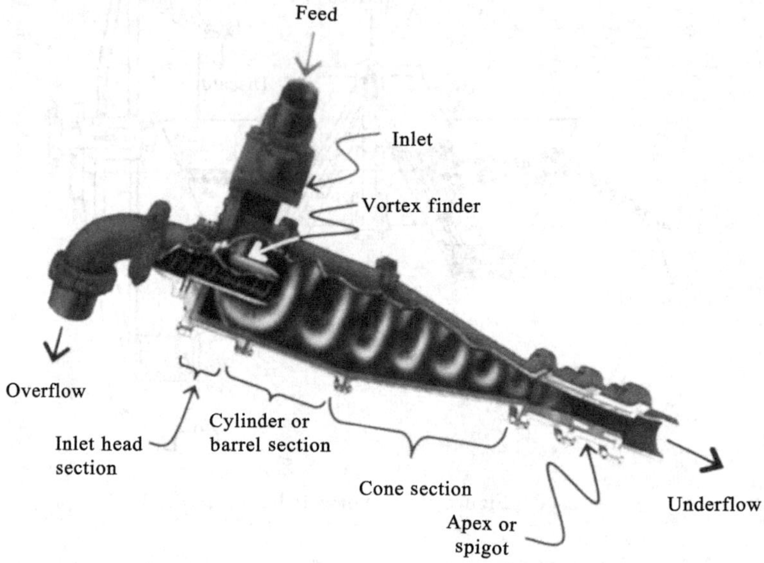

Figure 5. 8 Structure of a dense medium cyclone

The ore or coal is suspended in the medium and introduced tangentially to the cyclone either via a pump or it is gravity-fed. Gravity feeding requires a taller and therefore more expensive building but achieves a more consistent flow and less pump wear and ore degradation. The dense material (reject in the case of coal, product in the case of iron ore) is centrifuged to the cyclone wall and exits at the apex. The light product "floats" to the flow around the axis and exits via the vortex finder.

(2) Vorsyl separator

Vorsyl separator (Figure 5. 9) is used in many coal-preparation plants for the treatment of small coal sizes up to about 50 mm at feed rates of up to 120 t/h. The feed to the separator, consisting of de-slimed raw coal, together with the separating medium of magnetite, is introduced tangentially, or more recently by an involute entry, at the top of the separating chamber, under pressure. Material of specific gravity less than that of the medium passes into the clean coal outlet via the vortex finder, while the near gravity material and the heavier shale particles move to the wall of the vessel due to the centrifugal acceleration induced. The particles move in a spiral path down the chamber towards the base of the vessel where the drag caused by the proximity of the orifice plate reduces the tangential velocity and creates a strong inward flow towards the throat.

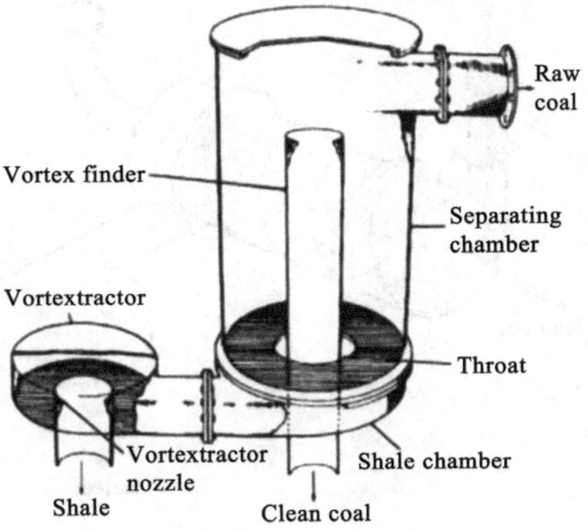

Figure 5.9 Vorsyl separator

This carries the shale, and near gravity material, through zones of high centrifugal force, where a final precise separation is achieved. The shale, and a proportion of the medium, discharge through the throat into the shallow shale chamber, which is provided with a tangential outlet, and is connected by a short duct to a second shallow chamber known as the vortextractor. This is also a cylindrical vessel with a tangential inlet for the medium and reject and an axial outlet. An inward spiral flow to the outlet is induced, which dissipates the inlet pressure energy and permits the use of a large outlet nozzle without the passing of an excessive quantity of medium.

(3) The LARCODEMS

The LARCODEMS (large coal dense medium separator) was developed to treat a wide size range of coal (−100 mm) at high capacity in one vessel. It has also been used in concentrating iron ore. The unit (Figure 5.10) consists of a cylindrical chamber which is inclined at approximately 30° to the horizontal. Feed medium at the required relative density is introduced under pressure, either by pump or static head, into the involute tangential inlet at the lower end. At the top end of the vessel is another involute tangential outlet connected to the vortextractor. Raw coal of 0.5~100 mm is fed into the separator by a chute connected to the top end, the clean coal after separation being removed through the bottom outlet. High relative density particles pass rapidly to the separator wall and are removed through the top involute outlet and the vortextractor.

The first installation of the device was as the main processor in the 250 t/h coal preparation plant at Point of Ayr Colliery in the United Kingdom. As the 250 t/h LARCODEMS is only 1.2 m diameter by 3 m long, it could have a dramatic effect on the design and construction of future coal preparation plants.

(4) Dyna Whirlpool separator

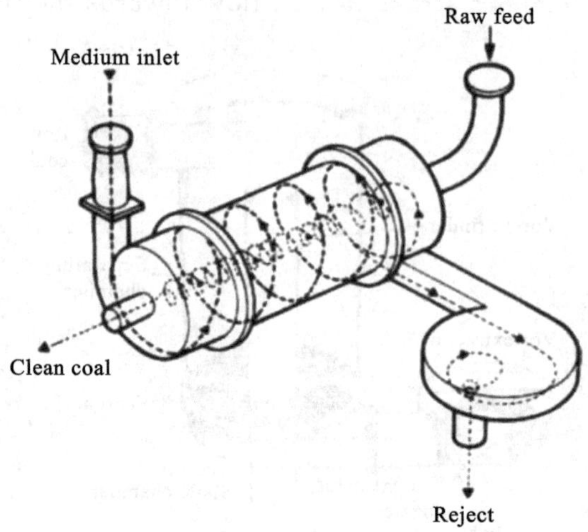

Figure 5.10 LARCODEMS

Dyna Whirlpool separator is similar to the LARCODEMS, and is used for treating fine coal, particularly in the Southern Hemisphere, as well as diamonds, fluorspar, tin, and lead-zinc ores, in the size range 0.5～30 mm.

It consists of a cylinder of predetermined length (Figure 5.11), having identical tangential inlet and outlet sections at either end. The unit is operated in an inclined position and medium of the required density is pumped under pressure into the lower outlet. The rotating medium creates a vortex throughout the length of the unit and leaves via the upper tangential discharge and the lower vortex outlet tube. Raw feed entering the upper vortex tube is sluiced into the unit by a small quantity of medium and a rotational motion is quickly imparted by the open vortex. Float material passes down the vortex and does not contact the outer walls of the unit, thus greatly reducing wear. The floats are discharged from the lower vortex outlet tube. The heavy sink particles of the feed penetrate the rising medium towards the outer wall of the unit and are discharged with medium through the sink discharge pipe. Since the sinks discharge is close to the feed inlet, the sinks are removed from the unit almost immediately, again reducing wear considerably. Only near-gravity particles which are separated further along the unit actually come into contact with the main cylindrical body. The tangential sink discharge outlet is connected to a flexible sink hose and the height of this hose may be used to adjust back pressure to finely control the cut-point. The capacity of the separator can be as high as 100 t/h, and it has several advantages over the DSM cyclone. Apart from the reduced wear, which not only decreases maintenance costs but also maintains performance of the unit, operating costs are lower, since only the medium is pumped. The unit has a much higher sinks capacity and can accept large fluctuations in sink/float ratios.

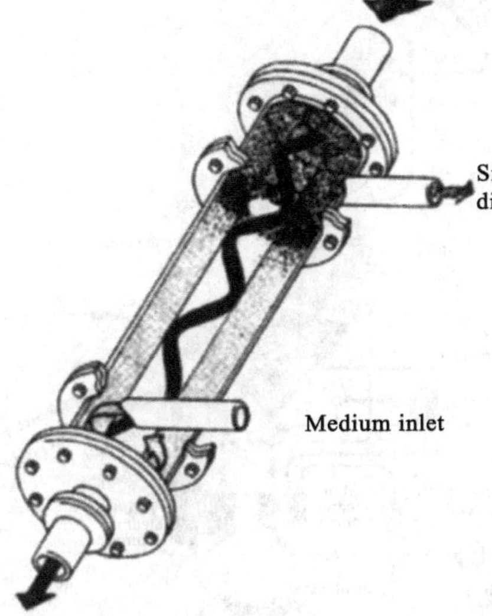

Figure 5.11 Dyna Whirlpool separator

3. Dense medium separation (DMS) circuits

Although the separating vessel is the most important element of a DMS process, it is only one part of a relatively complex circuit. Other equipment is required to prepare the feed, and to recover, clean, and recirculate the medium.

The greatest expense in any dense medium circuit is the provision for reclaiming and cleaning the medium which leaves the separator with the sink and float products. A typical circuit is shown in Figure 5.12.

In a typical coal preparation plant, the deslimed raw coal is fed to a head tank where it is mixed with the dense medium, and this mixture flows by gravity to the cyclone. The overflow (float), which contains the washed coal, flows to a sieve bend screen and then to a vibrating screen. Most of the medium drains through these screens and is returned to the medium sump for reuse. The coal is also rinsed to remove the remaining magnetite while passing over the last section of the vibrating screen. The dilute medium (predominately magnetite and rinse water, but containing some nonmagnetic material) flows to a magnetic separator which recovers and concentrates the magnetite. The magnetic separator also removes the nonmagnetic particles which would otherwise tend to accumulate in the dense medium. The magnetite flows from the magnetic separator back to the medium sump. The underflow (sink) from the cyclone, which contains the refuse, is processed similarly.

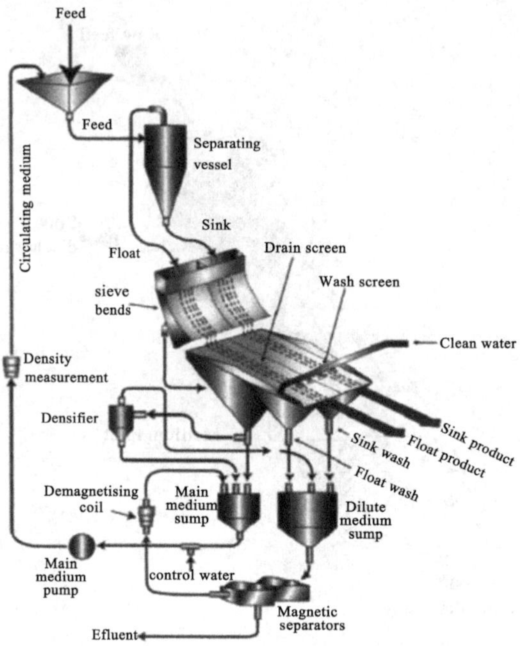

Figure 5.12　Typical dense medium separation circuit

Notes

1. Gravitational units comprise some form of vessel into which the feed and medium are introduced and the floats are removed by paddles, or merely by overflow.

重力沉降型分选机为某种形状的设备，给入的原料和介质完成分选后，浮物以刮板刮出，或以溢流形式排出。

2. Longitudinal partitions separate the float surface from the sink-discharge action of the revolving lifters.

纵向隔板将浮物表面与旋转提升的沉物隔开。

3. The comparatively shallow pool depth in the drum compared with the cone separator minimises settling out of the medium particles giving a uniform gravity throughout the drum.

与圆锥选矿机相比，滚筒选矿机的分选池深度较小，最大限度地减少了介质颗粒的沉降，使整个滚筒具有均匀的密度。

4. By far the most widely used centrifugal DM separator is the cyclone (Figure 5.8) whose principle of operation is similar to that of the conventional hydrocyclone.

迄今为止，使用最广泛的离心式重介质分选机是旋流器(图 5.8)，这种旋流器的工作原理与传统水力旋流器相似。

5. Vorsyl separator (Figure 5.9) is used in many coal-preparation plants for the treatment of small coal sizes up to about 50 mm at feed rates of up to 120 t/h.

沃西尔重介质旋流器(图 5.9)用于许多选煤厂，以高达 120 t/h 的进料速度处理最大粒度约 50 mm 的细粒煤。

6. In a typical coal preparation plant, the deslimed raw coal is fed to a head tank where it is mixed with the dense medium, and this mixture flows by gravity to the cyclone.

在一个典型的选煤厂中,脱泥的原煤被送入第一个槽中,在那里与重介质混合,并在重力作用下流向旋流器。

Exercises

1. Answer the following questions according to the text.

(1) How does Wemco cone separator work in separation?

(2) What are the heavy medium separation vessels introduced in the text?

(3) What is the requirement of the feed to these cyclone dense medium separators?

(4) What is the greatest expense in any dense medium circuit?

2. Translate the following phrases into English.

(1) 圆筒分选机 (2) 重介质旋流器 (3) 选煤厂 (4) 稀介质

(5) 沉砂口 (6) 重介质分选流程 (7) 溢流管 (8) 介质回收与净化

3. Translate the following paragraphs into Chinese.

Paragraphs 1 and 2 in section 2 (1) "By far the most widely used centrifugal DM ...".

4. Translate the following sentences into English.

(1) 沉物的排出是分选机设计中最困难的部分。其目的是在不移除大量介质的情况下排放沉物颗粒,避免在分选机中引起向下的介质流的扰动。

(2) 给料通过自由落体的方式进入介质,使其能够进入介质中几厘米。

(3) 重介质旋流器现已广泛应用于矿石和煤炭的处理。它们在介质中提供高离心力和低黏度的分选环境,使其能够实现比重力分选机更好的分选效果。

(4) 虽然分选设备是重介质分选工艺中最重要的部分,但它只是相对复杂流程的一部分。需要其他设备来准备进料,并回收、净化和循环介质。

6 MAGNETIC AND ELECTROSTATIC SEPARATION (磁电分选)

6.1 Principles of Magnetic Separation(磁选原理)

1. Concept of magnetic separation

Magnetic separation is the process of separating ore minerals exploiting the difference in magnetic properties between them. Magnetic separators are used to separate either valuable minerals from non-magnetic gangue, e. g. , magnetite from quartz, or magnetic contaminants or other valuable minerals from the non-magnetic values. An example of this is the tin-bearing mineral cassiterite, which is often associated with traces of magnetite or wolframite which can be removed by magnetic separators.

Magnetic separation has been used for more than 200 years in the concentration of iron ores, and is still used extensively today. A wide variety of devices have been used. Since the start of this century there has been a steady expansion in both the equipment available and the range of ores for which magnetic separation is applicable. The removal of small amounts of iron and iron-bearing minerals from industrial minerals has become an important application, as has the concentration of various ferrous and nonferrous minerals.

2. Magnetic properties of materials

All materials display certain magnetic properties, regardless of their composition and state. According to their magnetic properties, materials can be divided into five basic groups: diamagnetic, paramagnetic, ferromagnetic, antiferromagnetic and ferrimagnetic. The last three groups have generally higher magnetic susceptibility than the other groups and are frequently termed "ferromagnetic".

Materials can also be classified into two broad groups, according to whether they are attracted or repelled by a magnet: ① Diamagnetics are repelled along the lines of magnetic force to a point where the field intensity is smaller. The forces involved here are very small and diamagnetic substances cannot be concentrated magnetically. ② Paramagnetics are attracted along the lines of magnetic force to points of greater field intensity. Paramagnetic materials can be concentrated in high-intensity magnetic separators. Examples of paramagnetics which are separated in commercial magnetic separators are ilmenite ($FeTiO_3$), rutile (TiO_2), wolframite ($(Fe, Mn)WO_4$), monazite (rare earth phosphate), siderite ($FeCO_3$), pyrrhotite (FeS), chromite ($FeCr_2O_4$), hematite (Fe_2O_3), and manganese minerals.

Some elements are themselves paramagnetic, such as Ni, Co, Mn, Cr, Ce, Ti, O, and the Pt group metals, but in most cases the paramagnetic properties of minerals are due to the presence of iron in some ferromagnetic form. Ferromagnetism can be regarded as a special case of paramagnetism, involving very high forces.

Ferromagnetic materials have very high susceptibility to magnetic forces and retain some magnetism when removed from the field (remanence). They can be concentrated in low-intensity magnetic separators and the principal ferromagnetic mineral separated is magnetite (Fe_3O_4), although hematite (Fe_2O_3) and siderite ($FeCO_3$) can be roasted to produce magnetite and hence give good separation. The removal of "tramp" iron from ores can also be regarded as a form of low-intensity magnetic separation.

The shapes of magnetization curves of various classes of materials are shown in Figure 6. 1. A simple linear relationship exists for the magnetization curves of paramagnetic and diamagnetic materials. Such a relationship does not suffer for ferromagnetic materials; their behavior is much more complicated.

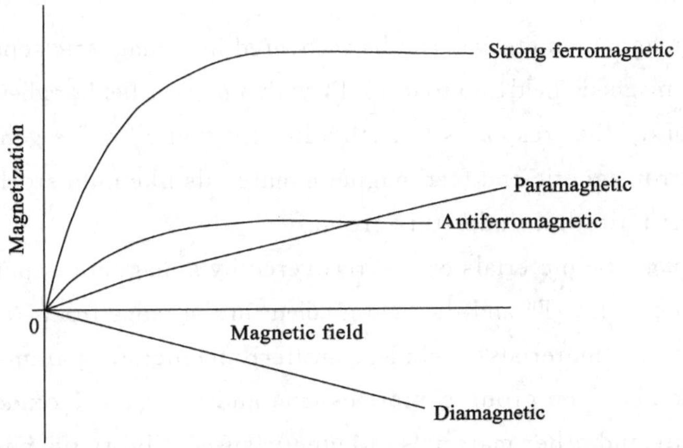

Figure 6. 1 Magnetization curves of various materials

The slope of the magnetization curve is the magnetic susceptibility of the material. The magnetic susceptibility is a small positive constant for paramagnetic materials, a very small negative constant for diamagnetic materials, and for ferromagnetic materials it is variable and depends on the applied magnetic field and the previous history of the sample.

Figure 6. 2 shows a typical magnetization curve of ferromagnetic materials. In an applied magnetic field (H), the magnetization of ferromagnetic materials increases until the saturation magnetization is reached (B_m). If the applied field intensity is reduced, the magnetization decreases but does not come back to its original value. Such an irreversible process of magnetization is called hysteresis. When the intensity H decreases to 0, there is a residual magnetization of the material (B_r), which is called its magnetic remanence. In order to reduce the value of B to 0, it is necessary to inversely increase the intensity H to the value of H_c. The value of H_c is called coercive force.

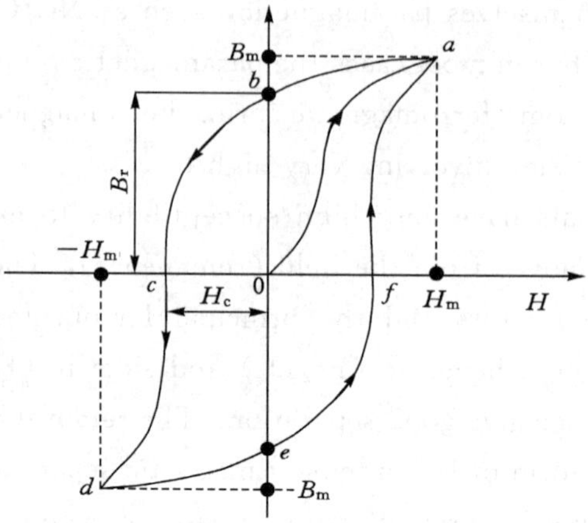

Figure 6.2 Magnetization curve of ferromagnetic materials

For the purpose of magnetic separation, materials can be divided into three basic groups:

(1) Strongly magnetic materials can be recovered by a magnetic separator with the use of a relatively weak magnetic field, up to 0.15 T, with a modest field gradient of 0.5 T/m. The magnetic susceptibility that responds to such a low magnetic field is generally greater than 10^{-4} m^3/kg and ferromagnetic and ferrimagnetic materials like mild steel, iron, magnetite, maghemite and pyrrhotite fall under this group.

(2) Weakly magnetic materials can be recovered by a magnetic separator that generates magnetic induction up to 1.0 T, and the field gradient in the range from 50 T/m to 500 T/m. This broad group of materials includes antiferromagnetic, paramagnetic and some ferromagnetic minerals. The group comprises iron and manganese oxides and carbonates, limonite, wolframite and other materials. Magnetic susceptibility of these materials ranges from 10^{-7} m^3/kg to 5×10^{-6} m^3/kg.

(3) Non-magnetic materials are those materials that cannot be easily recovered by conventional magnetic separators. They include very weakly paramagnetic and antiferromagnetic materials with magnetic susceptibility smaller than, for example, 10^{-7} m^3/kg. Austenitic stainless steel, aluminium, rutile, pyrite, garnet and red blood cells belong to this group. Diamagnetic materials, with their negative magnetic susceptibility are also included.

3. Magnetic field and magnetic force

(1) Forces on particles

The separation is achieved by passing the suspensions or the mixtures of particles through a non-homogeneous magnetic field. In magnetic separation the separating external force is the magnetic force. The separation of one material from another or the removal of magnetizable particles from streams depend upon their motion in response to the magnetic force and to other competing external forces, namely gravitational, inertial, hydrodynamic and centrifugal forces. Magnetic separation is a physical separation of discrete particles based on a three-way competition between: ① Magnetic forces; ② Gravitational, centrifugal, frictional, or inertial forces; ③ Attractive or repulsive interparticle forces.

These magnetic, competing, and interparticle forces determine separator performance; they are dependent on the nature of the feed and the character of the separator. The nature of the feed includes its size distribution, magnetic susceptibility, and other physical and chemical properties that may affect the various forces involved.

Thus a necessary condition for a successful separation of more strongly magnetic from less strongly magnetic particles in a magnetic field is that the magnetic force F^{m}_{mag} acting on more strongly magnetic particles must be greater than the sum of all the competing forces F_{comp}. Simultaneously, the magnetic force acting on less strongly magnetic particles F^{n}_{mag} must be smaller than the sum of the corresponding competing forces. Therefore, the following conditions must be met in a magnetic separator:

$$F^{m}_{mag} \geqslant F_{comp} \quad \text{and} \quad F^{n}_{mag} \leqslant F_{comp} \tag{6.1}$$

In general, a mixture of particles introduced into the magnetic separator will be split into two or more components. However, in any real separation, both magnetic and non-magnetic particles can be found in the magnetic fraction, non-magnetic fraction and the middling fractions. The efficiency of separation is usually expressed by the recovery of the magnetic component, and by the grade of the magnetic product.

(2) Magnetic field and magnetization

When we describe a magnetic field, we employ two different quantities, namely magnetic field strength H and magnetic flux density (or magnetic induction) B. In vacuum B and H are not independent and are related by the equation:

$$B = \mu_0 H \tag{6.2}$$

where the magnetic permeability of vacuum μ_0 is numerically equal to $4\pi \times 10^{-7}$ H/m. In a magnetic material of magnetization M, the total magnetic induction becomes:

$$B = \mu_0 (H + M) \tag{6.3}$$

In practical situations the magnetic flux density B is frequently used, rather than the magnetic field strength H.

The magnetization of a material, in general, depends on the magnetic field acting on it. For many materials, M is proportional to H (at least when H is not too large):

$$M = kH \tag{6.4}$$

where k, the volume magnetic susceptibility, is a property of the material. Since M and H have the same dimensions, k is dimensionless. Then

$$B = \mu_0 (1 + k) H = \mu_0 \mu_r H = \mu H \tag{6.5}$$

where $\mu_r = (1+k)$ and $\mu = \mu_0 (1+k)$.

The quantity μ_r is called relative magnetic permeability and is dimensionless, while μ

is called magnetic permeability and has a unit of H/m. For materials that have very small susceptibilities, it is much more convenient to use κ than μ_r. For instance, for $\kappa = 10^{-6}$, $\mu_r = 1.000,001$, and for $\kappa = -10^{-6}$, $\mu_r = 0.999,999$.

Another property of a material that determines its response to a magnetic field is the mass, or specific susceptibility χ. It has a relationship with κ as following:

$$\chi = \kappa/\rho \tag{6.6}$$

where ρ is the density of the material.

Since a sample mass is usually easy to known than its volume, mass magnetic susceptibility is very commonly employed.

(3) Magnetic force

The magnetic force on a particle which is placed in a magnetic field is dependent not only on the value of the field intensity, but also on the magnetic field gradient, i. e., the rate at which the field intensity increases along the magnetic flux. It can be shown that

$$F = \mu_0 kVH \text{ grad } H \tag{6.7}$$

where F is the magnetic force on particle, μ_0 is the magnetic permeability in vacuum, κ is the magnetic susceptibility of the particle, V is the volume of the particle, H is the magnetic field intensity, grad H is the magnetic field gradient.

Thus in order to generate a given magnetic force, there are an infinite number of combination of field and gradient which will give the same effect. Production of a high field gradient as well as high intensity is therefore an important aspect of separator design.

4. Magnetic separator design

Certain elements of design are incorporated in all magnetic separators, whether they are low or high intensity, wet or dry. The prime requirement, as has already been mentioned, is the provision of a high-intensity field in which there is a steep field strength gradient. In a field of uniform flux, such as in Figure 6.3(a), magnetic particles will orient themselves, but will not move along the lines of flux. The most straightforward method for producing a converging field is by providing a V-shaped pole above a flat pole, as in Figure 6.3(b). The tapering of the upper pole concentrates the magnetic flux into a very small area giving high intensity. The lower fiat pole has the same total magnetic flux distributed over a larger area. Thus there is a steep field gradient across the gap by virtue of the different intensity levels. Another method of producing a high field gradient is by using a pole which is constructed of alternate magnetic and non-magnetic laminations.

Provision must be incorporated in the separator for regulating the intensity of the magnetic field so as to deal with various types of material. This is easily achieved in electromagnetic separators by varying the current, while with permanent magnets the interpole distance can be varied.

Commercial magnetic separators are continuous process machines and separation is

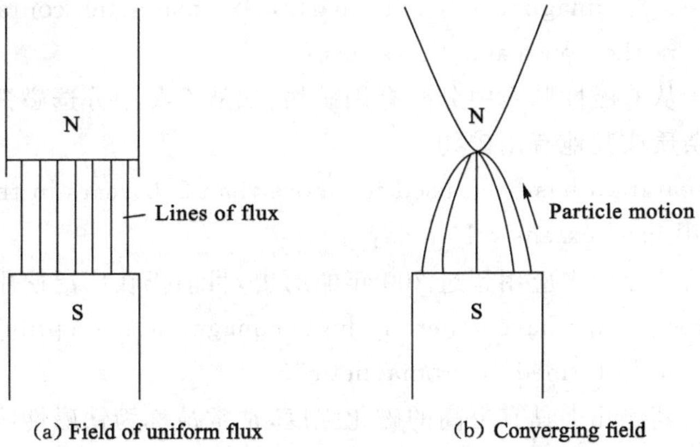

Figure 6. 3

carried out on a moving stream of particles passing into and through the magnetic field. Close control of the speed of passage of the particles through the field is essential, which rules out free fall as a means of feeding.

Since the lines of force converge to the particles, a high field gradient is produced, which causes the particles themselves to behave as magnets, thus attracting each other. Flocculation, or agglomeration, of the particles can occur if they are small and highly susceptible and if the field is intense. This has great importance as these magnetic "flocs" can entrain gangue and can bridge the gaps between magnetic poles, reducing the efficiency of separation. Flocculation is especially serious with dry separating machines operating on fine material. If the ore can be fed through the field in a monolayer, this effect is much less serious, but, of course, the capacity of the machine is drastically reduced. Flocculation is often minimised by passing the material through consecutive magnetic fields, which are usually arranged with successive reversal of the polarity. This causes the particles to turn through 180°, each reversal tending to free the entrained gangue particles. The main disadvantage of this method is that flux tends to leak from pole to pole, reducing the effective field intensity.

Provision for collection of the magnetic and non-magnetic fractions must be incorporated into the design of the separator. Rather than allow the magnetics to contact the pole-pieces, which would cause problems of detachment, most separators are designed so that the magnetics are attracted to the pole-pieces, but come into contact with some form of conveying device, which carries them out of the influence of the field, into a bin or a belt. Non-magnetic disposal presents no problems, free fall from a conveyor into a bin often being used. Middlings are readily produced by using a more intense field after the removal of the highly magnetic fraction.

Notes

1. Magnetic separators are used to separate either valuable minerals from non-

magnetic gangue, e. g., magnetite from quartz, or magnetic contaminants or other valuable minerals from the non-magnetic values.

磁选机可以用于从非磁性脉石中分离有用矿物,如从石英中分选磁铁矿,或从非磁性有用矿物中分离磁性杂质或其他有用矿物。

2. Magnetic separation has been used for more than 200 years in the concentration of iron ores, and is still used extensively today.

磁选已在铁矿石的分选中应用超过 200 年的历史,目前仍在广泛使用。

3. The last three groups have generally higher magnetic susceptibility than the other groups and are frequently termed "ferromagnetic".

最后三类通常比其他几类具有更高的磁化强度,常常被称为铁磁性物质。

4. Diamagnetics are repelled along the lines of magnetic force to a point where the field intensity is smaller. The forces involved here are very small and diamagnetic substances cannot be concentrated magnetically.

逆磁性物质沿着磁力线被排斥到场强较小的点。这里涉及的力非常小,逆磁性物质不能被磁力富集。

5. Paramagnetics are attracted along the lines of magnetic force to points of greater field intensity.

顺磁性物质沿着磁力线被吸引到场强较大的点。

6. In an applied magnetic field (H), the magnetization of ferromagnetic materials increases until the saturation magnetization is reached (B_m). If the applied field intensity is reduced, the magnetization decreases but does not come back to its original value. Such an irreversible process of magnetization is called hysteresis.

在外加磁场中(磁场强度为 H),铁磁性物料的磁化强度一直升高直到达到磁饱和(磁通量密度达到峰值 B_m)。若外磁场强度降低,磁化强度也降低,但并不会降低到原来的值。

7. The separation of one material from another or the removal of magnetizable particles from streams depend upon their motion in response to the magnetic force and to other competing external forces, namely gravitational, inertial, hydrodynamic and centrifugal forces.

一种物料与另一种物料的分离,或者从物料流中去除磁性颗粒取决于它们在磁力和其他外在阻力作用下的运动情况,这些阻力包括重力、惯性力、流体动力以及离心力。

8. Thus a necessary condition for a successful separation of more strongly magnetic from less strongly magnetic particles in a magnetic field is that the magnetic force F_{mag}^{m} acting on more strongly magnetic particles must be greater than the sum of all the competing forces F_{comp}.

因此,在磁场中成功分离强磁性颗粒和弱磁性颗粒的必要条件是作用在强磁性颗粒上的磁力必须大于所有阻力的总和。

9. In general, a mixture of particles introduced into the magnetic separator will be split into two or more components. However, in any real separation, both magnetic and non-magnetic particles can be found in the magnetic fraction, non-magnetic fraction and

the middling fractions.

一般地，进入磁选机的颗粒混合物会分成两个或多个产品，但在实际分选中，磁性颗粒和非磁性颗粒都会出现在磁性物产品、非磁性物产品以及中矿中。

Exercises

1. Answer the following questions according to the text.

(1) What are the principles of magnetic separation?

(2) What are the applications of magnetic separation?

(3) Describe the classification of materials based on magnetic property and the differences among them.

(4) What is hysteresis and remanence?

(5) What are the elements that determine the magnetic force on a mineral?

(6) How to express the efficiency of magnetic separation?

2. Translate the following phrases into English.

(1) 磁选　(2) 磁场　(3) 磁力　(4) 重力

(5) 剩磁　(6) 磁选机　(7) 非磁性颗粒　(8) 顺磁性物质

(9) 逆磁性物质　(10) 永久磁铁

3. Translate the following paragraphs into Chinese.

Paragraphs 1 and 2 in section 1 "Magnetic separation is the process of separating ore minerals ...".

4. Translate the following sentences into English.

(1) 磁选是利用矿物之间磁性质的差异进行分选的过程。

(2) 所有的物料都表现出一定的磁性，无论其成分和状态如何。物料按其磁性能可分为五大类：抗(逆)磁性、顺磁性、铁磁性、反铁磁性和亚铁磁性。

(3) 置于磁场中的颗粒上的磁力不仅取决于磁场强度的大小，还取决于磁场梯度。

(4) 分选机中必须包含调节磁场强度的装置，以便处理不同类型的原料。

6.2 Magnetic Separators(磁选机)

Magnetic separators can be classified into low- and high-intensity machines, which may be further classified into dry-feed and wet-feed separators.

Low-intensity separators are used to treat ferromagnetic materials and some highly paramagnetic minerals. These separators can operate either on dry or wet modes. High-intensity separators are used for treatment of weakly magnetic materials, coarse or fine, in wet or dry modes. Very fine, feebly magnetic materials can be treated in high-gradient magnetic separators.

1. Low-intensity magnetic separators

Dry low-intensity magnetic separators are usually used to remove tramp iron, to concentrate coarse strongly magnetic iron ores, and to recover iron values form blast furnace and steel mill slag. Removal of strongly magnetic impurities from a variety of materials is also accomplished with dry low-intensity separators.

(1) Dry low-intensity magnetic separators—Magnetic pulleys

An easy and simple way to remove tramp iron from material on a conveyor belt is by means of a magnetic head pulley. These are available both in permanent magnetic and electromagnetic construction. They are relatively inexpensive and easy to install, and accomplish continuous tramp iron removal. The magnetic pulley, shown in Figure 6. 4, usually replaces the head pulley of a belt conveyor, converting the conveyor into a magnetic separator. Material carried on the conveyor passes over the magnetic pulley which holds the magnetic particles until they leave the region of the magnetic field while the non-magnetic material is discharged over the pulley.

(2) Wet low-intensity magnetic separators—Drum separators

Drum separators are the most common machines in current use to recover ferromagnetic solids from a slurry feed. The major areas for their use are in media recovery in dense medium separation plants and in beneficiation of magnetite ore.

Drum separators consist essentially of a rotating non-magnetic drum (Figure 6. 5) containing three to six stationary magnets of alternating polarity, which improves the agitation at the drum surface and so reduces the entrapment of nonmagnetics. The magnets cover an angle up to 150°. The drum is usually made of stainless steel and often protected by an anti-wear cover. The size of the feed treated in these separators rarely exceeds 6 mm while the bottom size of the feed can be as small as 20 μm. Field intensities of up to 0. 7 T at the pole surfaces can be obtained in this type of separator.

The design of the tanks for the drum separator is very important and is determined by the objectives of the separation process. There are three basic tank design, based on the slurry flow: concurrent, counter-current, and counter-rotation.

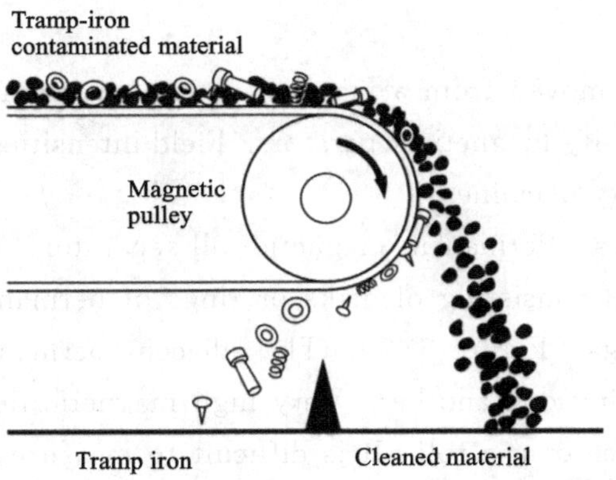

Figure 6.4 Principle of operation of a magnetic pulley

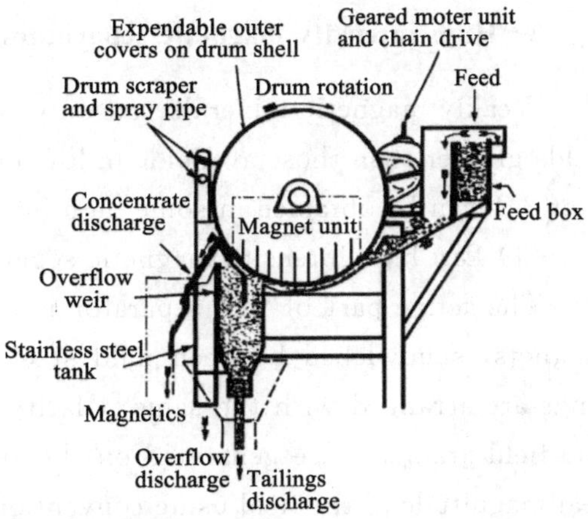

Figure 6.5 Drum separator

The concurrent tank design is shown schematically in Figure 6.6(a). The feed is introduced at one end of the separator through the feed box and flows in the direction of drum rotation. The concurrent design of the tank is used mainly for treating coarse material in the size range 5 mm and below. High throughput and high quality of the magnetic concentrate are characteristic features of this design.

In the counter-rotation design of the tank [Figure 6.6(b)], the feed is introduced through a special feed box to the drum, which rotates in the opposite direction to the slurry flow. Very high recoveries can thus be achieved with this design, while the grade of the concentrate is of secondary importance. Therefore, it is suitable for roughing operation. The particle size can be up to 4 mm, but is preferably smaller than 0.5 mm.

The counter-current tank [Figure 6.6(c)] is used mostly for finishing operation where very thorough cleaning must be required. The term, counter-current, refers to the fact that the tailings are forced to travel in the opposite direction to the drum rotation when leaving the separator. This ensures a high recovery of magnetic material and a high quality of the concentrate. These tanks are most suited to finely ground material of 100 μm or 250 μm and less.

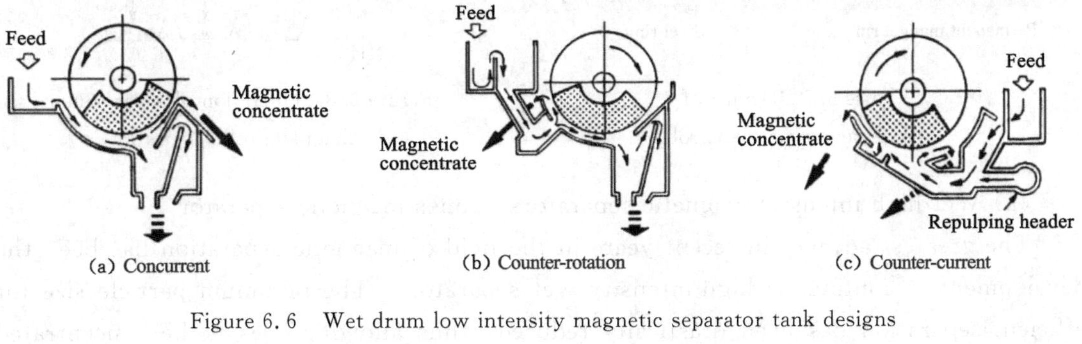

Figure 6.6 Wet drum low intensity magnetic separator tank designs

2. High intensity magnetic separators

Weakly magnetic minerals can only be removed from a material stream in magnetic fields greater than those available in low-intensity magnetic separators. Field intensities of over 2 T can be obtained in some high-intensity machines.

(1) Dry high intensity magnetic separators—Permanent magnetic roll separator

The active part of this separator is a roll consisting of disks or rings of permanent magnets, sandwiched between mild steel rings (Figure 6.7). The adjacent permanent rings are arranged with the same polarity facing one another. Very high magnetic fields and field gradients are generated on the surface of the roll. It is difficult to measure the real magnitude of the field using conventional measuring equipment because the field varies over a scale of several hundreds of microns.

The values of the magnetic field strength and its gradient on the roll surface and around the roll are strongly dependent on the configuration of the magnetic and steel rings. Thin magnet and steel rings generally produce a very high field, field gradient and magnetic force at the surface of the roll, while a long reach of the magnetic force can be achieved using thicker steel rings. The most common ratio of the widths of the magnet to the steel is 4 : 1.

Operation of permanent magnetic roll separators is schematically shown in Figure 6.8. For easy removal of magnetic particles, the roll is covered by a belt, which can be as thin as 0.12 mm, to allow the feed material to be as close to the magnet as possible. Below the conveyor is a hopper which collects the discharging material, while adjustable splitters divert the different fractions into collection bins placed beneath the hopper.

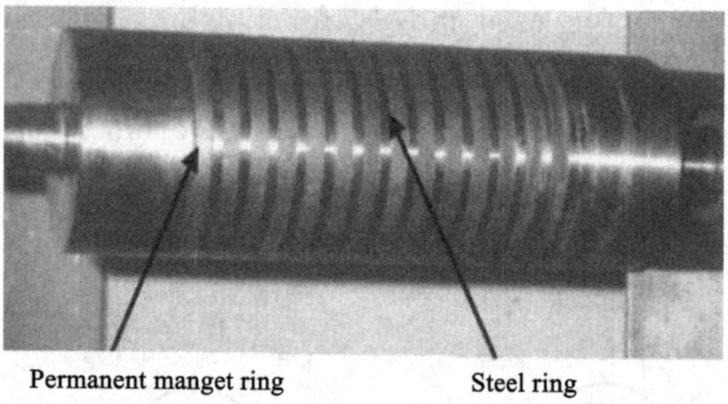

Figure 6.7 Details of a permanent magnetic roll

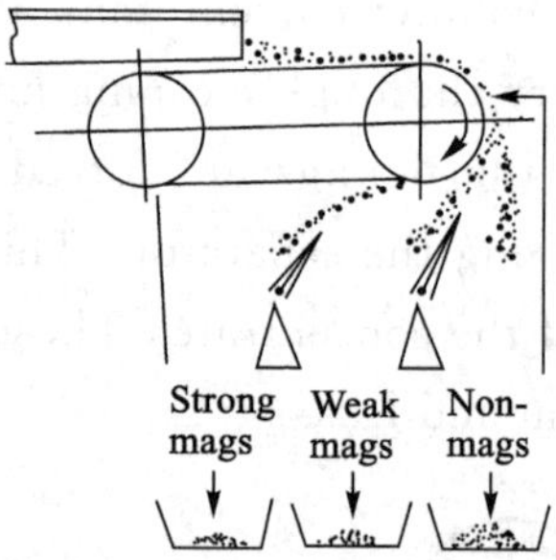

Figure 6.8 Operation of a permanent magnetic roll separator

(2) Wet high intensity magnetic separators—Jones magnetic separator

The greatest advance in recent years in the field of magnetic separation has been the development of continuous high-intensity wet separators. The minimum particle size for efficient separation has been drastically reduced, thus allowing ores to be concentrated magnetically that cannot be concentrated effectively by dry high-intensity methods, because fine grinding is necessary to ensure complete liberation of the magnetic fraction. In some flow sheets, expensive drying operations can be eliminated by using a wet concentration system.

The most well-known separator now being used to treat fine hematite ores is the Jones wet high-intensity magnetic separator, the design principle of which is utilized in many other types of wet separator used today.

The machine consists of a strong main frame (Figure 6.9), made of structural steel. The magnet yokes are welded to this frame, with the electromagnetic coils enclosed in air-cooled cases. The actual separation takes place in the plate boxes which are on the periphery of the one or two rotors attached to the central roller shaft and are grooved to concentrate the magnetic field at the tip of the ridges (Figure 6.10). Figure 6.11 shows a perspective view of Jones separator.

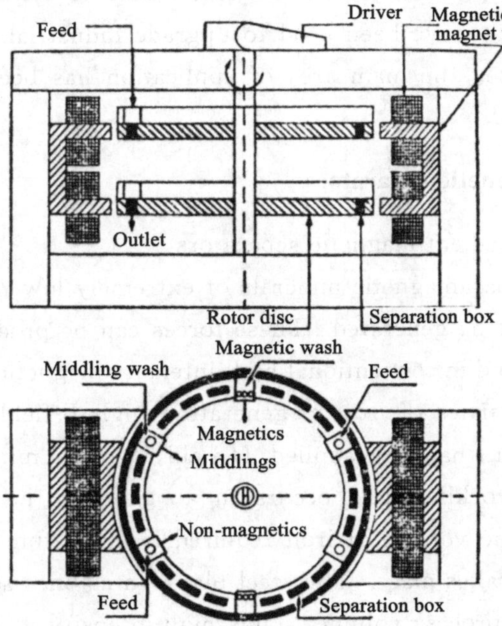

Figure 6.9 Operating principle of the Jones high-intensity wet magnetic separator

The slurry is gravity-fed onto the plate box where the feebly magnetic particles are held by the plates, whereas the remaining non-magnetic pulp passes straight through the plate boxes and is collected in a launder. Before leaving the field any entrained non-magnetics are washed out by low-pressure water and are collected as a middling product. When the plate boxes reach a point midway between the two magnetic poles, where the magnetic field is essentially zero, the magnetic particles are washed out with high pressure scour water sprays. Feeding is continuous due to the rotation of the plate boxes on the rotors and the feed points are at the leading edges of the magnetic fields. Each rotor has two symmetrically disposed feed points.

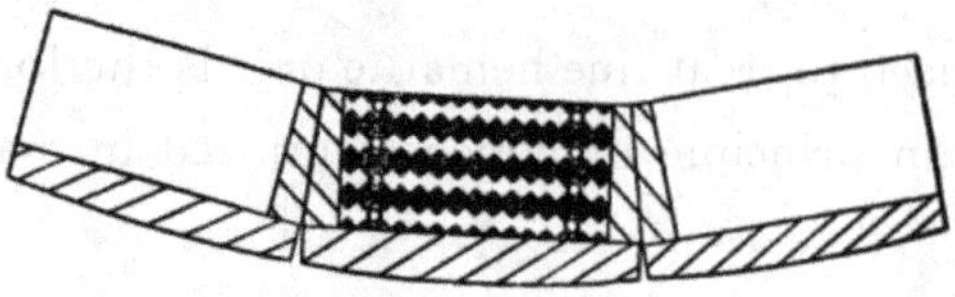

Figure 6.10 Plan of Jones box showing grooved plates and spacer bars

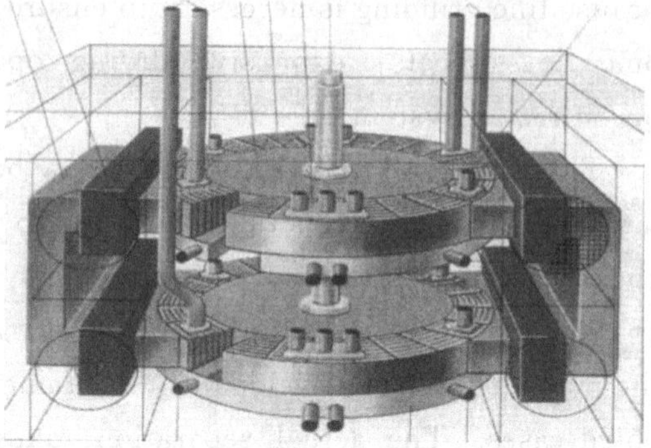

Figure 6.11 Perspective view of a Jones continuous wet high-intensity magnetic separator

The Jones magnetic separators have been widely used to beneficiate a wide spectrum of minerals. Although they have been used to upgrade industrial minerals such as glass sand, feldspar and bauxite, the main area of application has been the concentration of weakly magnetic iron ores.

3. High gradient magnetic separators

(1) Design of high gradient magnetic separators

In order to separate paramagnetic minerals of extremely low magnetic susceptibility, high magnetic forces must be generated. These forces can be produced by increasing the magnetic field strength, and in conventional high-intensity magnetic separators use is made of the ferromagnetic properties of iron to generate a high B-field (induced field) many hundreds of times greater than the applied H-field, with a minimum consumption of electrical energy. The working field occurs in air-gaps in the magnetic circuit, the disadvantage being that the volume of iron required is many times greater than the gap volume where separation takes place. The steel plates in a Jones separator, for example, occupy up to 60% of the process volume. Thus high-intensity magnetic separators using conventional iron circuits tend to be very massive and heavy in relation to their capacity. A large separator may contain over 200 t of iron to carry the flux, hence capital and installation costs are extremely high.

As iron saturates at around 2~2.5 T, conventional iron circuits are of little value for generating fields above about 2 T. Such fields can only be generated by the use of high H-fields produced in solenoids, but the energy consumption is extremely high and there are problems in cooling the solenoid.

An alternative is to increase the magnetic force by increasing the value of the magnetic field gradient. Instead of using one large convergent field in the gap of a magnetic circuit, the uniform field of a solenoid is used (Figure 6.12). The core, or working volume, is filled with a matrix of secondary poles, such as ball bearings, or wire wool, the latter filling only about 10% of the working volume. Each secondary pole, due to its high permeability, can produce a maximum field strength of the order of 2 T, but more importantly, each pole produces, in its immediate vicinity, high field gradients of up to 14 T/mm. Thus a multi-rode of high gradients across numerous small gaps, centred around each of the secondary poles, is achieved.

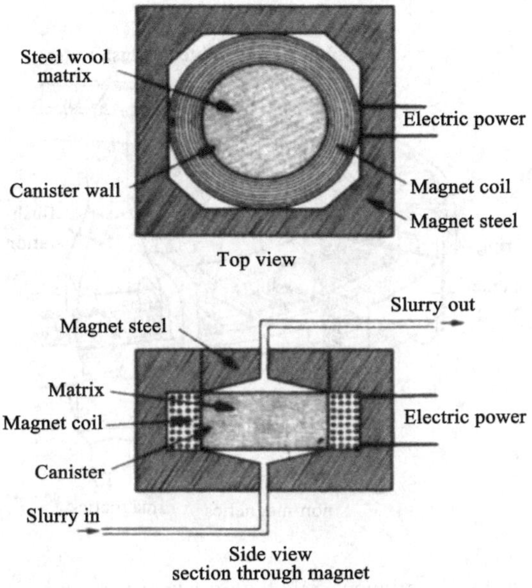

Figure 6. 12 High-gradient magnetic separator

The solenoid can be clad externally with an iron frame to form a continuous return path for the magnetic flux, thus reducing the energy consumption for driving the coil. The matrix is held in a canister into which the slurry is fed. As particles are captured, the ability of the matrix to extract particles is reduced. Periodically the magnetic field can be removed and the matrix flushed with water to remove the captured material.

An inherent disadvantage of high gradient separators is that an increase in field gradient necessarily reduces the working gap between secondary poles, the magnetic force having only a short reach, usually not more than 1 mm. It is therefore necessary to use gaps of only about 2 mm between poles, such that the matrix separators are best suited to the treatment of very fine particles. They are used mainly in the kaolin industry for removing micron-sized particles which contain iron. Several large separators, with the ferromagnetic matrix contained in baskets approximately 2 m in diameter are in commercial use in the United States and in Cornwall, England. They operate with fields of 2 T, and have capacities ranging between 10 t/h and 80 t/h depending on the final clay quality desired.

(2) Sala-HGMS carousel separator

The Sala-HGMS carousel separator was developed from the cyclic separator to meet the need for a continuous industrial machine capable of processing ores containing a high proportion of magnetics. As can be seen in Figure 6. 13, a double-saddle iron-clad coil surrounds the carousel ring. Seals mounted on the ring allow independent control of feed and flush velocity to enhance the separation and to increase the flushing efficiency of the magnetics. The iron return frame reduces the magnetic flux leakage to places outside the magnetic head.

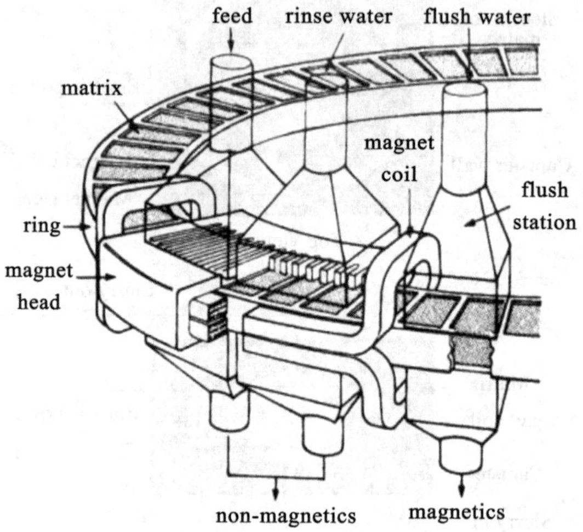

Figure 6. 13 Continuous Sala high-gradient magnetic separator

4. Superconducting separators

Undoubtedly the future developments and applications of magnetic separation in the mineral industry will lie in the use of high magnetic forces. Matrix separators with very high field gradients and multiple small working gaps can draw little advantage from field strengths above the saturation levels of the secondary poles. However, "open-gradient" separators, with large working volumes to deflect coarser particles at high capacity, rather than capture particles, as in high-gradient separators, need to use the highest possible field strengths in order to generate the high magnetic forces required to treat feebly paramagnetic particles.

Field strengths in excess of 2 T can only be generated economically by the use of superconducting magnets. Certain alloys have the property of presenting no resistance to electric currents at extremely low temperatures. An example is niobium-titanium at 4. 2 K, the temperature of liquid helium. Once a current commences to flow through a coil made from such a superconducting material, it will continue to flow without being connected to a power source, and the coil will become, in effect, a permanent magnet. Superconducting magnets can produce extremely intense and uniform magnetic fields, of up to 15 T.

In 1986 a superconducting high-gradient magnetic separator was designed and built by Eriez Magnetics to process kaolinite clay in the United States. This machine uses only about 0.007 kW in producing 5 T of flux, the ancillary equipment needed requiting another 20 kW. In comparison, a conventional 2 T high-gradient separator of similar throughput would need about 250 kW to produce the flux, and at least another 30 kW to cool the magnet windings.

The 5 T machine is an assembly of concentric components (Figure 6.14). A removable processing canister is installed in a processing chamber located at the centre of the assembly. This is surrounded by a double-walled, vacuum-insulated container that accommodates the super-conductive magnet's niobium/titanium-tantalum winding, and the liquid helium coolant. A thermal shield, cooled with liquid nitrogen to 77 K, limits radiation into the cryostat. In operation, the supply of slurry is periodically cut off, the magnetic field is shut down, and the canister backwashed with water to clear out accumulated magnetic contaminants.

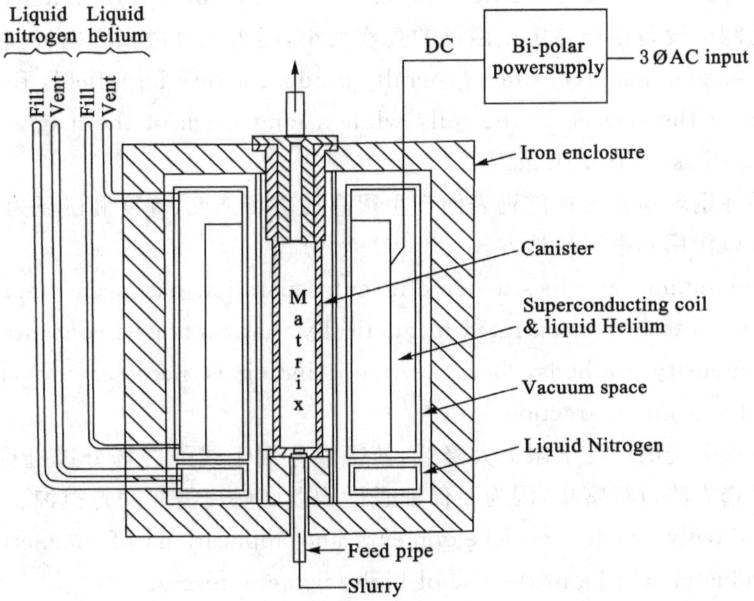

Figure 6.14 5 T superconducting magnetic separator

Notes

1. Dry low-intensity magnetic separators are usually used to remove tramp iron, to concentrate coarse strongly magnetic iron ores, and to recover iron values form blast furnace and steel mill slag. Removal of strongly magnetic impurities from a variety of materials is also accomplished with dry low-intensity separators.

干式弱磁场磁选机通常用于去除杂铁,富集粗粒强磁性铁矿石,以及从高炉和钢厂渣中回收含铁的有用物质。还可以使用干式弱磁场磁选机从各种物料中去除强磁性杂质。

2. Drum separators are the most common machines in current use to recover ferromagnetic solids from a slurry feed. The major areas for their use are in media recovery in dense medium separation plants and in beneficiation of magnetite ore.

筒式磁选机是目前用于从矿浆中回收铁磁性固体的最常用设备。它们的主要应用领域是重介质分选厂的介质回收和磁铁矿的分选。

3. The active part of this separator is a roll consisting of disks or rings of permanent magnets, sandwiched between mild steel rings (Figure 6.7). The adjacent permanent rings are arranged with the same polarity facing one another. Very high magnetic fields and field gradients are generated on the surface of the roll.

永磁辊式磁选机的运动部分是一个辊,它由夹在低碳钢圆环之间的永磁磁铁圆盘或圆环组成(图 6.7)。相邻永磁环同极相对布置,磁辊表面产生的磁场强度和磁场梯度很高。

4. It is difficult to measure the real magnitude of the field using conventional measuring equipment because the field varies over a scale of several hundreds of microns.

用传统的测试设备很难测定该磁场的真实大小,因为几百微米之内场强就会发生变化。

5. Thin magnet and steel rings generally produce a very high field, field gradient and magnetic force at the surface of the roll, while a long reach of the magnetic force can be achieved using thicker steel rings.

薄的磁环和钢环通常会在磁辊表面产生很高的磁场强度、梯度和磁场力,而用厚点的钢环时磁场力可以作用到很远的地方。

6. The minimum particle size for efficient separation has been drastically reduced, thus allowing ores to be concentrated magnetically that cannot be concentrated effectively by dry high-intensity methods, because fine grinding is necessary to ensure complete liberation of the magnetic fraction.

磁选能有效分选的粒度下限大大降低,使得那些用干式强磁场分选机难以有效分离的矿石也能用磁选方法进行富集,因为要保证磁性物彻底解离需要磨得很细。

7. Undoubtedly the future developments and applications of magnetic separation in the mineral industry will lie in the use of high magnetic forces.

毫无疑问,未来磁选技术在矿业中的发展和应用将取决于强磁力的使用。

Exercises

1. Answer the following questions according to the text.

(1) What are the classification of magnetic separators and their applications?

(2) What are the classification of drum separators and the differences among them?

(3) What is the structure of a permanent magnetic roll separator?

(4) What are the disadvantages of conventional high intensity separators and the advantages of high gradient magnetic separators?

(5) What will the future developments and applications of magnetic separation in the

mineral industry lie in?

2. Translate the following phrases into English.

(1) 磁滑轮　　(2) 顺流型筒式磁选机　　(3) 半逆流型筒式磁选机

(4) 高梯度磁选机　　(5) 超导磁选机　　(6) 逆流型筒式磁选机

(7) 交替极性　　(8) 弱磁场磁选机

3. Translate the following paragraphs into Chinese.

Paragraphs 1, 2 and 3 in section 1 (2) "Drum Separators are the most common machines . . . ".

4. Translate the following sentences into English.

(1) 磁选机可分为弱磁场磁选机和强磁场磁选机,又可分为干式磁选机和湿式磁选机。

(2) 筒式磁选机的槽体设计非常重要,由分选工艺目的决定。根据矿浆流向,有三种基本的设计:顺流型、逆流型和半逆流型。

(3) 只有使用超导磁体才能经济地产生超过 2 T 的场强。

6.3 Electrostatic Separation(静电分选)

1. Principle and application of electrostatic separation

(1) True electrostatic separation and high-tension separation

Electrical separation utilizes the difference in electrical conductivity between the various minerals in the ore feed. Since almost all minerals show some difference in conductivity it would appear to represent the universal concentrating method. In practice, however, the method has fairly limited application, and its greatest use is in separating some of the minerals found in heavy sands from beach or stream placers. The fact that the feed must be perfectly dry imposes limitations on the process, but it also suffers from the same great disadvantage as dry magnetic separation—the capacity is very small for finely divided material. For most efficient operation, the feed should be in a layer, one particle deep, which severely restricts the throughput if the particles are as small as, say, 75 μm.

The first mineral separation processes utilizing high voltage were virtually true electrostatic processes employing charged fields with little or no current flow. High-tension separation, however, makes use of a comparatively high rate of electrical discharge, with electron flow and gaseous ionisation having major importance.

When a particle is placed on the grounded rotor in the presence of the electric field, the particle rapidly develops a surface charge by induction. Whether the particle is a conductor or a nonconductor, it can be considered to have become polarized. However, as indicated in Figure 6.15, a conductor particle rapidly becomes an equipotential surface and has the same potential as the grounded rotor; therefore it is attracted toward the electrode. Thus, the conductor particle is drawn from the surface by attraction to the electrode while the nonconductor particle continues to adhere to the rotor surface until gravity causes it to fall.

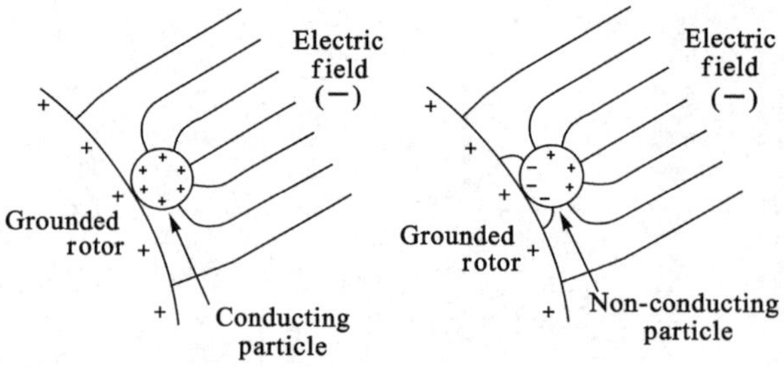

Figure 6.15 Charging particles by induction

The attraction of particles carrying one kind of charge towards an electrode of the opposite charge is known as the "lifting effect", as such particles are lifted from the separating surface towards the electrode. Materials which have a tendency to become charged with a definite polarity may be separated from each other by the use of the lifting effect even though their conductivities may be very similar. As an example, quartz assumes a negative charge very readily and may be separated from other poor conductors by an electrode which carries a positive charge. Pure electrostatic separation is relatively inefficient, even with very clean mineral, and is sensitive to changes of humidity and temperature.

A large percentage of the commercial applications of high-tension separation has been made using the "pinning effect", in which non-conducting mineral particles, having received a surface charge from the electrode, retain this charge and are pinned to the oppositely charged separator surface by positive-negative attraction. Figure 6. 16 shows the principle of high-tension separation diagrammatically.

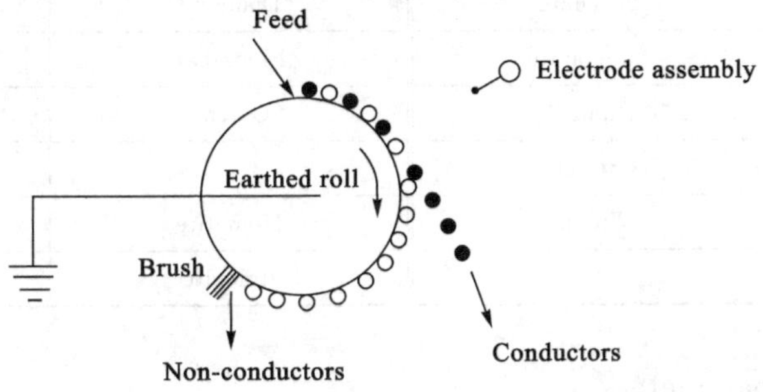

Figure 6. 16 Principle of high-tension separation

The mixture of ore minerals, of varying susceptibilities to surface charge, is fed on to a rotating drum made from mild steel, or some other conducting material, which is earthed through its support bearings. An electrode assembly, comprising a brass tube in front of which is supported a length of fine wire, spans the complete length of the roll, and is supplied with a fully rectified DC supply of up to 50 kV, usually of negative polarity. The voltage supplied to the assembly should be such that ionisation of the air takes place. This can often be seen as a visible corona discharge. Arcing between the electrode and the roll must be avoided, as this destroys the ionisation. When ionisation occurs, the minerals receive a spray discharge of electricity which gives the poor conductors a high surface charge, causing them to be attracted to and pinned to the rotor surface. The particles of relatively high conductivity do not become charged as rapidly, as the charge rapidly dissipates through the particles to the earthed rotor. These particles of higher conductivity follow a path, when leaving the rotor, approximating to the one which they would assume if there were no charging effect at all.

The electrode assembly is designed to create a very dense high-voltage discharge. The fine wire of the assembly is placed adjacent to and parallel to the large diameter electrode and is mechanically and electrically in contact with it. This fine wire tends to discharge readily, whereas the large tube tends to have a short, dense, non-discharging field. This combination creates a very strong discharge pattern which may be "beamed" in a definite direction and concentrated to a very narrow arc. The effect on the minerals passing through the beam is very strong and is due largely to gaseous ions which are created due to the high-voltage gradient in the field of the corona.

Table 6.1 shows typical minerals which are either pinned to or thrown from the rotor during high-tension separation.

Table 6.1 Typical behaviour of minerals in high-tension separators

Minerals pinned to rotor		Minerals thrown from rotor	
Apatite	Monazite	Cassiterite	Limonite
Barite	Quartz	Chromite	Magnetite
Calcite	Scheelite	Diamond	Pyrite
Corundum	Sillimanite	Fluorspar	Rutile
Garnet	Spinel	Galena	Sphalerite
Gypsum	Tourmaline	Gold	Stibnite
Kyanite	Zircon	Hematite	Tantalite
		Ilmenite	Wolframite

(2) Operational parameters

To cater for such an extensive range of minerals, all the parameters influencing separation must be readily adjusted while the separator is performing. These variables include the roll speed, the position of the electrode wire with respect to the electrode tube, the position of the electrode assembly with respect to the roll, variation of the direct current (DC) voltage and polarity, the splitter plate position, the feed rate, and heating of the feed. Heating the feed is important, since best results are generally obtained only with very dry material. This is particularly difficult in high humidity regions. It is not often that a single pass will sufficiently enrich an ore and Figure 6.17 shows a typical flowsheet, where the falling particles are deflected to lower sets of rollers and electrodes until the required separation has taken place.

2. Electrostatic separators

(1) High-tension separators

A combination of the effects of pinning and lifting can be created by using a static electrode large enough to preclude corona discharge, following the ionizing electrode. The conducting particles, which are flung from the rotor, are attracted to this electrostatic electrode, and the compound process produces a very wide and distinct separation between the conducting and the non-conducting particles (Figure 6.18).

High-tension separators operate on feeds containing particles of between 60 μm and 500 μm in diameter. Particle size influences separation behaviour, as the surface charges

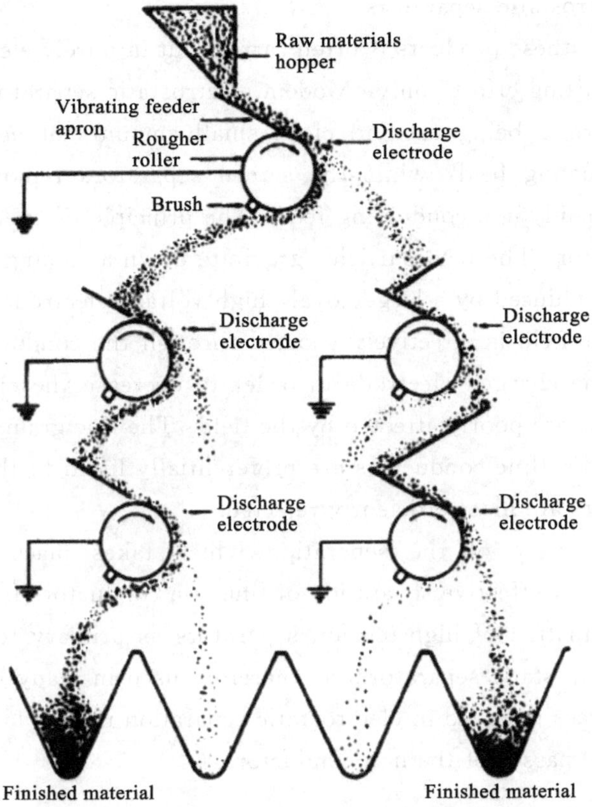

Figure 6.17 Arrangement of separators in practice

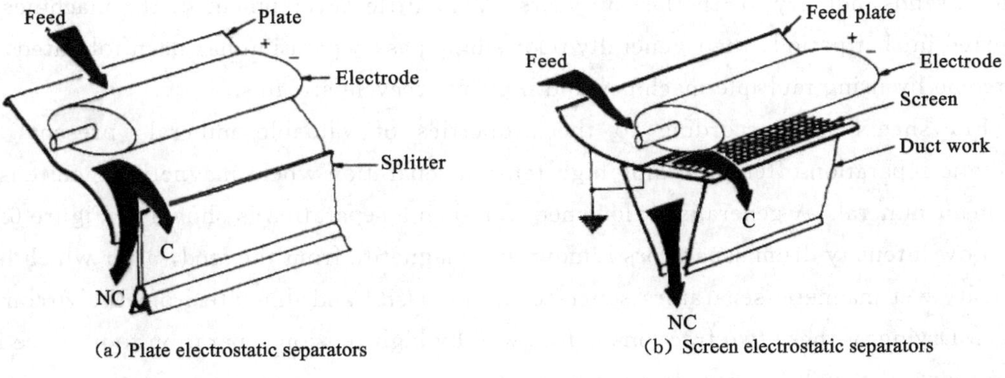

Figure 6.18

on a coarse grain are lower in relation to its mass than on a fine grain. Thus, a coarse grain is more readily thrown from the roll surface, and the conducting fraction often contains a small proportion of coarse non-conductors. Similarly, the finer particles are most influenced by the surface charge, and the non-conducting fraction often contains some fine conducting particles.

(2) Purely electrostatic separators

Final cleaning of these products is often carried out in purely electrostatic separators, which employ the "lifting effect" only. Modem electrostatic separators are of the plate or screen type, the former being used to clean small amounts of non-conductors from a predominantly conducting feed, while the screen separators remove small amounts of conductors from a mainly non-conducting feed. The principle of operation is the same for both types of separator. The feed particles gravitate down a sloping, grounded plate into an electrostatic field induced by a large, oval, high-voltage electrode (Figure 6.18).

The electrostatic field is effectively shorted through the conducting particles, which are lifted towards the charged electrode in order to decrease the energy of the system. Non-conductor grains are poorly affected by the field. The fine grains are most affected by the lifting force, and so fine conductors are preferentially lifted to the electrode, whereas coarse non-conductors are most efficiently rejected.

This is the converse of the separation which takes place in the high-tension separators, where most effective separation of fine non-conductors from coarse conductors takes place; a combination of high-tension separators as primary roughers, followed by final cleaning in electrostatic separators, is therefore used in many flowsheets. Since the magnitude of the forces involved in electrostatic separation is very low, the separators are designed for multiple passes of the non-conductors.

3. Typical flowsheet

High tension roll and electrostatic plate separators have been the mainstay of the mineral sands industry for the last 50 years. Very little development of the machines has occurred in that period, their generally poor single pass separation has been tolerated, and overcome by using multiple machines and multiple recycle streams.

Flowsheets vary according to the properties of valuable minerals present, wet magnetic separation often preceding high-tension separation where magnetic ilmenite is the dominant mineral. A generalised flowsheet for such a separation is shown in Figure 6.19.

Low-intensity drum separators remove any magnetite from the feed, after which high-intensity wet magnetic separators separate the monazite and ilmenite from the zircon and rutile. Drying of these two fractions is followed by high-tension separation to produce final separation, although further cleaning is sometimes carried out by electrostatic separators. For example, screen electrostatic separators may be used to clean the zircon and monazite concentrates, removing fine conducting particles from these fractions. Similarly, plate electrostatic separators could be used to reject coarse non-conducting particles from the rutile and ilmenite concentrates.

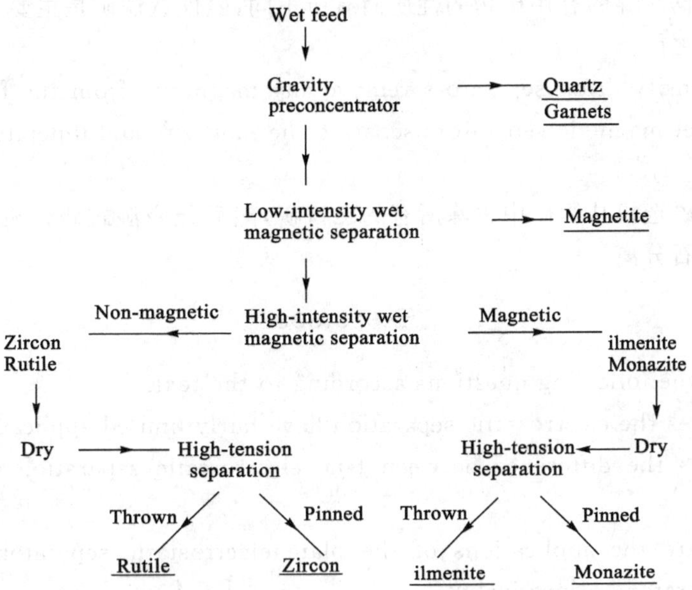

Figure 6.19 Typical beach sand treatment flowsheet

Notes

1. High-tension separation, however, makes use of a comparatively high rate of electrical discharge, with electron flow and gaseous ionisation having major importance.

然而,高压电选利用了相对较高的放电速率,电子流和气态电离具有重要意义。

2. Thus, the conductor particle is drawn from the surface by attraction to the electrode while the nonconductor particle continues to adhere to the rotor surface until gravity causes it to fall.

因此,导体颗粒被吸引到电极表面,而非导体颗粒继续附着在转子表面,直到重力使其下落。

3. An electrode assembly, comprising a brass tube in front of which is supported a length of fine wire, spans the complete length of the roll, and is supplied with a fully rectified DC supply of up to 50 kV, usually of negative polarity.

一个电极组件包括一根黄铜管及其前面支撑的一段金属细丝,长度跨越整个辊长,并提供高达 50 kV 的全整流直流电,通常为负极。

4. This combination creates a very strong discharge pattern which may be "beamed" in a definite direction and concentrated to a very narrow arc.

这种电极组合能产生一种非常强的放电模式,放电可以向着某个特定方向发射,聚集成一条很窄的圆弧。

5. Flowsheets vary according to the properties of valuable minerals present, wet magnetic separation often preceding high-tension separation where magnetic ilmenite is the dominant mineral.

分选流程根据赋存的有用矿物的性质而有所不同，磁性钛铁矿是主要矿物时，湿式磁选通常在高压分选之前。

6. Low-intensity drum separators remove any magnetite from the feed, after which high-intensity wet magnetic separators separate the monazite and ilmenite from the zircon and rutile.

弱磁场筒式磁选机从给矿中去除所有的磁铁矿，然后强磁湿式磁选机将独居石和钛铁矿与锆石和金红石分离。

Exercises

1. Answer the following questions according to the text.

(1) Why does the electrostatic separation have fairly limited application?

(2) What is the difference between true electrostatic separation and high-tension separation?

(3) What are the applications of the plate electrostatic separator and the screen electrostatic separator, respectively?

(4) How does a high-tension separator achieve separation?

(5) Describe the typical beach sand treatment flowsheet in this text.

2. Translate the following phrases into English.

(1) 静电分选　(2) 电导率　(3) 导电颗粒　(4) 静电场

(5) 非电离电极　(6) 静电电选机　(7) 高压电选机　(8) 电晕放电

3. Translate the following paragraphs into Chinese.

Paragraph 1 in section 1 (1) “Electrical separation utilizes the difference in ...”.

4. Translate the following sentences into English.

(1) 电选是利用给矿中各种矿物之间的电导率差异进行分选的过程。

(2) 携带一种电荷的粒子对另一种电荷的电极的吸引力被称为“提升效应”。

(3) 导体颗粒被吸引到电极表面，而非导体颗粒继续附着在辊筒表面，直到重力使其下落。

7 FLOTATION(浮选)

7.1 Principles of Flotation(浮选原理)

1. Introduction

(1) Concept of flotation

Flotation is undoubtedly the most important and versatile mineral-processing technique, and its application is being expanded to treat greater tonnages and cover new areas. Originally patented in 1906, flotation has permitted the mining of low-grade and complex ore bodies which would have otherwise been regarded as uneconomic. In earlier practice the tailings of many gravity plants were of a higher grade than the ore treated in many modern flotation plants.

Flotation is a selective process and can be used to achieve specific separations from complex ores such as lead zinc, copper-zinc, etc. Initially developed to treat the sulphides of copper, lead, and zinc, the field of flotation now has been expanded to include the oxidized minerals, non-metallic ores, and fine coal.

Formerly the term flotation with descriptive adjective was used for all processes of concentration in which levitation in water of particles heavier than water was obtained. Thus, if some particles were retained in an oil layer or at the interface between an oil layer and a water layer, the process was spoken of as bulk-oil flotation; if the particles were retained at a free water surface as a layer of one particle deep, the process was spoken of as skin flotation; and if the particles were retained in a foamy layer several inches thick, the process was froth flotation.

Nowadays, both bulk-oil flotation and skin flotation have been sifted out, but froth flotation has survived the test of time and has been developing rapidly. The term flotation is now used universally to describe froth flotation. The theory of froth flotation is complex, involving three phases (solids, water, and froth) with many sub-processes and interactions, and is not completely understood.

(2) Basic principles

Froth flotation utilizes the differences in physico-chemical surface properties of particles of various minerals. After treatment with reagents, such differences in surface properties between the minerals within the flotation pulp become apparent and, for flotation to take place, an air bubble must be able to attach itself to a particle, and lift it to the water surface. Figure 7.1 illustrates the principles of flotation in a mechanical flotation cell. The agitator provides enough turbulence in the pulp phase to promote collision of particles and bubbles which results in the attachment of valuable particles to bubbles and their transport into the froth phase for recovery.

The process can only be applied to relatively fine particles, as if they are too large the adhesion between the particle and the bubbles will be less than the particle weight and the bubble will therefore drop its load.

In flotation concentration, the mineral is usually transferred to the froth, or float fraction, leaving the gangue in the pulp or tailing. This is direct flotation as opposed to reverse flotation, in which the gangue is separated into the float fraction.

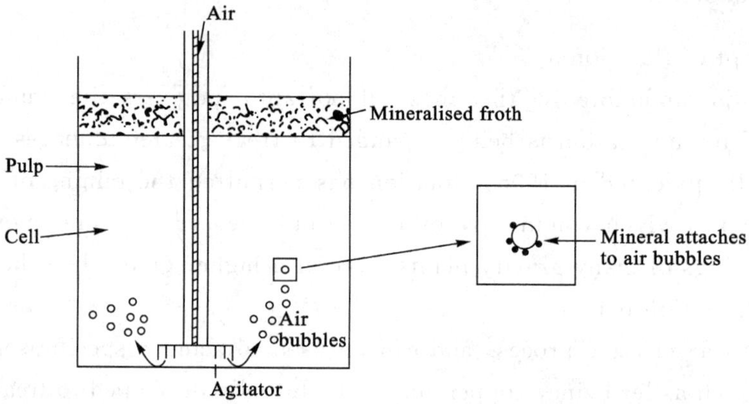

Figure 7.1　Principle of froth flotation

The air-bubbles can only stick to the mineral particles if they can displace water from the mineral surface, which can only happen if the mineral is to some extent water repellent or hydrophobic. Having reached the surface, the air bubbles can only continue to support the mineral particles if they form a stable froth, otherwise they will burst and drop the mineral particles. To achieve these conditions it is necessary to use the numerous chemical compounds known as flotation reagents.

2. Wetting phenomena and contact angle

(1) Wetting phenomena

Among the various properties of flotation separation, the wetting phenomena on the mineral surface is the most important.

Flotation separation of dissimilar minerals is accomplished through the selective affinity of their surfaces which properly modified by reagents if necessary, for air bubbles, or for water. Of various phases the gas phase is the simplest and the solid phase the most complex. The liquid phase, while in appearance very simple, is actually complex and important. Both the selectivity and the behavior of flotation separation are determined by the properties of the three phases and interfaces made up of them.

Some mineral surfaces have a higher affinity for water, some lower. For example, quartz has a high affinity for water, and sulfur lower. Water can spread out through quartz surface, otherwise, water can hardly spread out through sulfur surface, the former is called wetting phenomena and the latter called non-wetting phenomena.

(2) Contact angle

In general, the contact angle is used to describe the wetting phenomena on the mineral surface. Figure 7. 2 represents the contact angle θ at a solid-water-air contact line. The activity of a mineral surface in relation to flotation reagents in water depends on the forces which operate on that surface. At equilibrium, the surface and interfacial tensile forces acting at O (Figure 7. 2) must balance exactly.

$$\gamma_{SA} = \gamma_{SW} + \gamma_{WA} \cdot \cos\theta \tag{7.1}$$

where, γ_{SA}, γ_{SW} and γ_{WA} are the surface tensile forces between solid-air, solid-water and water-air, respectively, and θ is the contact angle between the mineral surface and the bubble.

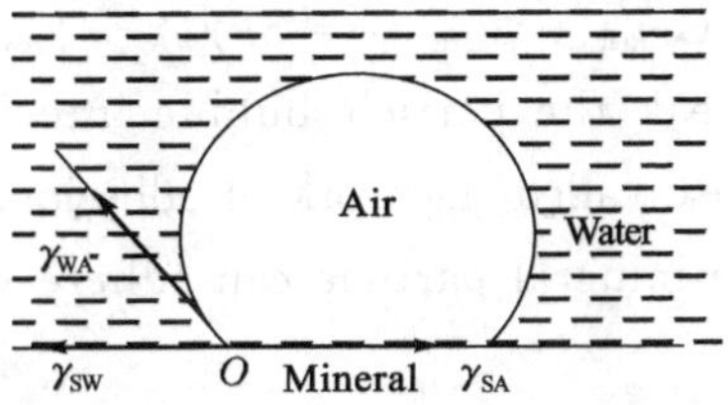

Figure 7. 2 Contact angle between bubble and particle in an aqueous medium

(3) Work of adhesion

The force required to break the particle-bubble interface is called the work of adhesion, W_{SA}, is equal to the work required to separate the solid-air interface and produce air-water and solid-water interfaces, i. e.

$$W_{SA} = \gamma_{SW} + \gamma_{WA} - \gamma_{SA} \tag{7.2}$$

Combining with Equation (7. 1) gives

$$W_{SA} = \gamma_{WA}(1 - \cos\theta) \tag{7.3}$$

It can be seen that the greater the contact angle the greater is the work of adhesion between particle and bubble, and the more resilient the system is to disruptive forces. The floatability of a mineral therefore increases with the contact angle; minerals with a high contact angle are said to be aerophilic, i. e., they have a higher affinity for air than for water. Most minerals are not water repellent in their natural state and flotation reagents must be added to the pulp.

The contact angle is of great significance in flotation. If the contact angle of mineral is nil, this means that water wets the mineral in preference to air and air-mineral contact is impossible. A contact angle of 0 represents nonflotability.

Conversely a contact angle of 180° would represent wetting of the mineral by air to exclusion of water. No solid is known to give an air-water contact angle larger than about 110°, so the condition of a 180° angle can't be realized experimentally.

3. Thermodynamics and kinetics of flotation

(1) Thermodynamics of flotation

It is called the mineralization of air bubbles that mineral particles collide with air bubbles and adhere to them. The mineralization of air bubbles is the primary act in froth flotation.

Assume that a fine particle comes in contact with an air bubble relatively large that bubble surface can be assumed to be plane. Adhesion does not change the bubble volume, so that effectively the air-water interface is decreased by S, the mineral-water interface is decreased by S, and the mineral-air interface is increased by S, S being, of course, the area of contact.

The decrease in Gibbs free energy (Figure 7.3) is the sum of initial interfacial energies less the sum of the terminal surface energies:

$$\Delta W = G_{\text{initial}} - G_{\text{terminal}} = S(\gamma_{WA} + \gamma_{SW} - \gamma_{SA}) \tag{7.4}$$

The work required to break the particle-bubble interface is equal to the Gibbs free energy change ΔW, which is also called the work of adhesion. According to Equation (7.4), if $\Delta W > 0$, the bubble and the mineral particle can adhere spontaneously to each other; if $\Delta W < 0$, they can not.

Combining with the equation:

$$\gamma_{SA} = \gamma_{SW} + \gamma_{WA}\cos\theta \tag{7.5}$$

gives

$$\Delta W = S \cdot \gamma_{WA}(1 - \cos\theta) \tag{7.6}$$

It can be seen that the greater the contact angle the greater is the work of adhesion between particle and bubble (ΔW).

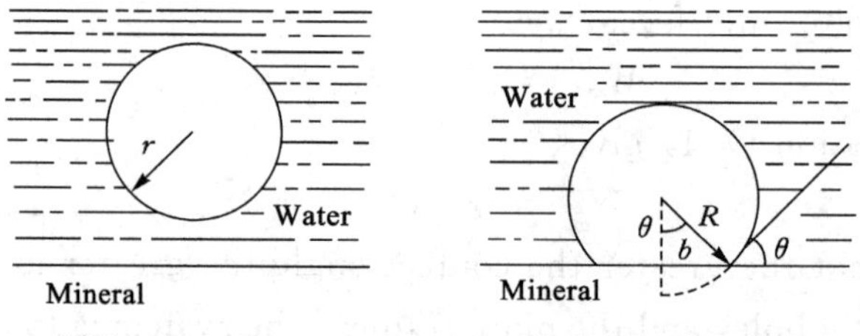

Figure 7.3 Change of Gibbs free energy between before and after particle-bubble attachment

Most minerals are not water repellent in their natural state and flotation reagents must be added to the pulp. The most important reagents are the collectors, which adsorb on mineral surfaces, rendering them hydrophobic and facilitating bubble attachment. Frothers help maintain a reasonably stable froth. Regulators are used to control the flotation process; either activate or depress mineral attachment to air-bubbles and also used to control the pH of the system.

(2) Kinetics of flotation

The flotation operation involves not only the conditioning of solid surfaces with agents to induce the adhesion of the selected solids to air bubble, but also the mechanical separation of the gassy solids from the wetted solids which depends on a number of factors. Among these are the grade of flotation feed, the extent of liberation and non-liberation, the distribution on sizes, the properties and types of reagents, the type of equipment used for flotation with its specific characteristics, and the type of aeration and froth removal. For a quantitative evaluation of selectivity in the separation, it is necessary to deal with the rates of progressive removal of the floated mineral. In a batch test the rate of flotation can be measured as the rate at which the concentration (c) of the given mineral decreases, hence:

$$-\mathrm{d}c/\mathrm{d}t = Kc \quad \text{or} \quad -\mathrm{d}c/\mathrm{d}t = K'c^{n} \tag{7.7}$$

where $n = 1.2$ or another empirical exponent. Another approach is to apply a simply probability theory, resulting in an expression:

$$W = W_0(1-P)^{m} \quad \text{or} \quad \lg W/W_0 = m\lg(1-P) \tag{7.8}$$

where W_0 is the weight of selectively floated mineral in the flotation feed, W is the weight of the same mineral remaining in the pulp after passage through m cells, P is the probability of flotation.

4. Classification of minerals

All minerals are classified into polar or non-polar types according to their surface characteristics. The surfaces of non-polar minerals are characterized by relatively weak molecular bonds. The minerals are composed of covalent molecules held together by van der Waals forces, and the non-polar surfaces do not readily attach to the water dipoles, and in consequence are hydrophobic. Minerals of this type, such as graphite, sulphur, molybdenite, diamond, coal, and talc, thus have high natural floatabilities with contact angles between 60° and 90°. Although it is possible to float these minerals without the aid of chemical agents, it is universal to increase their hydrophobicity by the addition of hydrocarbon oils or frothing agents. Creosote, for example, is widely used to increase the floatability of coal.

Minerals with strong covalent or ionic surface bonding are known as polar types, and exhibit high free energy values at the polar surface. The polar surfaces react strongly with water molecules, and these minerals are naturally hydrophilic. In general, the degree of polarity increases from sulphide minerals, through sulphates, to carbonates, halides, phosphates, etc., then oxides-hydroxides, and, finally, silicates and quartz.

Notes

1. Nowadays, both bulk-oil flotation and skin flotation have been sifted out, but froth flotation has survived the test of time and has been developing rapidly.

现在,全油浮选和表层浮选已被淘汰,而泡沫浮选则经受住了时间的考验,并得到迅速发展。

2. Froth flotation utilizes the differences in physico-chemical surface properties of particles of various minerals.

浮选利用的是矿物颗粒之间表面物化性质的不同进行分选的。

3. The liquid phase, while in appearance very simple, is actually complex and important.

液相虽然表面上很简单,实际上非常复杂也非常重要。

4. Both the selectivity and the behavior of flotation separation are determined by the properties of the three phases and interfaces made up of them.

浮选分离的选择性和过程都是由这三相以及它们组成的界面的特性决定的。

5. The activity of a mineral surface in relation to flotation reagents in water depends on the forces which operate on that surface.

在水中矿物表面与浮选药剂的反应活性,取决于该表面的各种作用力。

6. It can be seen that the greater the contact angle the greater is the work of adhesion between particle and bubble and the more resilient the system is to disruptive forces.

可以看出,接触角越大,颗粒与气泡间的黏附功就越大,体系(矿化气泡)更不易被破坏。

7. Any systematic change in one of the factors listed at the beginning of this section that produces a change in the slope of the overall position of curve *C*, provides a more reliable and informative assessment of flotation behavior than that inferred from the recovery and grade data.

在本节开始所列的任何一个因素的系统性变化,都会使曲线 *C* 整体位置的斜率发生改变,这样就能够比使用回收率和品位数据更为可靠和更为全面地对浮选行为进行评价。

8. All minerals are classified into polar or non-polar types according to their surface characteristics.

所有矿物都根据其表面特征分为极性和非极性两种类型。

Exercises

1. Answer the following questions according to the text.

(1) What is froth flotation?

(2) What is wetting phenomena?

(3) What is the work of adhesion?

(4) What is the mineralization of air bubbles?

(5) How to classify minerals according to their surface characteristics, and what are their characteristics?

2. Translate the following phrases into English.

(1) 浮选药剂	(2) 复杂矿体	(3) 固气界面	(4) 接触角
(5) 界面张力	(6) 黏附功	(7) 气泡矿化	(8) 亲水性颗粒
(9) 浮选速度	(10) 润湿现象	(11) 表面物化性质	(12) 泡沫浮选

3. Translate the following paragraphs into Chinese.

Paragraphs 1, 2 and 3 in section 1 (2) "Froth flotation utilizes the differences in physico-chemical surface properties ...".

4. Translate the following sentences into English.

(1) 浮选是一种选择性强的选矿方法,可用于从复杂硫化矿中分离某一种特定的硫化矿物。

(2) 接触角是一个用来描述矿物表面润湿现象的参数。

(3) 如果某种矿物的接触角是零,这就意味着在矿浆中空气与矿物表面的接触是不可能的。

(4) 气泡的矿化是指矿物颗粒与气泡碰撞并附着于气泡上。

(5) 气泡表面的接触角越大,与空气的亲和力越强,其可浮性越好。

(6) 大多数矿物不是天然疏水性矿物,浮选过程中需要添加浮选药剂来改善其表面性质。

7.2 Collectors(捕收剂)

1. Classification of collectors

(1) Concept of collectors

Hydrophobicity has to be imparted to most minerals in order to float them. In order to achieve this, surfactants known as collectors are added to the pulp and time is allowed for adsorption during agitation in what is known as the conditioning period.

Collectors are organic compounds which render selected minerals water-repellent by adsorption of molecules or ions on to the mineral surface, reducing the stability of the hydrated layer separating the mineral surface from the air bubble to such a level that attachment of the particle to the bubble can be made on contact.

(2) Types of collectors

Collector molecules may be ionizing compounds, which dissociate into ions in water, or non-ionizing compounds, which are practically insoluble, and render the mineral water-repellent by covering its surface with a thin film.

Ionizing collectors have found very wide application in flotation. They have complex molecules which are asymmetric in structure and are heteropolar, i. e. , the molecule contains a non-polar hydrocarbon group and a polar group which may be one of a number of types. The non-polar hydrocarbon radical has pronounced water-repellent properties, whereas the polar group reacts with water. Ionizing collectors are classed in accordance with the type of ion, anion or cation that produces the water-repellent effect in water. This classification is given in Figure 7. 4.

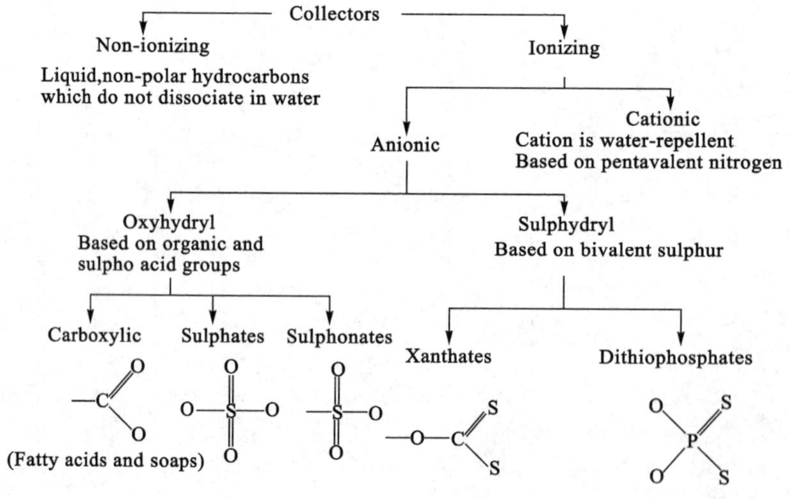

Figure 7. 4 Classification of collectors

(3) Principles

Because of chemical, electrical, or physical attraction between the polar portions and surface sites, collectors adsorb on the particles with their non-polar ends orientated towards the bulk solution, thereby imparting hydrophobicity to the particles (Figure 7.5). They are usually used in small amounts, substantially those necessary to form a monomolecular layer on particle surfaces (starvation level), as increased concentration, apart from the cost, tends to float other minerals, reducing selectivity. It is always harder to eliminate a collector already adsorbed than to prevent its adsorption.

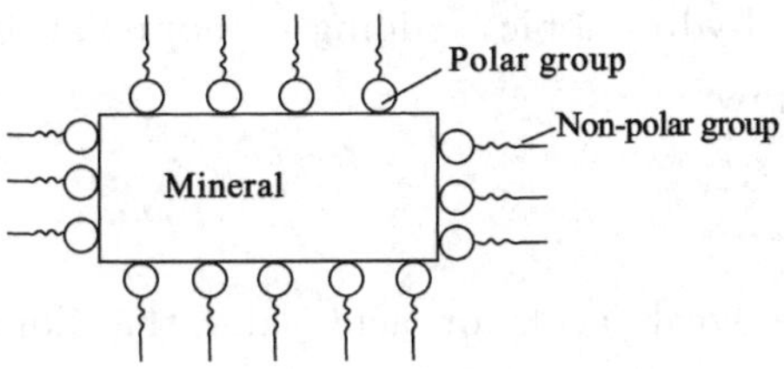

Figure 7.5 Collector adsorption on mineral surface

An excessive concentration of a collector can also have an adverse effect on the recovery of the valuable minerals, possibly due to the development of collector multi-layers on the particles, reducing the proportion of hydrocarbon radicals orientated into the bulk solution. The hydrophobicity of the particles is thus reduced, and hence their floatability.

The flotation limit may be extended without loss of selectivity by using a collector with a longer hydrocarbon chain, thus producing greater water repulsion, rather than by increasing the concentration of a shorter chain collector. However, chain length is usually limited to two to five carbon atoms, since the solubility of the collector in water rapidly diminishes with increasing chain length. Not only the chain length but also the chain structure, affects solubility and adsorption; branched chains have higher solubility than straight chains.

It is common to add more than one collector to a flotation system. A selective collector may be used at the head of the circuit, to float the highly hydrophobic minerals, after which a more powerful, but less selective one, is added to promote recovery of the slower floating minerals.

2. Nonionic collectors

The nonionic collectors are simple hydrocarbon oils. Hydrocarbon oils, and similar compounds, have an affinity for surfaces that are already partially hydrophobic. They selectively adsorb on these surfaces, and increase their hydrophobicity. The most commonly floated naturally-hydrophobic material is coal. Addition of collectors such as fuel oil and kerosene significantly enhances the hydrophobicity of the coal particles without affecting the surfaces of the associated ash-forming minerals. This improves the recovery of the coal, and increases the selectivity between coal particles and mineral matter. Fuel oil and kerosene have the following advantages over specialized collectors for froth flotation: ① they have low enough viscosity to disperse in the slurry and spread over the coal particles easily, and ② they are very low-cost compared to other compounds which can be used as coal collectors.

In addition to coal, it is also possible to float naturally-hydrophobic minerals such as molybdenite, elemental sulfur, and talc with nonionic collectors. Nonionic collectors can also be used as "extenders" for other collectors. If another, more-expensive collector makes a surface partially hydrophobic, adding a nonpolar oil will often increase the hydrophobicity further at low cost.

3. Anionic collectors

Anionic collectors are weak acids or acid salts that ionize in water, producing a collector that has a negatively-charged end that will attach to the mineral surfaces, and a hydrocarbon chain that extends out into the liquid.

(1) Anionic collectors for sulfide minerals

The most common collectors for sulfide minerals are the sulfhydryl collectors, such as the various xanthates and dithiophosphates.

The xanthates are the most important for sulphide mineral flotation. They are prepared by reacting an alkali hydroxide, an alcohol and carbon disulphide:

$$ROH + CS_2 + KOH = ROCSSK + H_2O \tag{7.9}$$

where R is the hydrocarbon group and contains normally one to six carbon atoms, the most widely used xanthates being ethyl, isopropyl, isobutyl, amyl, and hexyl.

Sodium ethyl xanthate is typical and has the structure shown in Figure 7.6. The anion consists of a hydrocarbon non-polar radical and a connected polar group. Xanthates are highly selective collectors for sulfide minerals, as they chemically react with the sulfide surfaces and do not have any affinity for the common non-sulfide gangue minerals. The $OCSS^-$ group attaches irreversibly to the sulfide mineral surface. Using xanthates with longer hydrocarbon chains tends to increase the degree of hydrophobicity when they adsorb onto the surface.

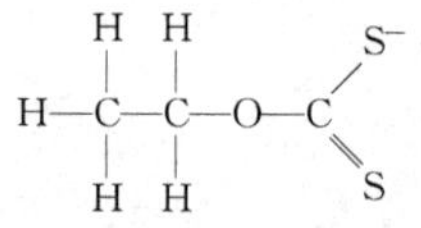

Figure 7.6 Structure of a typical xanthate collector (ethyl xanthate)

The solubilities of the hydrophobic xanthates of copper, lead, silver, and mercury are very low, but the xanthates of zinc and iron are much more soluble. Typically, ethyl xanthates are only weak collectors of pure sphalerite, but replacement of the crystal lattice zinc atoms by copper improves the flotation properties of the mineral. The alkali earth metal xanthates (calcium, barium, magnesium) are very soluble and xanthates have no collector action on the minerals of such metals, nor on oxides, silicates, or aluminosilicates, which permits extremely selective flotation of sulphides from gangue minerals.

The dithiophosphates have pentavalent phosphorus in the polar group, rather than tetravalent carbon (Figure 7. 7). Dithiophosphates are not as widely used as the xanthates, but are still important reagents in practice. They are comparatively weak collectors, but give good results in combination with xanthates. They are often used in the separation of copper from lead sulphides, as they are effective selective collectors for copper sulphide minerals.

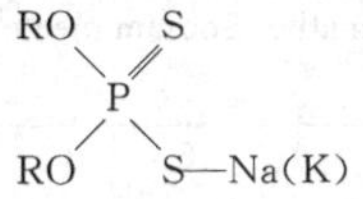

Figure 7. 7 Dithiophosphates

(2) Anionic collectors for oxide minerals

Oxyhydryl collectors are commonly used in the flotation of oxide minerals. A typical anionic collector for oxide mineral flotation is sodium oleate, the sodium salt of oleic acid, which has the structure shown in Figure 7. 8. The anionic group responsible for attaching it to the mineral surface is the carboxyl group, which dissociates in water to develop a negative charge. The negatively-charged group is then attracted to positively-charged mineral surfaces.

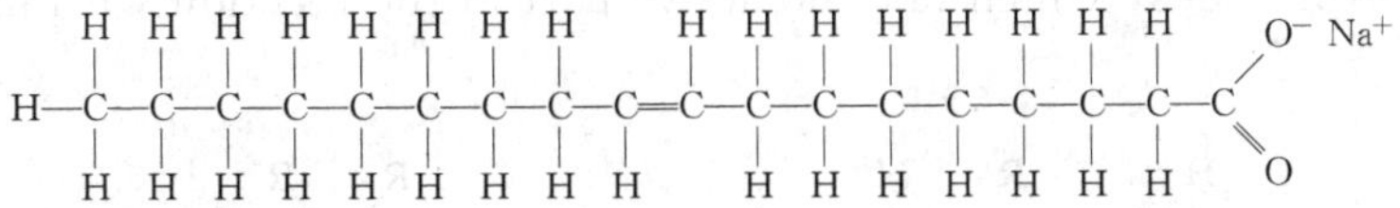

Figure 7. 8 The structure of oleic acid, a very commonly-used anionic collector.

Typically, oxyhydryl collectors are organic acids or soaps. The carboxylates are known as fatty acids, and occur naturally in vegetable oils and animal fats from which they are extracted by distillation and crystallisation. The salts of oleic acid, such as sodium oleate and linoleic acid, are commonly used. As with all ionic collectors, the longer the hydrocarbon chain length, the more powerful is the water-repulsion produced, but solubility decreases. Soaps (the salts of fatty acids), however, are soluble even if the chain length is long. The carboxylates are strong collectors, but have relatively low selectivity. They are used for the flotation of minerals of calcium, barium, strontium, and magnesium, the carbonates of non-ferrous metals, and the soluble salts of alkali metals and alkaline earth metals.

Since particles that are immersed in water develop a net charge due to exchanging ions with the liquid, it is often possible to manipulate the chemistry of the solution so that one mineral has a strong positive charge while other minerals have a charge that is either only weakly positive, or negative (Figure 7.9). In these conditions, the anionic collector will preferentially adsorb onto the surface with the strongest positive charge and render them hydrophobic.

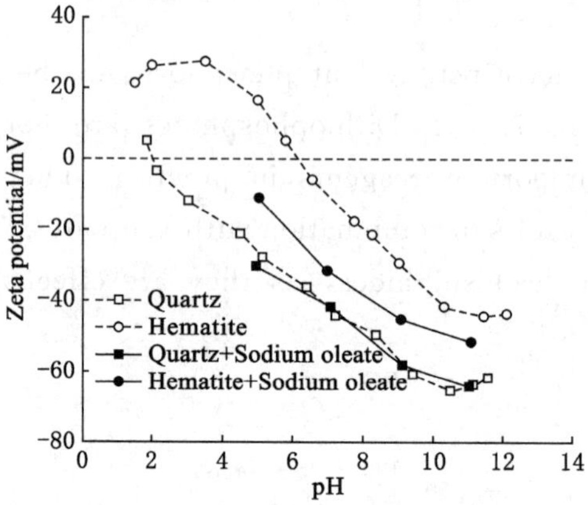

Figure 7.9 Relationship between Zeta potential and pH of minerals (sodium oleate, 30 mg/L)

4. Cationic collectors

Cationic collectors use a positively-charged amine group to attach to mineral surfaces. The characteristic property of this group of collectors is that the water-repulsion is produced by the cation where the polar group is based on pentavalent nitrogen, the amines (Figure 7.10) being the most common. The anions of such collectors are usually halides, or more rarely hydroxides, which take no active part in the reaction with minerals.

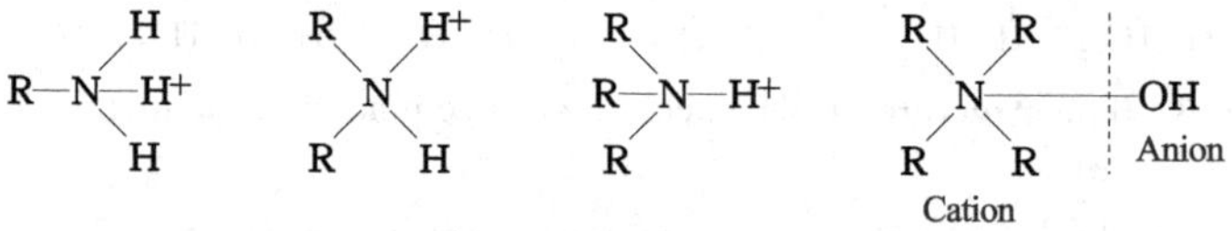

Figure 7.10 Primary, secondary, tertiary and quaternary amine groups that can be used for cationic collectors

Unlike the xanthates, the amines are considered to adsorb on mineral surfaces primarily due to electrostatic attraction between the polar head of the collector and the charged electrical double layer on the mineral surface. Such forces are not as strong or irreversible as the chemical forces characteristic of anionic collectors, so these collectors tend to be relatively weak in collecting power.

Cationic collectors are very sensitive to the pH of the medium, being most active in slightly acid solutions and inactive in strongly alkaline and acid media. They are used for floating oxides, carbonates, silicates, and alkali earth metals such as barite, carnallite, and sylvite. The primary amines (i.e., those where only one hydrocarbon group is present with two hydrogen atoms) are strong collectors of apatite and they can selectively float sedimentary phosphates from calcareous ores. The collector requirement can be reduced by adding a non-polar reagent such as kerosene, which is co-adsorbed on the mineral surface.

Notes

1. Collectors are organic compounds which render selected minerals water-repellent by adsorption of molecules or ions on to the mineral surface, reducing the stability of the hydrated layer separating the mineral surface from the air bubble to such a level that attachment of the particle to the bubble can be made on contact.

捕收剂是有机化合物,通过将分子或离子吸附到矿物表面,使目标矿物具有疏水性,从而将矿物与气泡分开的水化膜的稳定性降低到一定程度,以便实现矿粒对气泡的接触性黏附。

2. Collector molecules may be ionizing compounds, which dissociate into ions in water, or non-ionizing compounds, which are practically insoluble, and render the mineral water-repellent by covering its surface with a thin film.

捕收剂分子可以是在水中解离成离子的离子型化合物,或者是实际上不溶的非离子型化合物,通过在其表面覆盖一层薄膜而使矿物疏水。

3. Ionizing collectors have found very wide application in flotation. They have complex molecules which are asymmetric in structure and are heteropolar.

离子型捕收剂已在浮选中得到非常广泛的应用。它们具有复杂的分子,其结构不对称并且是异极性的。

4. Because of chemical, electrical, or physical attraction between the polar portions and surface sites, collectors adsorb on the particles with their non-polar ends orientated towards the bulk solution, thereby imparting hydrophobicity to the particles.

由于捕收剂的极性部分与颗粒表面部位之间的化学的、电学的及物理方面的吸引力,捕收剂以非极性基朝向本体溶液(极性基朝向矿物颗粒)的方式吸附到颗粒上赋予其疏水性。

5. They are usually used in small amounts, substantially those necessary to form a monomolecular layer on particle surfaces (starvation level), as increased concentration, apart from the cost, tends to float other minerals, reducing selectivity.

通常捕收剂用量较小,能够在矿物表面形成一个单分子层即可。用量过大时,除了成本以外,还可能使其他矿物上浮,降低浮选的选择性。

6. The nonionic collectors are simple hydrocarbon oils.

非离子型捕收剂是简单的烃类油。

7. The most common collectors for sulfide minerals are the sulfhydryl collectors, such as the various xanthates and dithiophosphates.

硫化矿物最常见的捕收剂是巯基捕收剂,例如各种黄原酸盐(黄药)和二硫代磷酸盐(黑药)。

8. Cationic collectors use a positively-charged amine group to attach to mineral surfaces.

阳离子捕收剂使用带正电荷的胺基附着在矿物表面。

Exercises

1. Answer the following questions according to the text.

(1) What is collector?

(2) What is the classification of collectors?

(3) What is the function of collectors such as fuel oil and kerosene in coal flotation?

(4) What are the most common collectors for sulfide minerals?

(5) How do the cationic collectors attach to mineral surface and what is their characteristic property?

2. Translate the following phrases into English.

(1) 非离子型捕收剂	(2) 黄药	(3) 水偶极	(4) 碳氢链长度
(5) 静电吸附	(6) 直链烷烃	(7) 极性基团	(8) 脂肪酸
(9) 胺类	(10) 阳离子		

3. Translate the following paragraphs into Chinese.

Paragraphs 1 and 2 in section 1 (1) "Hydrophobicity has to be imparted to most..." and paragraph 2 in section 1 (2) "Ionizing collectors have found very wide application in flotation ...".

4. Translate the following sentences into English.

(1) 为了使矿物浮起，必须赋予它们疏水性。

(2) 捕收剂是有机化合物，通过分子或离子吸附在矿物表面上，使目标矿物具有疏水性。

(3) 非极性烃基具有明显的疏水特性，而极性基团会与水发生反应。

(4) 捕收剂浓度过高也会对有用矿物的回收产生不利影响，这可能是由于捕收剂在颗粒表面形成了多层结构，降低了烃类自由基在溶液中的比例。

(5) 离子型捕收剂是根据在水中产生疏水效果的离子是阴离子或阳离子来分类的。

7.3 Frothers and Regulators(起泡剂和调整剂)

1. Frothers

(1) Action of the frother

When mineral surfaces have been rendered hydrophobic by the use of a collector, stability of bubble attachment, especially at the pulp surface, depends to a considerable extent on the efficiency of the frother.

Ideally the frother acts entirely in the liquid phase and does not influence the state of the mineral surface. In practice, however, interaction does occur between the frother, mineral, and other reagents, and the selection of a suitable frother for a given ore can only be made after extensive test work.

Frothers are in many respects chemically similar to ionic collectors, and, indeed, many of the collectors, such as the oleates, are powerful frothers, being in fact too powerful to be used as efficient frothers, since the froths which they produce can be too stable to allow efficient transport to further processing. Froth build-up on the surfaces of thickeners and excessive frothing of flotation cells are problems occurring in many mineral processing plants. A good frother should have negligible collecting power, and also produce a froth which is just stable enough to facilitate transfer of floated mineral from the cell surface to the collecting launder.

Frothers are generally heteropolar surface-active organic reagents, capable of being adsorbed on the air-water interface. When surface-active molecules react with water, the water dipoles combine readily with the polar groups and hydrate them, but there is practically no reaction with the non-polar hydrocarbon group, the tendency being to force the latter into the air phase. Thus the heteropolar structure of the frother molecule leads to its adsorption, i. e. , the molecules concentrate in the surface layer with the non-polar groups orientated towards the air and the polar groups towards the water (Figure 7. 11).

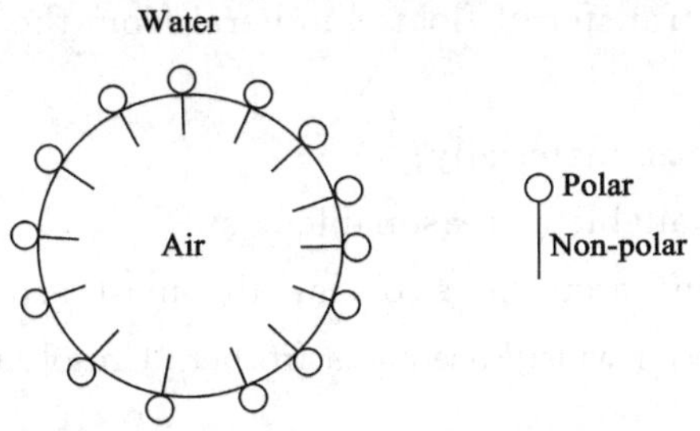

Figure 7. 11 Action of the frother

Frothing action is thus due to the ability of the frother to adsorb on the air-water interface because of its surface activity and to reduce the surface tension, thus stabilising the air-bubble.

(2) Types and characteristics

Frothers must be to some extent soluble in water, otherwise they would be distributed very unevenly in an aqueous solution and their surface active properties would not be fully effective. The most effective frothers include in their composition one of the following groups:

Hydroxyl	—OH
Carboxyl	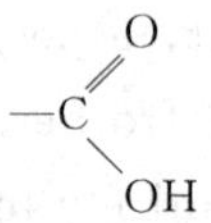
Carbonyl	=C=
Amino group	$—NH_2$
Sulpho group	$—OSO_2OH$, $—SO_2OH$

The acids, amines, and alcohols are the most soluble of the frothers. The alcohols (—OH) are the most widely used, since they have practically no collector properties, and in this respect are preferable to other frothers, such as the carboxyls, which are also powerful collectors; the presence of collecting and frothing properties in the same reagent may make selective flotation difficult. Frothers with an amino group, and certain sulpho group frothers also have weak collector properties.

The principal quality sought in a frothing agent is found in many substances, but choice is somewhat restricted by certain considerations, of which cost, availability, and lack of collecting properties are the most important.

To sum up, a practical frother must be:

① An organic substance;

② Whose molecules are heteropolar and consist of one more hydrocarbon radicals attached to one polar group;

③ Whose solubility is neither large nor very small;

④ Which should have negligible collecting power and also produces froth which is just stable enough to facilitate transfer of floated mineral from the cell surface to the collecting launder;

⑤ Which does not ionize materially;

⑥ Which is readily available at reasonable cost.

Pine oil, which contains aromatic alcohols, the most active frothing component being terpineol, $C_{10}H_{17}OH$, has been widely used as a frother. Cresol (cresylic acid), $CH_3C_6H_4OH$, has also had wide use.

A wide range of synthetic frothers, based mainly on high molecular-weight alcohols, are now in use in many plants. They have the important advantage over industrial products such as pine oil and cresol in that their compositions are much more stable, which makes it easier to control the flotation process and improves performance. A widely used synthetic alcohol frother is methyl isobutyl carbinol (MIBC) (Figure 7. 12). The alcohol groups provide a selective, often brittle, froth, which allows good control and materials transfer through the launders and pumps.

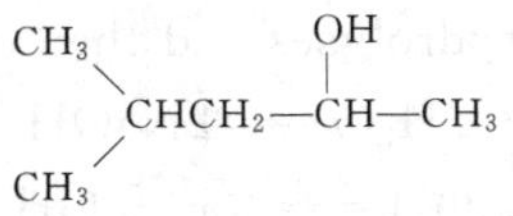

Figure 7. 12 Structural formula of MIBC

2. Regulators

Regulators, or modifiers, are used extensively in flotation to modify the action of the collector, either by intensifying or by reducing its water-repellent effect on the mineral surface. They thus make collector action more selective towards certain minerals. Regulators can be classed as activators, depressants, or pH modifiers.

(1) Activators

The process by which collection is increased is generally termed activation, these reagents which can alter the chemical nature of mineral surfaces so that they become hydrophobic due to the action of the collector are called activators, they are generally soluble salts which ionize in solution, and the ions reacting with the mineral surface.

A classical example is the activation of sphalerite by copper in solution. Sphalerite is not floated satisfactorily by a xanthate collector, since the collector products formed, such as zinc xanthate, are relatively soluble in water, and so do not provide a hydrophobic film around the mineral.

Floatability can be improved by the use of large quantities of long-chain xanthates, but a more satisfactory method is to use copper sulphate as an activator, which is readily soluble and dissociates into copper ions in solution. Activation is due to the formation of molecules of copper sulphide at the mineral surface, due to the fact that copper is more electronegative than zinc and therefore ionizes less readily:

$$ZnS + Cu^{+} = CuS + Zn^{+} \quad (7.10)$$

The copper sulphide deposited on the sphalerite surface reacts readily with xanthate to form insoluble copper xanthate, which renders the sphalerite surface hydrophobic.

Very careful control of reagent feeding must be observed when copper sulphate is used as an activator in conjunction with a xanthate collector. It is essential that the minerals are conditioned with the activator separately, before the collector is added, and that when the conditioned pulp enters the collector conditioner, there is little residual copper sulphate in solution. Xanthates react readily with copper ions, needlessly consuming both reagents.

The main use of copper sulphate as an activator is in the differential flotation of lead-zinc ores, where after lead flotation the sphalerite is activated and floated. To some extent, copper ions can also activate galena, calcite, and pyrite.

Oxidised minerals of lead, zinc, and copper, such as cerussite, smithsonite, azurite, and malachite, float very inefficiently with sulphydryl collectors and require an extremely large amount. Such minerals are activated by the use of sodium sulphide or sodium hydrosulphide.

In solution, sodium sulphide hydrolyses and then dissociates:

$$Na_2S+2H_2O=2NaOH+H_2S \quad (7.11)$$

$$NaOH=Na^++OH^- \quad (7.12)$$

$$H_2S=H^++HS^- \quad (7.13)$$

$$HS^-=H^++S^{2-} \quad (7.14)$$

Since the dissociation constants of Equations (7.13) and (7.14) are extremely low and that of Equation (7.12) is high, the concentration of OH^- ions increases at a faster rate than that of H^+ ions and the pulp becomes alkaline. Hydrolysis and dissociation of sodium sulphide release OH^-, S^{2-}, and HS^- ions into solution and these can react with and modify the mineral surfaces. Sulphidation causes sulphur ions to pass into the crystal lattice of the oxidised minerals, giving them a relatively insoluble pseudo-sulphide surface coating and allowing them to be floated by sulphydryl collectors.

(2) Depressants

The process by which collection is decreased is generally termed depression, those reagents which alter the chemical nature of mineral surfaces so that they become hydrophilic, thus preventing their flotation, are called depressants.

There are many types of depressant and their actions are complex and varied, and in most cases little understood, making depression more difficult to control than the application of other types of reagent.

Slime coating is an example of a naturally occurring form of depression. Slimes in a crushed and ground ore make flotation difficult, as they coat the mineral particles, retarding collector adsorption. They can sometimes be removed from the surfaces by vigorous agitation, but a more reliable method is to use slime dispersant.

Sodium silicate in solution increases the double-layer charge on particles, so that the slime layers which have formed readily disperse. The clean mineral surface can then interact with the collector. In this respect, therefore, sodium silicate is used as an activator, preventing depression by the slimes. Sodium silicate is also used as a depressant in some systems, depressing silicates, and calcite in fluorite-calcite and scheelite-calcite separations.

Cyanides are widely used in the selective flotation of lead-copper-zinc and copper-zinc ores, as depressants for sphalerite, pyrite and certain copper sulphides.

It is fairly well established that pure clean sphalerite does not adsorb short-chain xanthates until its surface is activated by copper ions. However, copper ions resulting from very slight dissolution of copper minerals present in the ore may cause unintentional activation and prevent selective separation. Cyanide is added to the pulp to desorb the surface copper and to react with copper in solution forming soluble cyanide complexes.

Sodium cyanide is most commonly used, which hydrolyses in aqueous solution to form free alkali, and relatively insoluble hydrogen cyanide:

$$NaCN + H_2O \rightleftharpoons HCN + NaOH \quad (7.15)$$

The hydrogen cyanide then dissociates:

$$HCN \rightleftharpoons H^+ + CN^- \quad (7.16)$$

The dissociation constant of Equation (7.16) is extremely low compared with that of Equation (7.15), so that an increase in pulp alkalinity reduces the amount of free HCN, but increases the concentration of CN^- ions. An alkaline pulp is essential, as free hydrogen cyanide is extremely dangerous.

Since the production of even small traces of hydrogen cyanide is extremely dangerous, cyanides must always be used in an alkaline medium, since free alkali in solution will force the hydrolysis reaction to the left.

Cyanides are, of course, extremely toxic and must be handled with great care. They also have the disadvantage of being expensive. And, they depress and dissolve gold and silver. Despite these disadvantages, the cyanides are widely used due to their high degree of selectivity in flotation.

Potassium dichromate ($K_2Cr_2O_7$) is used to depress galena in copper-lead separations. The depressive action is due to the chemical reaction between the galena surface and CrO_4^{2-} anions, which produces insoluble lead dichromate which increases wettability and prevents flotation.

Organic reagents such as starch, tannin and dextrin do not ionize in solution, but form colloidal particles in the pulp which can be deposited on the mineral surfaces, preventing flotation in a similar manner to a slime coating. Large quantities of these reagents will depress all minerals and they are not as selective as the electrolytic depressants. They are used in small amounts to depress talc, graphite, and calcite. Starch is also used to depress molybdenite in copper-molybdenum separation.

(3) pH modifiers

Flotation is usually carried out in an alkaline medium, as most collectors are stable under these conditions and corrosion of cells, pipework, etc., is minimised. It is evident from the foregoing that pulp alkalinity plays a very important, though very complex, role in flotation, and, in practice, selectivity in complex separations is dependent on a delicate balance between reagent concentrations and pH.

Alkalinity is controlled by the addition of lime, sodium carbonate (soda ash), and to a lesser extent sodium hydroxide or ammonia. Sulphuric or sulphurous acids are used where a decrease in pH is required.

These chemicals are often used in very significant amounts in almost all flotation operations. Although they are cheaper than collectors and frothers, the overall cost is generally higher with pH regulators per tonne of ore treated than with any other processing chemical.

Lime, being cheap, is very widely used to regulate pulp alkalinity and is used in the form of milk of lime, a suspension of calcium hydroxide particles in a saturated aqueous solution. Lime prevents the adverse effects of soluble salts on flotation, by precipitating them as metal hydroxides. It is also used extensively as a depressant for pyrite and arsenopyrite. Both the hydroxyl ion and the calcium ion participate in the depressive effect of lime on pyrite by the formation of mixed films of Fe(OH), FeO(OH), $CaSO_4$, and $CaCO_3$ on the surface, reducing the adsorption of xanthate. Lime has no depressant effect with copper minerals, but does depress galena to some extent. In the flotation of galena, therefore, pH control is often effected by the use of soda ash (Na_2CO_3), pyrite and sphalerite being depressed by cyanide.

As was shown earlier, the effectiveness of sodium cyanide and sodium sulphide is governed to such a large extent by the value of pH that these reagents are of scarcely any value in the absence of alkalis. Since, where cyanide is used as a depressant, the function of the alkali is to control the cyanide ion concentration, there is, for each mineral and given concentration of collector, a "critical cyanide ion concentration" above which flotation is impossible.

Notes

1. Froth build-up on the surfaces of thickeners and excessive frothing of flotation cells are problems occurring in many mineral processing plants.

浓缩机表面的泡沫堆积和浮选槽中的过量泡沫是许多选矿厂存在的问题。

2. When surface-active molecules react with water, the water dipoles combine readily with the polar groups and hydrate them, but there is practically no reaction with the non-polar hydrocarbon group, the tendency being to force the latter into the air phase.

当表面活性分子与水发生反应时，水的偶极子极易与极性基耦合并与其水合，但是与非极性的烃基几乎不发生反应，这种倾向使得后者进入气相。

3. They have the important advantage over industrial products such as pine oil and cresol in that their compositions are much more stable, which makes it easier to control the flotation process and improves performance.

与松油和甲酚等工业产品相比，它们的重要优势在于成分更加稳定，这使得合成起泡剂更容易控制浮选的过程和提高浮选质量。

4. Sphalerite is not floated satisfactorily by a xanthate collector, since the collector products formed, such as zinc xanthate, are relatively soluble in water, and so do not provide a hydrophobic film around the mineral.

对于闪锌矿而言，黄药类捕收剂浮选效果并不令人满意，因为形成的捕收产品如黄原酸锌相对易溶于水，因此在矿物的周围就不易形成疏水性膜。

5. Very careful control of reagent feeding must be observed when copper sulphate is used as an activator in conjunction with a xanthate collector.

当硫酸铜与黄药类捕收剂一起作为活化剂时，必须谨慎控制加入药剂的量。

6. Organic reagents such as starch, tannin and dextrin do not ionize in solution, but form colloidal particles in the pulp which can be deposited on the mineral surfaces, preventing flotation in a similar manner to a slime coating.

淀粉、丹宁和糊精等有机药剂,在溶液中不发生电离,而是在矿浆中以胶体颗粒的形式存在,它们可以沉积在矿物表面,以类似矿泥罩盖的方式抑制浮选。

7. Lime has no depressant effect with copper minerals, but does depress galena to some extent.

石灰对铜矿物没有抑制作用,但在某种程度上对方铅矿有抑制作用。

Exercises

1. Answer the following questions according to the text.

(1) What properties should a good frother have?

(2) What is the action of the frother?

(3) Why do synthetic frothers have important advantage over industrial products?

(4) What is the function of regulators?

(5) What is the definition of activators?

2. Translate the following phrases into English.

(1) 表面活性剂　(2) 起泡能力　(3) 合成起泡剂　(4) 捕收能力

(5) 表面张力　(6) 异极性物质　(7) 矿泥罩盖　(8) 疏水性薄膜

(9) 硫化铜　(10) 胶体颗粒

3. Translate the following paragraphs into Chinese.

Paragraphs 3 and 4 in section 1 (1) "Frothers are in many respects chemically similar ...".

4. Translate the following sentences into English.

(1) 起泡剂一般为异极性表面活性有机药剂,能够吸附在空气-水界面上。

(2) 锌、铝、铜的氧化矿物可采用硫化钠或硫氢化钠来活化。

(3) 抑制剂的作用是改变矿物表面的化学性质,使其表面亲水。

(4) 氰化物是闪锌矿、黄铁矿的有效抑制剂,常用于铜-锌矿石和铅-锌矿石的选择性浮选。

7.4 Flotation Circuits(浮选流程)

1. Flotation banks

Commercial flotation is a continuous process. Many flotation machines (cells) are arranged in series forming a bank (Figure 7.13). Pulp enters the first cell of the bank and gives up some of its valuable mineral as a froth; the overflow from this cell passes to the second cell, where more mineralized froth is removed, and so on down the bank, until barren tailings overflow the last cell in the bank.

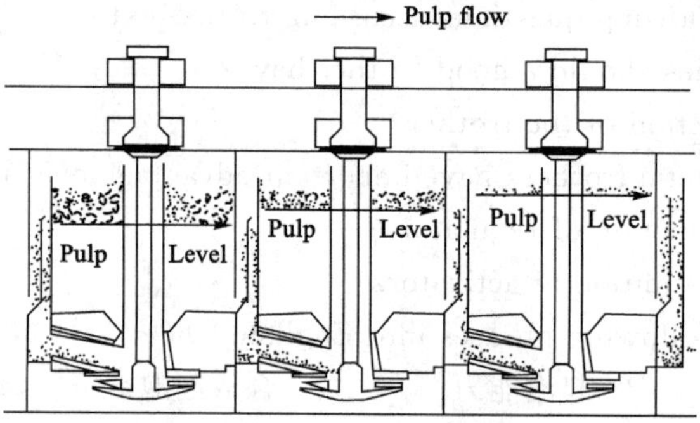

Figure 7.13 Bank of flotation cells

In the case of flotation cells that use weir-type level control, the height of the froth column for each cell is determined by adjusting the height of the tailings overflow weir, the difference in height between this and the cell overflow lip determining the froth depth. In modern tank cells, pulp level is often maintained by adjusting the cell's tailings discharge with a rubber sleeve pinch valve.

New feed enters the first cell of the bank, the froth column in the first few cells being kept high, since there are plenty of hydrophobic mineral particles to sustain it. The pulp level is raised from cell to cell, as the pulp becomes depleted in floatable minerals, by progressively raising the cell tailings weir height. The last few cells in the bank contain relatively low-grade froths, comprising weakly aerophilic particles. These are the scavengers, usually containing middlings particles, which are often recirculated to the head of the system.

In earlier cell designs, the scavenger cells, having little mineral to sustain a deep froth, have their tailings weirs raised so that pulp is almost overflowing the cell lip. This policy, which is used to remove all weakly floating material, ensures maximum recovery from the bank of cells. Excessive circulating loads should be avoided, however, as the rougher feed may be diluted, and the flotation time reduced.

2. Basic flotation circuits

(1) Simple rougher-scavenger flotation circuit

The flowsheet for this basic system is shown in Figure 7.14. This flowsheet can be successfully operated only when the gangue is relatively unfloatable, and it requires extremely careful control to produce an even grade of concentrate if there are any fluctuations in the head grade.

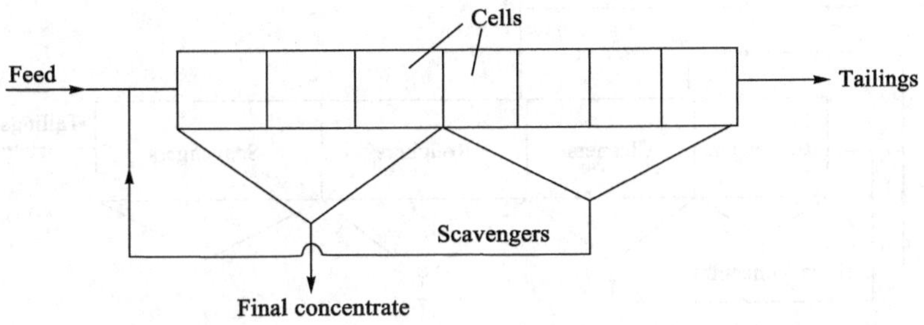

Figure 7.14 Simple flotation circuit

(2) Rougher-scavenger-cleaner system

A preferable system is to dilute the concentrate from the first few cells of the bank, known as rougher concentrates, and refloat them in cleaner cells, where the weirs are kept low to maintain a deep froth and produce a high-grade concentrate. In this rougher-scavenger-cleaner system (Figure 7.15), the cleaning cells receive a comparatively high-grade feed, whilst the scavenging section can be run with an excess of air so as to obtain maximum recovery. Tailings from the cleaner cells, usually containing aerophilic mineral particles, are generally recirculated to the rougher cells, along with the scavengers. This type of circuit, besides being useful for ores that need the maximum amount of aeration at the end of the bank to produce profitable recovery, is often employed when the gangue has a tendency to float and is difficult to separate from the mineral.

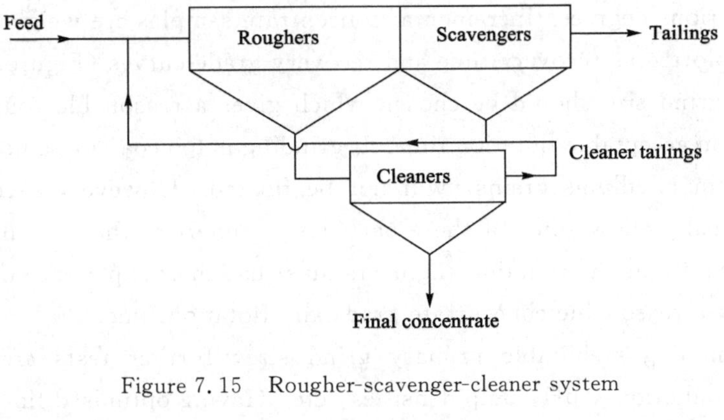

Figure 7.15 Rougher-scavenger-cleaner system

In such cases, it may be necessary to utilize one or more recleaning banks of cells

(Figure 7.16). It is worth noting that the diluting water used to lower the pulp density of the cleaning bank passes to the roughing cells and dilutes the primary feed, which should therefore contain a correspondingly smaller portion of water as it leaves the grinding section in order that the dilution of the cleaner tailing may bring it to the correct pulp ratio in the roughing cells.

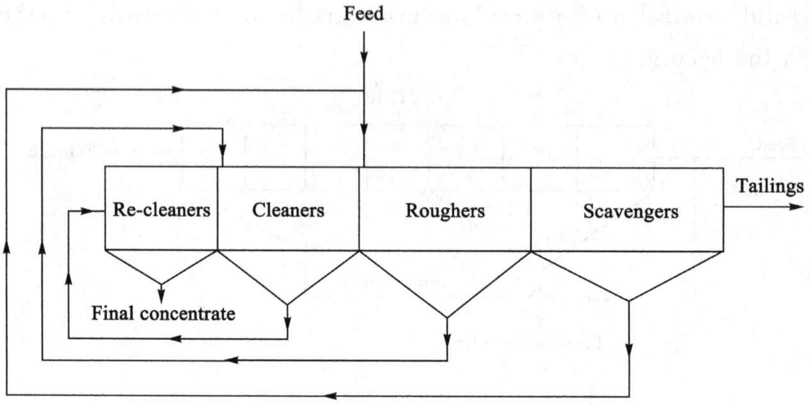

Figure 7.16 Circuit employing recleaning

3. Complex flotation circuits

(1) Flowsheet design

In designing a suitable flowsheet for a flotation plant, the primary grind size is of major consideration. This is mainly due to the fact that the flotation response depends on the level of liberation of the minerals in the ore. The target grind size can be estimated based on past experience and from mineralogical evaluation, but laboratory grind-flotation tests must be conducted to determine optimum conditions.

The purpose of the primary grind is to promote economic recovery of the valuable ore minerals. Batch tests are performed, utilizing various reagent combinations, on samples of ore ground to various degrees. Incremental concentrate samples are weighed and assayed, and the results plotted as recovery-time and recovery-grade curves (Figure 7.17).

Initially, a grind size should be chosen which gives a reasonable rougher grade and recovery within an acceptable flotation time. If grinding is too coarse, some of the valuable mineral, locked in middlings grains, will not be floated. However, excessive flotation times may eventually allow some of these particles to report to the concentrate, reducing its grade. It is here that the flotation engineers must use their experience in deciding what is, at this stage, a reasonable concentrate grade and flotation time.

After determining a suitable primary grind size, further tests are performed to optimise reagent additions, pH, pulp densities, etc. Having optimised flotation recovery, testwork is then aimed at producing the required concentrate grade, and determining the flowsheet which must be utilized to achieve this.

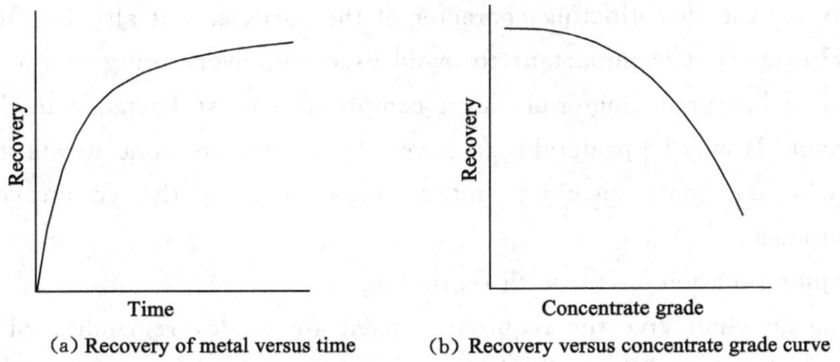

Figure 7.17

(2) Complex flotation circuit with low-grade cleaners

In a complex flotation circuit, the rougher flotation may be divided into stages, each delivering its concentrate into the cleaning circuit at a different location according to its grade. Thus, the basic flowsheet consisting of cleaners and recleaners may be supplemented by a low-grade cleaning circuit (Figure 7.18).

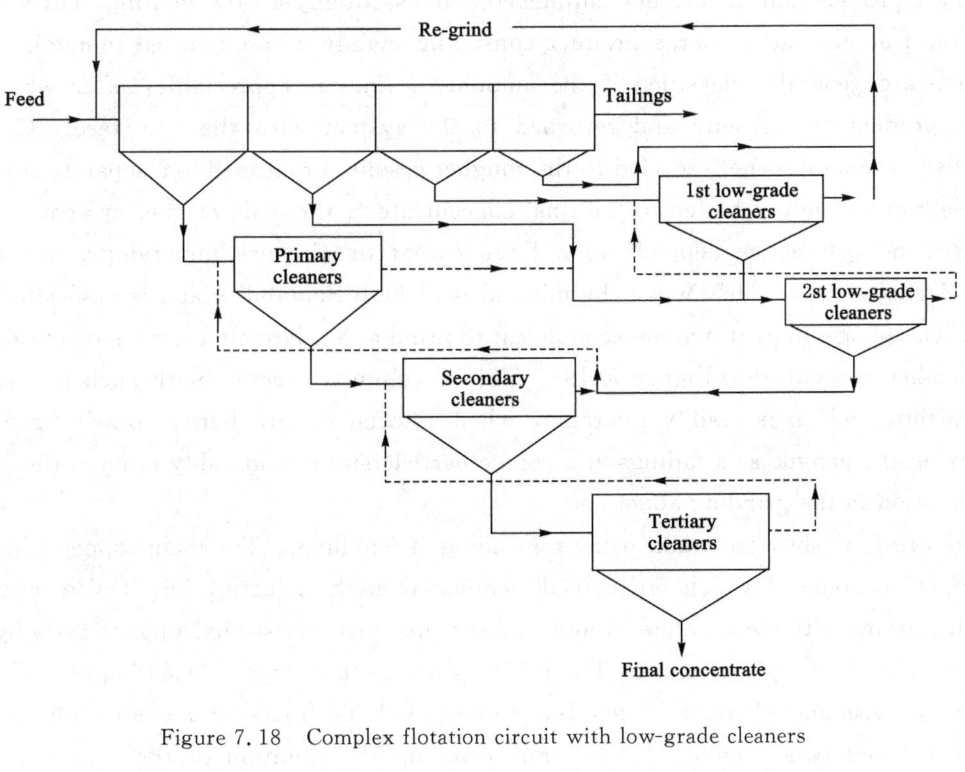

Figure 7.18 Complex flotation circuit with low-grade cleaners

To ensure the recovery of the weakly aerophilic particles which are passed to the particular section of the cleaning plant, it is essential that the retention time in the cleaning stage is at least that of the corresponding roughing section.

Since the object of the scavenging section is to promote maximum recovery by minimising the losses to tailings, it is advisable to dimension the scavengers generously, to allow not only for the slow-floating character of the particles but also for fluctuations in the circuit. However, it is important to avoid excessive overloading of the system with large volumes of low-grade material, so a compromise must be made in designing the scavenger circuit. It may be preferable to have a lower rougher concentrate grade (longer flotation time), and more cleaning stages, thus reducing the volume of scavenger concentrate produced.

(3) Complex flotation circuit with regrinding

If cleaning does not give the required concentrate grade, regrinding of the rougher concentrate may be needed, usually being necessary to at least regrind the scavenger concentrate, and sometimes the primary cleaner tailings, before recirculation to the rougher circuit. The purpose of primary grinding is to promote maximum recovery, by rendering most of the valuable mineral floatable so that the bulk of the gangue can be discarded, thus reducing the amount of material that must be further processed. In secondary grinding, or regrinding, the major consideration is the grade of the concentrate.

Regrinding of the middlings products is common practice in flotation plants. Both the scavenger product and the cleaner tailings contain essentially a slow floating, fairly metal-rich fine fraction and a coarse product consisting mainly of unliberated mineral. These products are generally classified if the amount of fines is appreciable, after which the coarse product is reground and returned to the system with the new feed. The fine classified product is either recycled to the rougher circuit, or cleaned in a separate circuit to a grade high enough to be fed to the final concentrate or the main cleaner system.

Regrinding practice depends to a large extent on the ore mineralogy. In certain circumstances, particularly when the mineral is of high floatability and is associated with an unfloatable gangue, it may be economical to grind at a relatively coarse size and regrind the rougher concentrate (Figure 7.19). This is common practice with such minerals as molybdenite, which is readily floatable, when associated with hard, abrasive gangues. Removing the gangue as a tailings at a coarse particle size considerably reduces the energy consumption in the grinding stage.

Figure 7.20 shows a circuit using regrinding of middlings. The main copper mineral is chalcopyrite, some of which is finely disseminated in the gangue, but it also occurs as complex grains with pyrite. The circuit removes the fairly coarse chalcopyrite early by one-stage rougher-scavenger-cleaning. The middlings from this stage, consisting essentially of the finely disseminated copper minerals, are reground and floated in a completely separate circuit utilizing two cleaning stages, thus isolating the flotation of the coarse material (80% $-188\ \mu m$) from the flotation of the very fine particles (80% $-40\ \mu m$).

Figure 7.21 shows a circuit which has been used to treat a complex ore containing copper, zinc, and iron sulphides, and cassiterite disseminated in a siliceous gangue.

A relatively coarse primary grind is used in order to recover as much cassiterite at as coarse a size as possible in the subsequent gravity process. After conditioning with copper

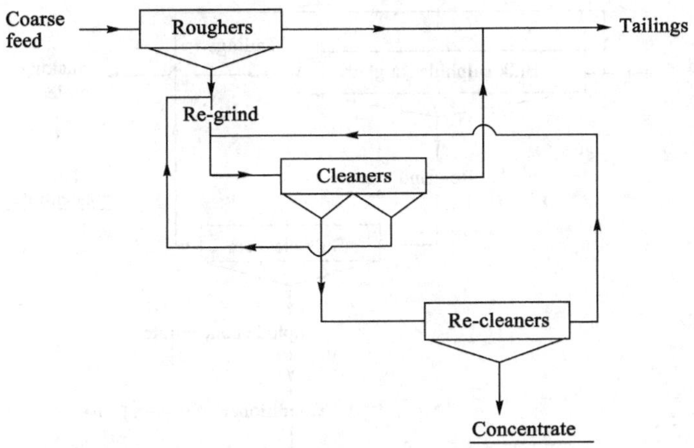

Figure 7.19 Regrinding of rougher concentrates

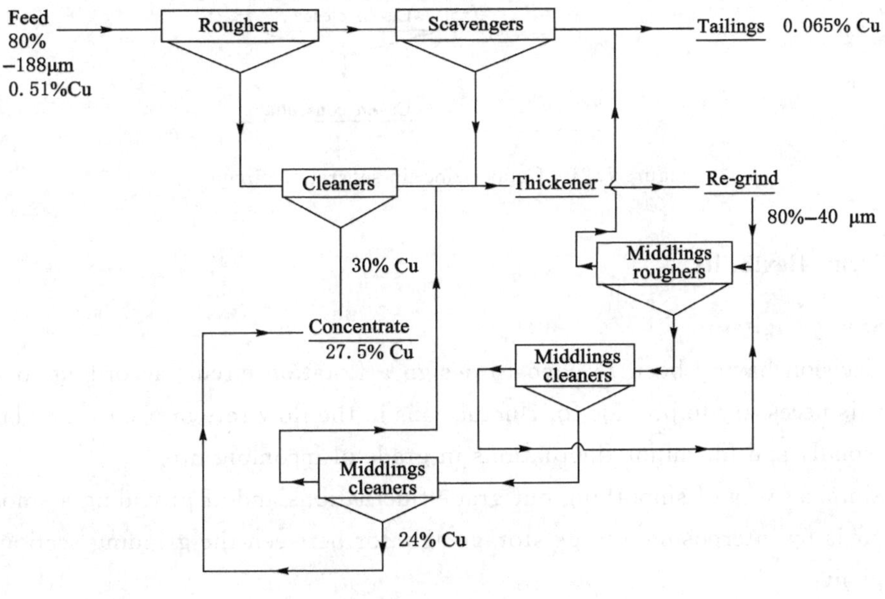

Figure 7.20 Regrinding of middlings

sulphate to activate the sphalerite, the relatively large amount of sulphide minerals, which would interfere with the recovery of the cassiterite, is removed by bulk flotation at neutral pH. The bulk rougher concentrate is reground to release finely disseminated cassiterite, after which cleaning is undertaken, the cleaner tailings being recirculated to the head of the system. The bulk cleaner concentrate is conditioned with lime to about pH 11, which depresses the pyrite, and the copper and zinc sulphides are floated and cleaned, leaving the pyrite in the rougher tailings.

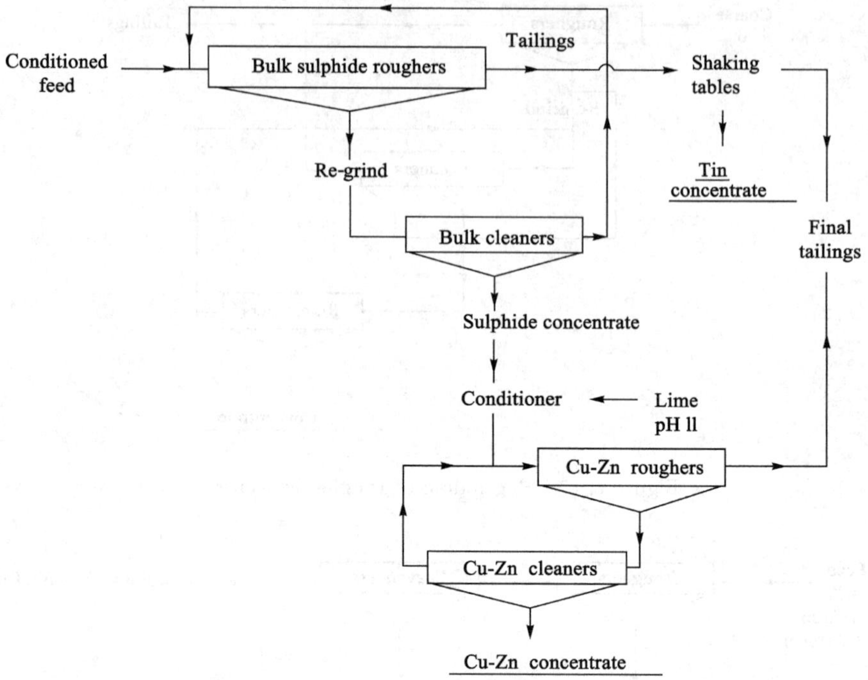

Figure 7.21 Copper-zinc-tin separation circuit

4. Circuit flexibility

(1) Storage agitator

The decision having been reached to design a flotation circuit according to a certain scheme, it is necessary to provide for fluctuations in the flow rate of ore to the plant, both large and small, and for minor fluctuations in grade of incoming ore.

The simplest way of smoothing out grade fluctuations and of providing a smooth flow to the plant is by interposing a large storage agitator between the grinding section and the flotation plant:

Crind ⟶ Storage Agitator ⟶ Flotation Plant

Any minor variations in grade and tonnage are smoothed out by the agitator, from which material is pumped at a controlled rate to the flotation plant. The agitator can also be used as a conditioning tank, reagents being fed directly into it. It is essential to precondition the pulp sufficiently with the reagents before feeding to the flotation banks, otherwise the first few cells in the bank act as an extension of the conditioning system, and poor recoveries result.

Control systems can be installed to maintain the flow rate of the slurry as constant as possible. The control system starts in the grinding circuit where the feed rate of ore to the grinding mills is maintained constant using variable speed feeders. The level of slurry in pump boxes is also maintained constant by automatically adjusting pump speed using variable speed drives. Levels of slurry in flotation cells are maintained constant by using automatic cell level control systems.

(2) Number of cells

In designing each flotation bank, the number of cells required must be assessed: should a few large cells be incorporated or many small cells giving the same total capacity?

This is determined by many factors. If a small cell in a bank containing many such cells has to be shut down, then its effect on production and efficiency is not as large as that of shutting down a large cell in a bank consisting of only a few such cells. Maintenance costs, however, tend to be lower with large cells, since there are relatively fewer parts to change in a particular bank.

Designing a bank with many small cells gives particles which have short-circuited in one or more cells the chance to float in succeeding cells. The designer, therefore, must decide between small cells for greater flexibility and metallurgical performance, and large cells, which have a smaller total capital cost, less floor area per unit volume, and lower operating costs. In eastern Europe, it has been common to install 30 or more machines in a single bank, while in the West the trend is to install very large cells, especially in the roughing stage.

In the 1990s and 2000s the flotation cell suppliers that produced Outokumpu and Wemco cells developed the tank cell designs. These cells can be as big as 150 m^3 or more in volume and can treat more that 100,000 tonnes per day of ore, particularly in large copper operations in South America and Asia. Tank cells of even greater volumes are under development from flotation cell suppliers. These cells are able to minimise short-circuiting of slurry by using big tanks with a single tailings discharge that is controlled by rubber sleeved pinch valves. A bank is typically designed with about eight to ten cells in the rougher section.

Notes

1. Commercial flotation is a continuous process.

工业浮选是一个连续的过程。

2. In the case of flotation cells that use weir-type level control, the height of the froth column for each cell is determined by adjusting the height of the tailings overflow weir, the difference in height between this and the cell overflow lip determining the froth depth.

对于使用堰式液位控制的浮选机,每个浮选室的泡沫层厚度是通过调整尾矿溢流堰的高度来确定的,尾矿溢流堰与浮选室溢流边缘之间的高度差决定了泡沫层的厚度。

3. A preferable system is to dilute the concentrate from the first few cells of the bank, known as rougher concentrates, and refloat them in cleaner cells, where the weirs are kept low to maintain a deep froth and produce a high-grade concentrate.

一个更可取的浮选系统是稀释浮选系列中最初几个浮选槽的精矿(称为粗精矿),然后将它们在精选槽中再浮,在这些浮选槽中,溢流堰保持在低位以保持厚的泡沫层并生产高品位的精矿。

4. In designing a suitable flowsheet for a flotation plant, the primary grind size is of major consideration.

在为浮选厂设计合适的流程时，粗磨粒度是主要考虑因素。

5. In a complex flotation circuit, the rougher flotation may be divided into stages, each delivering its concentrate into the cleaning circuit at a different location according to its grade.

在复杂的浮选流程中，粗浮选可分为多个阶段，每个阶段根据精矿品位将其输送到不同位置的精选流程。

6. The simplest way of smoothing out grade fluctuations and of providing a smooth flow to the plant is by interposing a large storage agitator between the grinding section and the flotation plant.

消除品位波动和为设备提供平稳流的最简单方法是在研磨段和浮选设备之间插入一个大的储存式搅拌机。

7. Control systems can be installed to maintain the flow rate of the slurry as constant as possible.

可以安装控制系统以保持矿浆的流速尽可能恒定。

Exercises

1. Answer the following questions according to the text.

(1) What is the flotation process of the flotation banks?

(2) How to maintain the pulp level in modem tank cells?

(3) What is the application of simple rougher-scavenger flotation circuit?

(4) Briefly describe the flotation process of the rougher-scavenger-cleaner system?

(5) What is the simplest way of smoothing out grade fluctuations and providing a smooth flow to the plant?

2. Translate the following phrases into English.

(1) 浮选流程	(2) 粗选	(3) 精选	(4) 扫选
(5) 浮选槽	(6) 溢流堰	(7) 品位波动	(8) 粗磨

3. Translate the following paragraphs into Chinese.

Paragraphs 2 and 3 in section 1 "In the case of flotation cells that use weir-type level ...".

4. Translate the following sentences into English.

(1) 许多浮选槽串联布置，形成一个浮选系列。

(2) 流程中的最后几个浮选槽含有相对低品位泡沫，包括弱亲气颗粒。

(3) 应避免过大的循环负荷，因为粗选槽的进料可能被稀释，浮选时间也会减少。

(4) 在设计每个浮选系列时，必须评估所需的浮选槽数量：是采用几个大的浮选槽，还是采用总体积相同的许多小浮选槽。

7.5 Flotation Machines (浮选机)

Although many different machines are currently being manufactured and many more have been developed and discarded in the past, it is fair to state that two distinct groups have arisen: pneumatic and mechanical machines.

1. Pneumatic machines

Pneumatic machines either use air entrained by turbulent pulp addition, or more commonly air either blown in or induced, in which case the air must be dispersed either by baffles or some form of permeable base within the cell. Generally, pneumatic machines give a low-grade concentrate and little operating trouble. Since air is used not only to produce the froth and create aeration but also to maintain the suspension and to circulate it, an excessive amount is usually introduced.

(1) Davcra cell

One of the early developments in the pneumatic field was the Davcra cell (Figure 7. 22), which has been claimed to yield equivalent or better performance than a bank of mechanical machines.

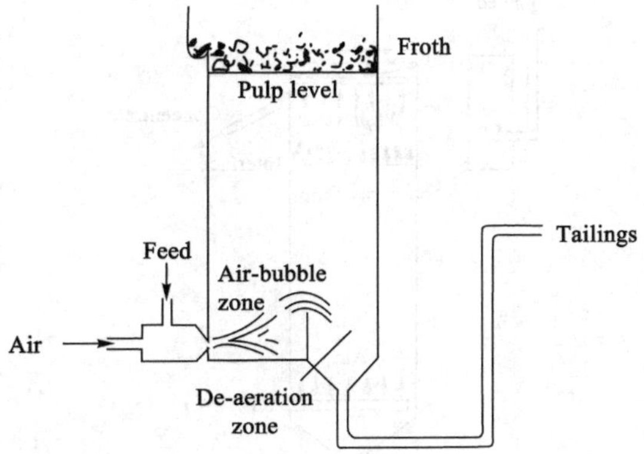

Figure 7. 22 Davcra cell

The cell consists of a tank segmented by a vertical baffle. Air and feed slurry are injected into the tank through a cyclone-type dispersion nozzle, the energy of the jet of pulp being dissipated against the vertical baffle. Dispersion of air and collection of particles by bubbles allegedly occurs in the highly agitated region of tank confined by the baffle. The pulp flows over the baffle into a quiescent region designed for bubble-pulp disengagement. The cell can be used for roughing or cleaning applications on a variety of minerals. Although not widely used, Davcra cells replaced some mechanical cleaner machines at Chambishi copper mine in Zambia, with reported lower operating costs, reduced floor area, and improved metallurgical performance.

(2) Flotation columns

A significant development in recent years has been the increasing industrial use of flotation columns. The main advantages of columns include improved separation performance, particularly on fine materials, low capital and operational cost, less plant space demand, and adaptability to automatic control.

A typical configuration of a column is shown in Figure 7. 23. It consists of two distinct sections. In the section below the feed point (the recovery section), particles suspended in the descending water phase contact a rising swarm of air bubbles produced by a sparger in the column base. Floatable particles collide with and adhere to the bubbles and are transported to the washing section above the feed point. Non-floatable material is removed from the base of the column as tailing. Gangue particles that are loosely attached to bubbles or are entrained in bubble slipstreams are washed back into the recovery section, hence reducing contamination of the concentrate. The wash water also serves to suppress the flow of feed slurry up the column towards the concentrate outlet. There is a downward liquid flow in all parts of the column preventing bulk flow of feed material into the concentrate.

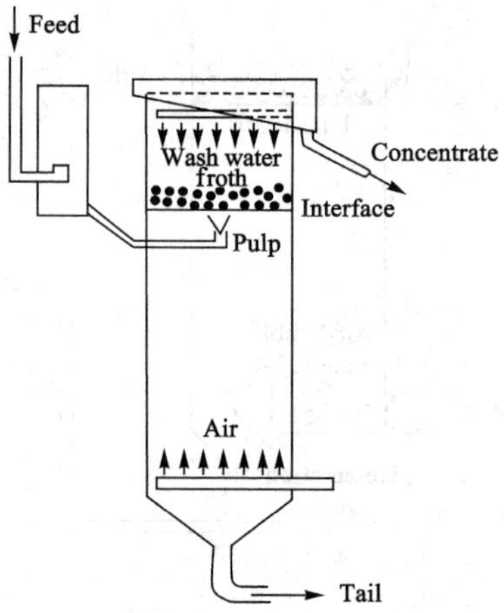

Figure 7. 23 The flotation column

Columns were originally developed in Canada, and were used initially for cleaning molybdenum concentrates. Two-column flotation units were installed in the molybdenum circuit at Mines Gaspe, Canada, in 1980, and excellent results were reported. The units replaced mechanical machines in the cleaner banks. Since then many of the copper-molybdenum producers in North America have installed columns for molybdenum cleaning, and their use has been expanded into the roughing, scavenging, and cleaning of a variety of ore types, in many parts of the world.

The United Coal Co. in the United States has also pioneered the use of columns for the flotation of fine coal. It is possible that, due to their froth washing capability, columns may find an increasing use in the future for treating ores that need extensive fine grinding, followed by de-sliming and multi-stage cleaning.

(3) Jameson cell

Flotation columns are usually about 12 m high, with diameters of up to about 3.5 m (round or square, the former being more popular). Several attempts have been made to develop column-type devices with much smaller height/diameter ratios, the Jameson cell being a successful example (Figure 7.24).

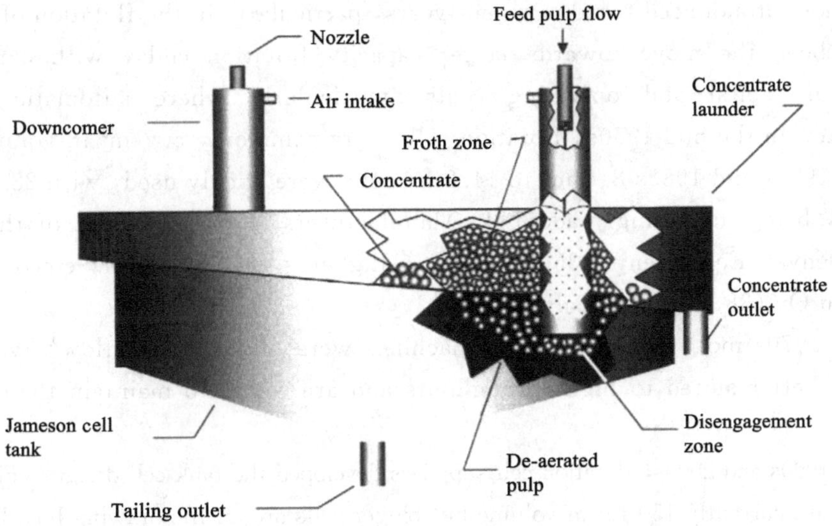

Figure 7.24 Principles of operation of the Jameson cell (Courtesy Xstrata Technology)

Contact between the feed and the air stream is made in a mixing device at the top of a vertical downcomer. The air-liquid mixture flows downwards to discharge into a shallow pool of pulp in the bottom of a short cylindrical column. The bubbles disengage and rise to the top of the column to overflow into a concentrate launder, while the tails are discharged from the bottom of the vessel. The main advantages of the device are that the overall height of the column is reduced to about 1 m, and the flotation column can be self-inducing with respect to the air supply.

The Jameson cell was developed jointly by Mount Isa Mines Ltd and the University of Newcastle, Australia. The cell was first installed for cleaning duties in base metal operations but it has also found uses in other duties including roughing and preconcentrating. The major advantage of the cell is its ability to produce clean concentrates in one stage of operation. The Jameson cell has also become widely used in the coal industry in Australia in the 1990s and 2000s.

2. Mechanical flotation machines

Mechanical flotation machines are the most widely used, being characterized by a mechanically driven impeller which agitates the slurry and disperses the incoming air into small bubbles. The machines may be self-aerating, the depression created by the impeller inducing the air, or "supercharged" where air is introduced via an external blower. In a typical flotation bank, there are a number of such machines in series, "cell-to-cell" machines being separated by weirs between each impeller, whereas "open-flow" or "free-flow" machines allow virtually unrestricted flow of the slurry down the bank.

The most pronounced trend in recent years, particularly in the flotation of base metal ores, has been the move towards larger capacity flotation cells, with corresponding reduction in capital and operating costs, particularly where automatic control is incorporated. In the mid-1960s, flotation cells were commonly 5.7 m^3 in volume, or less and in the 1970s and 1980s 8.5 m^3 to 14.2 m^3 cells were widely used, with 28.3 m^3 cells, and larger, being increasingly adopted. Manufacturers in the forefront of this industry included Denver Equipment (36.1 m^3), Galigher (42.5 m^3), Wemco (85 m^3), Outokumpu Oy (38 m^3), and Sala (44 m^3).

In the 1970s most of the flotation machines were of the "open flow" type, as they were much better suited to high throughputs and are easier to maintain than cell-to-cell types.

In the 1990s and 2000s, flotation cell suppliers developed the tank cell designs (Figure 7.25). These cells are currently 150 m^3 in volume but bigger cells are on the drawing boards. The cells are circular in shape, fitted with froth crowders, multiple froth launders and discharge and designed with effective level control systems.

Figure 7.25 160 m^3 Outokumpu tank cell (Courtesy JKMRC and JKTech Pty Ltd)

(1) Cell-to-cell machine

The Denver "Sub-A" was perhaps the most well-known cell-to-cell machine, having been used widely in the past in small plants, and in multi-stage cleaning circuits, where the pumping action of the impellers permitted the transfer of intermediate fluid without external pumps. They were manufactured with cell sizes of up to 14.2 m^3, and were used mostly as coal-cleaning devices, where the users reported a significant improvement in selectivity over open-flow designs.

The flotation mechanism is suspended in an individual square cell separated from the adjoining cell by an adjustable weir (Figure 7.26). A feed pipe conducts the flow of pulp from the weir of the preceding cell to the mechanism of the next cell, the flow being aided by the suction action of the impeller. The positive suction created by the impeller draws air down the hollow stand-pipe surrounding the shaft. This air stream is sheared into fine bubbles by the impeller action and is intimately mixed with the pulp which is drawn into the cell onto the rotating impeller.

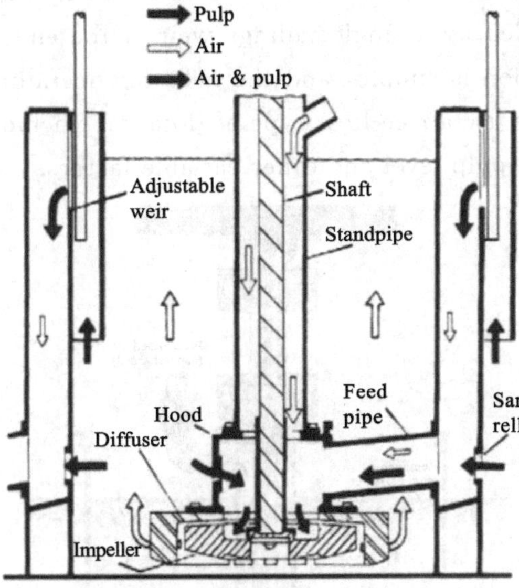

Figure 7.26 Denver sub-aeration cell

Directly above the impeller is a stationary hood, which prevented "sanding-up" of the impeller if the machine is shut down. Attached to this hood are four baffle vanes, which extend almost to the comers of the cell. These prevented agitation and swirling of the pulp above the impeller, thus producing a quiescent zone where bubbles can ascend with their mineral load without being subjected to scouting which may cause them to drop it. The mineral-laden bubbles separate from the gangue in this zone and pass upward to form a froth. As the bubbles move to the pulp level, they are carried forward to the overflow lip by the crowding action of succeeding bubbles, and quick removal of froth is accomplished by froth paddles which aid the overflow.

Pulp from the cell flowed over the adjustable tailings weir, and was drawn on to the impeller of the next cell where it was again subjected to intense agitation and aeration. Particles which are too heavy to flow over the tailings weir are by-passed through sand relief ports, which prevented the build-up of coarse material in the cell.

The amount of air introduced into the pulp depends upon the impeller speed, which is normally in the range of 7～10 m/s peripheral. More air may be obtained by increasing the impeller speed, but this may in certain circumstances over-agitate the pulp as well as increase impeller wear and energy consumption. In such cases, supercharging may be applied by introducing additional air down the stand-pipe by means of an external blower.

(2) Free-flow type machine

Supercharging is required with the Denver D-R machine (Figures 7.27), which ranges in size from 2.8 m^3 to 36.1 m^3, and which was developed as a result of the need for a machine to handle larger tonnages in bulk-flotation circuits. These units are characterized by the absence of intermediate partitions and weirs between cells. Individual cell feed pipes have been eliminated, and pulp is free to flow through the machine without interference. The pulp level is controlled by a single tailings weir at the end of the trough. Flotation efficiency is high, operation is simple, and the need for operator attention is minimised. Most high-tonnage plants use a free-flow type of flotation machine and many are equipped with automatic control of pulp level and other variable factors.

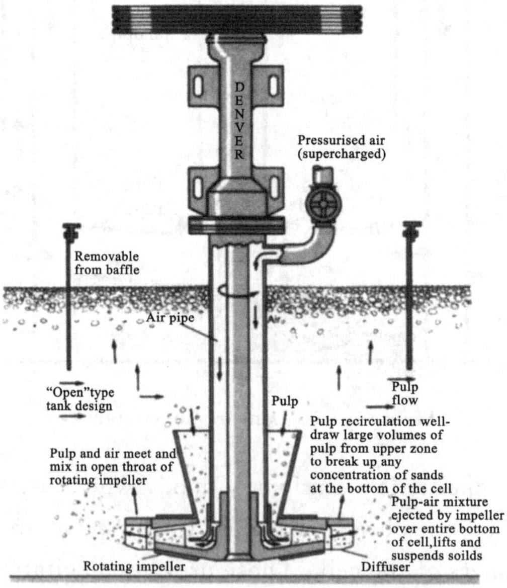

Figure 7.27 Denver D-R flotation machine

(3) Wemco Fagergren cell

A widely used flotation machine is the Wemco Fagergren (Figure 7.28) manufactured in sizes up to 85 m^3. The modem 1+1 design consists of a rotor-disperser assembly, rather than an impeller, and the unit usually comprises a long rectangular trough, divided into sections, each containing a rotor-disperser assembly. The feed enters below the first partition, and tails go over partitions from one section to the next, the pulp level being adjusted at the end tailings weir.

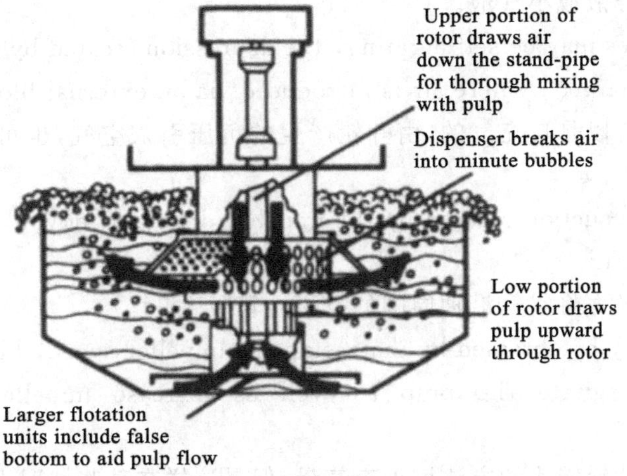

Figure 7.28 Wemco Fagergren cell

Pulp passing through each cell, or compartment, is drawn upwards into the rotor by the suction created by the rotation. The rotor also draws air down the standpipe, no external blower being needed. The air is thoroughly mixed with the pulp before being broken into small, firm bubbles by the disperser.

Notes

1. Although many different machines are currently being manufactured and many more have been developed and discarded in the past, it is fair to state that two distinct groups have arisen: pneumatic and mechanical machines.

尽管目前正在制造许多不同的浮选机，过去还研发和淘汰了更多的浮选机，但可以说已经出现了两种不同类型的浮选机：充气式浮选机和机械搅拌式浮选机。

2. Pneumatic machines either use air entrained by turbulent pulp addition, or more commonly air either blown in or induced, in which case the air must be dispersed either by baffles or some form of permeable base within the cell.

充气式浮选机可以利用涡流矿浆给矿的夹带作用充气，或者更为普遍地利用充气的方法导入空气，当采用充气方法时需要利用槽体中的折流板或某种多孔的底板使气体分散。

3. The main advantages of columns include improved separation performance, particularly on fine materials, low capital and operational cost, less plant space demand, and adaptability to automatic control.

浮选柱的主要优点包括：分选性能提高(特别是对细物料)，投资和生产成本低，占地面积小，适合采用自动控制。

4. Mechanical flotation machines are the most widely used, being characterized by a mechanically driven impeller which agitates the slurry and disperses the incoming air into small bubbles.

机械搅拌式浮选机是应用最广泛的浮选机，其特点是有一个机械驱动的叶轮，它搅动矿浆并将进入的空气分散成小气泡。

5. The machines may be self-aerating, the depression created by the impeller inducing the air, or "supercharged" where air is introduced via an external blower.

机械式浮选机可以是自充气的,由叶轮产生的负压引入空气,也可以是"增压"的,空气通过外部鼓风机引入。

6. The positive suction created by the impeller draws air down the hollow stand-pipe surrounding the shaft.

叶轮产生的正吸力将空气沿轴周围的空心立管吸入。

7. More air may be obtained by increasing the impeller speed, but this may in certain circumstances over-agitate the pulp as well as increase impeller wear and energy consumption.

可以通过提高叶轮转速的方法增加充气量,但是这样在某些情况下会过度搅拌矿浆,而且增加叶轮的磨损和能耗。

Exercises

1. Answer the following questions according to the text.

(1) What are the two distinct groups of flotation machines?

(2) Why have the pneumatic machines been little used?

(3) What are the advantages of flotation column?

(4) What is the characteristic of mechanical flotation machines?

2. Translate the following phrases into English.

(1) 充气式浮选机 (2) 机械搅拌式浮选机 (3) 叶轮 (4) 浮选柱

(5) 吸力作用 (6) 底部充气式浮选机 (7) 詹姆森浮选机 (8) 盖板

3. Translate the following paragraphs into Chinese.

Paragraphs 1 and 2 in section 1 (2) "A significant development in recent years has been the increasing ...".

4. Translate the following sentences into English.

(1) 由于空气不仅用于产生泡沫和充气,而且还用于维持悬浮液并使其循环,因此通常会引入过量的空气。

(2) 近年来浮选工艺变得越发重要的一个重要体现是浮选柱的工业应用越来越多。

(3) 机械搅拌式浮选机可以是自充气式的,由叶轮产生的低压吸入空气,也可以是"加压"式的,即通过外部鼓风机引入空气。

(4) 充气式浮选机可以利用涡流矿浆给矿的夹带作用充气。

8 HYDROMETALLURGY(湿法冶金)

8.1 Introduction to Hydrometallurgy(湿法冶金概述)

Hydrometallurgy originated in the 16th century, but its principal development took place in the 20th century, stimulated partly by the desire to extract gold from low-grade ores. The development of ion exchange, solvent extraction, and other processes has led to an extremely broad range of applications of hydrometallurgy. Nowadays, hydrometallurgy is used to produce more than 70 metallic elements. In addition to most gold and silver, large tonnages of copper and zinc are produced by hydrometallurgy.

Hydrometallurgy involves the use of aqueous solutions for the recovery of metals from ores, concentrates, and recycled or residual materials. This method is mainly used in low-grade, hard-to-melt or micro-powdered ores. Typically, hydrometallurgy involves leaching, solid-liquid separation, solution concentration and purification, and metal or metal compound recovery.

1. Leaching

Leaching involves the process of extracting metal from metal bearing materials with chemical solvents. The first examples come from 17th century Germany and Spain where it was applied to the extraction of copper. According to the types of lixiviant solutions, it can be divided into acid leaching, alkali leaching and salt leaching. According to the leaching chemical process, it is divided into oxidation leaching and reduction leaching. According to the leaching method, it is divided into in-situ leaching, heap leaching, vat leaching, tank leaching and autoclave leaching, etc. According to the pressure of the leaching process, it can be divided into atmospheric leaching and pressure leaching.

The lixiviant solution conditions vary in terms of pH, oxidation-reduction potential, presence of chelating agents and temperature, to optimize the rate, extent and selectivity of dissolution of the desired metal component into the aqueous phase. Through the use of chelating agents, one can selectively extract certain metals. Such chelating agents are typically amines of schiff bases.

(1) Leaching methods based on lixiviant solutions

Acid leaching

Acid leaching is a method of leaching valuable metals with acid as a solvent. Commonly used acids are inorganic acids and organic acids. Sulfuric acid, hydrochloric acid, nitric acid, sulfurous acid, hydrofluoric acid and aqua regia are used in industry. As sulfuric acid has a high boiling point, a wide range of sources, low price, and weak corrosiveness, it is the most widely used. In non-ferrous metallurgy, sulfuric acid is often used for leaching of copper oxide ore, leaching of zinc calcine, and oxygen pressure leaching of nickel matte and zinc sulfide concentrate.

Hydrochloric acid has strong reactivity and can leach a variety of metals, metal oxides and certain sulfides. Such as leaching nickel matte and cobalt slag. However, hydrochloric acid and the generated chloride are more corrosive, which requires higher anti-corrosion for equipment. Nitric acid is a strong oxidant, it is expensive, and reacts to precipitate toxic nitrogen oxides. Therefore, it is only used in a few special cases.

Alkali leaching

Alkali leaching is a leaching method using alkaline solution as solvent. Commonly used alkalis are sodium hydroxide, sodium carbonate and sodium sulfide. Pressure alkaline leaching of bauxite is the most important application example of alkaline leaching. Alkaline leaching is also used for leaching wolframite, uranium ore (Na_2CO_3 leaches UO_3), sulfide and antimony oxide ore ($Na_2S+NaOH$ as leaching solution), etc. The leaching ability of alkaline solutions is generally weaker than that of acidic solutions; but the leaching selectivity is better, the leaching solution is purer, less corrosive to the equipment, no special corrosion protection is required, and the material of the manufacturing equipment is easier to solve.

Bioleaching

Bioleaching is the extraction of a metal from sulfide ores or concentrates using materials found native to the environment; namely, water, air and microorganisms. In other words, bioleaching is the commercialization of the ability of certain bacteria and archaea, found in nature, to catalyze the oxidation of sulfide minerals. It is the leaching of sulfide minerals that distinguishes bioleaching from conventional acid leaching wherein only oxidized minerals are leached. An associated process is biooxidation. This is the oxidation of sulfide minerals associated with but not necessarily part of the mineral of interest to be extracted. Bioleaching is used today in commercial operations to process ores of copper, nickel, cobalt, zinc and uranium, whereas, biooxidation is used in gold processing and coal desulfurization.

Other lixiviant solutions

Salt leaching is the process of leaching valuable metals with salt as a solvent. For example, sulfide ores use iron sulfate to leach copper, sodium chloride to extract lead, and sodium cyanide to extract gold and silver in the ore.

Oxidation leaching uses oxidant to cause valuable components in ore, concentrate or other solid materials to undergo a leaching method characterized by oxidation reaction during the leaching process. The oxidants commonly used in industry are air, oxygen, Fe, MnO_2 (manganese dioxide) and Cl_2 (chlorine) etc. For metals or low-valent metal oxides, the purpose of oxidative leaching is to oxidize the metal into ions or oxidize low-valent ions into highly soluble high-valent ions into the solution. One example of the former is the cyanide leaching of gold and silver ores.

Reduction leaching uses a reducing agent to make the valuable components in the leached solid material undergo a leaching method characterized by a reduction reaction during the leaching process. Commonly used reducing agents in industry are SO_2, $FeSO_4$, etc. A typical example of reduction leaching is the leaching of cobalt slag. Cobalt in the cobalt slag is in the form of $Co(OH)_3$ which is insoluble high-valent hydroxide.

Each variation of leaching plays an important role in mining and metal recovery, and are used in a variety of applications across the globe.

(2) Leaching methods based on leaching process

In-situ leaching

In-situ leaching is also called "underground leaching", it injects the leaching agent into the ore body under natural burial conditions through drilling and live blasting, and selectively leaches useful components, pumps the solution to the ground through the pumping hole. And then collects and transports the solution to the extraction plant for processing. The Beverley uranium deposit is an example of in-situ leaching and also Trojan Mine in Zimbabwe.

Heap leaching

In heap leaching processes, crushed (and sometimes agglomerated) ore is piled in a heap which is lined with an impervious layer. Leach solution is sprayed over the top of the heap, and allowed to percolate downward through the heap. The heap design usually incorporates collection sumps, which allow the "pregnant" leach solution (i. e., solution with dissolved valuable metals) to be pumped for further processing. An example is gold cyanidation, where pulverized ores are extracted with a solution of sodium cyanide, which, in the presence of air, dissolves the gold, leaving behind the nonprecious residue.

Vat leaching

Vat leaching involves contacting material, which has usually undergone size reduction and classification, with leach solution in large vats.

Tank leaching

Stirred tank, also called agitation leaching, involves contacting material, which has usually undergone size reduction and classification, with leach solution in agitated tanks. The agitation can enhance reaction kinetics by enhancing mass transfer. Tanks are often configured as reactors in series.

Autoclave leaching

Autoclave reactors are used for reactions at higher temperatures, which can enhance the rate of the reaction. Similarly, autoclaves enable the use gaseous reagents in the system.

2. Solid-liquid separation

It is the process of separating the leachate into liquid and solid phases. Commonly used solid-liquid separation methods include sedimentation separation and filtration.

(1) Sedimentation separation

It is a liquid-solid separation method that uses gravity to separate the leached pulp into an underflow with a large solid content and a clear overflow. The prerequisite is that there is a density difference between the solid phase and the overflow liquid. Due to the wide range of particle diameters in the slurry, ranging from several hundred microns to several microns, the sedimentation rate cannot be calculated by theoretical formulas, and is usually determined by experiments.

Industrial equipment that increases the concentration of the underflow is called a thickening tank or a clarification tank. Its clarification capacity is proportional to the sinking area of the tank, and the concentration of the underflow is related to the residence time of the leached pulp in the tank (that is, it is related to the tank depth). In order to increase its capacity and save floor space, a multi-layer thickening tank is always used.

When dealing with slurries containing very fine materials, centrifugal force can be used instead of gravity to accelerate particle settling. Such as hydrocyclones and spiral centrifuges to enhance the sedimentation process. Or with the help of chemical reagents-coagulant or flocculant to promote the transformation of dispersed, non-agglomerated particles in the slurry into clear overflow and dense underflow. Coagulants (such as lime) can make particles coagulate each other, and flocculants can make fine particles form flocs to strengthen the sedimentation process.

(2) Filtration

Filtration or Filter separation is a liquid-solid separation method in which porous media is used to intercept the solid particles in the leaching slurry, and the pressure difference or other external force is used as the driving force to make the liquid pass through the micropores. There are various media for intercepting solid particles, such as woven fabrics, porous ceramics, porous metals, pulp and asbestos, etc. However, regardless of the filter medium, the pores are usually larger than the diameter of the particles to be filtered. The effective filtration is achieved by bridging the particles on the medium or loose filter cake. Only the filtrate after forming the filter cake is clear. Filtration is divided into two types: constant pressure and constant speed. The former will slow down the filtration speed without changing the suction power after the filter cake is thickened, and the latter must be added when the filter cake is thickened in order to maintain the filtration speed. Large pumping force can ensure constant speed.

The most important elements of filter selection are the specific resistance of the filter cake, the solid content of the filter material, and the viscosity of the filtrate. Commonly used filters include rotary drum vacuum filter, belt filter, plate and frame filter and so on.

3. Solutionconcentration and purification

After leaching, the leach liquor must normally undergo concentration of the metal ions that are to be recovered. Additionally, undesirable metal ions sometimes require removal. To separate the target metal from the impurities, there are generally two methods: one is to first precipitate the target metal from the solution, and the other is to allow the target metal to remain in the solution after the impurities are separated out. Common methods for solution concentration and purification are: precipitation, cementation, solvent extraction and ion exchange.

(1) Precipitation

Precipitation refers to the separation of one or several chemical components from solution in the form of an insoluble compound, which is called the precipitate. Precipitation is one of the most fundamental and commonly encountered separation techniques in quantitative and qualitative analytical chemistry. The addition of a solution of silver nitrate to a solution containing chloride ions results in the formation of insoluble silver chloride: $AgNO_3 + Cl^- \longrightarrow NO_3^- + AgCl\downarrow$. The precipitate formed may settle out spontaneously or may be collected by filtration or centrifugation. It is often difficult to obtain a pure substance by a single precipitation, and a substance may be further purified by reprecipitation after it has been redissolved.

Precipitation is the selective removal of a compound of the targeted metal or removal of a major impurity by precipitation of one of its compounds. Copper is precipitated as its sulfide as a means to purify nickel leachates.

(2) Cementation

Cementation is the conversion of the metal ion to the metal by a redox reaction. A typical application involves addition of scrap iron to a solution of copper ions. Iron dissolves and copper metal is deposited. In nonferrous metallurgy, cementation is a hydrometallurgical process based on the displacement of the more electropositive metals from solutions of their compounds by less electropositive metals in the solid state. For example, the normal electrode potential of copper is +0.344 volts, and that of zinc is −0.762 volts; the difference between the potentials makes it possible to obtain the reaction $Cu^{2+}_{solution} + Zn_{metal} \longrightarrow Zn^{2+}_{solution} + Cu_{metal}$. The greater the difference in potentials, the lower the content of precipitated metal that will remain in the solution. Cementation is extensively used to remove impurities from solutions and to recover metals from solutions. The process can also be used to precipitate metals from molten slags.

(3) Solvent extraction

Solvent extraction is a technique, also called liquid extraction, for separating the components of a liquid solution. This technique depends upon the selective dissolving of one or more constituents of the solution into a suitable immiscible liquid solvent. It is particularly useful industrially for separation of the constituents of a mixture according to chemical type, especially when methods that depend upon different physical properties, such as the separation by distillation of substances of different vapor pressures, either fail entirely or become too expensive.

(4) Ion exchange

Ion exchange uses chelating agents, natural zeolite, activated carbon, resins, and liquid organics impregnated with chelating agents to exchange cations or anions with the solution. Typical ion exchangers are ion-exchange resins (functionalized porous or gel polymer), zeolites, montmorillonite, clay, and soil humus. Ion exchangers are either cation exchangers, which exchange positively charged ions (cations), or anion exchangers, which exchange negatively charged ions (anions). There are also amphoteric exchangers that are able to exchange both cations and anions simultaneously. However, the simultaneous exchange of cations and anions can be more efficiently performed in mixed beds, which contain a mixture of anion—and cation-exchange resins, or passing the treated solution through several different ion-exchange materials. Ion exchanges can be unselective or have binding preferences for certain ions or classes of ions, depending on their chemical structure. This can be dependent on the size of the ions, their charge, or their structure.

4. Metal recovery

Metal recovery is the final step in a hydrometallurgical process. Metals suitable for sale as raw materials are often directly produced in the metal recovery step. Sometimes, however, further refining is required if ultra-high purity metals are to be produced. The primary types of metal recovery processes are electrolysis and chemical extraction.

(1) Electrowinning

Electrowinning and electrorefining respectively involve the recovery and purification of metals using electrodeposition of metals at the cathode, and either metal dissolution or a competing oxidation reaction at the anode. In the process of electrowinning, metals are plated onto a cathode, based on the relative reduction potential of each metal in solution. The metal with the most positive reduction potential will plate out preferentially. Once that metal is depleted from solution, the next most positive reduction potential will plate out. The electrochemical series dictates what metals will plate out of solution and in what order, and if pre-treatment is required after leaching in order to isolate the particular metal of interest prior to electrowinning. A major target of copper ore hydrometallurgy is copper, which is conveniently obtained by electrolysis. Cu^{2+} ions reduce at mild potentials, leaving behind other contaminating metals such as Fe^{2+} and Zn^{2+}. The process of electrowinning is as follows:

The cathode process: Cu^{2+} obtains electrons on the pole piece to precipitate metallic copper

$$Cu^{2+} + 2e \longrightarrow Cu \quad (E = 0.34\ V)$$

The anode process: the water on the anode desorbs oxygen

$$H_2O - 2e \longrightarrow 1/2O_2 + 2H^+$$

(2) Chemical extraction

Chemical extraction is the process of reducing metal ions in aqueous solution to metal with a reducing agent. Commonly used reducing agents in industry include hydrogen, SO_2 gas, ferrous ion, iron, zinc, aluminum, copper and other metals, as well as oxalic acid and hydrazine.

Electrowinning and chemical extraction have their own advantages and disadvantages. Electrolytic extraction does not require a large amount of reagents, and has little environmental pollution. It is especially suitable for large-scale production. It is the main industrial method for extracting copper, nickel, and zinc from aqueous solutions. However, this method consumes a lot of electric energy and is not suitable for areas lacking electricity. In addition, the disposable equipment has a large investment, a large area and a long operation cycle. The chemical extraction method has the advantages of not consuming a large amount of electric energy, needing less equipment investment, small footprint, short operation cycle, etc.; the disadvantage is that it needs to consume reducing agent, and the waste liquid produced need to be treated before discharge.

Notes

1. Leaching involves the process of extracting metal from metal bearing materials with chemical solvents.

浸出是用化学溶剂从含金属物料中提取金属的过程。

2. It is the process of separating the leachate into liquid and solid phases. Commonly used solid-liquid separation methods include sedimentation separation and filtration.

固液分离指的是将浸出液分离成液相和固相的过程,常用的固液分离方法有沉降分离和过滤两种方法。

3. When dealing with slurries containing very fine materials, centrifugal force can be used instead of gravity to accelerate particle settling. Such as hydrocyclones and spiral centrifuges to enhance the sedimentation process.

当处理含极细物料的矿浆时,可利用离心力代替重力以加速颗粒沉降。如水力旋流器和螺旋离心机就是利用离心力来强化沉降过程的。

4. To separate the target metal from the impurities, there are generally two methods: one is to first precipitate the target metal from the solution, and the other is to allow the target metal to remain in the solution after the impurities are separated out.

为了从杂质中分离目标金属,通常有两种方法:一种是首先从溶液中沉淀目标金属,另一种是在分离出杂质之后使目标金属保留在溶液中。

5. Precipitation refers to the separation of one or several chemical components from solution in the form of an insoluble compound, which is called the precipitate.

沉淀是指一种或几种化学成分以不溶性化合物的形式从溶液中分离出来的过程,这种化合物被称为沉淀物。

6. Electrowinning and electrorefining respectively involve the recovery and purification of metals using electrodeposition of metals at the cathode, and either metal

dissolution or a competing oxidation reaction at the anode.

电沉积和电精炼分别涉及利用金属在阴极的电沉积以及在阳极上进行金属溶解或竞争性氧化反应来回收和提纯金属。

7. Chemical extraction is the process of reducing metal ions in aqueous solution to metal with a reducing agent.

化学提取是用还原剂将水溶液中的金属离子还原为金属的过程。

Exercises

1. Answer the following questions according to the text.

(1) What is hydrometallurgy?

(2) Which solution can be used as alkali leaching solvent?

(3) What is in-suit leaching?

(4) What is the process of heap leaching?

(5) What are the primary types of metal recovery processes?

2. Translate the following phrases into English.

(1) 湿法冶金　(2) 酸浸　(3) 堆浸　(4) 固液分离

(5) 溶液净化　(6) 金属回收　(7) 浸出液　(8) 电积

(9) 离子交换　(10) 溶剂萃取

3. Translate the following paragraphs into Chinese.

Paragraphs 1 and 2 in section 1 “Leaching involves the process of extracting metal from ...”.

4. Translate the following sentences into English.

(1) 湿法冶金通常包括:浸出、固液分离、溶液富集和净化以及金属或金属化合物的回收。

(2) 酸浸法是一种以酸为溶剂浸出有价金属的方法。常用的酸包括无机酸和有机酸。

(3) 沉降分离是借助于重力作用将浸出矿浆分离为含固体量较多的底流和澄清的溢流的固液分离方法。

(4) 过滤分离是利用多孔介质拦截浸出矿浆中的固体粒子,用压差或其他外力为推动力,使液体通过微孔的固液分离方法。

(5) 溶液富集和净化的常用方法是:沉淀、胶结、溶剂萃取和离子交换。

8.2 Copper Ore Leaching(铜矿浸出)

1. Introduction

(1) Properties of copper

Copper is a chemical element with the symbol Cu and atomic number 29. It is a soft, malleable, and ductile metal with very high thermal and electrical conductivity. Native copper was probably one of the early metals worked by ancient people. Nuggets of the metal could be found in streams in a few areas, and its properties allowed it to be easily worked without a required processing step. Today most copper is produced from sulfide ores. Copper is used as a conductor of heat and electricity, as a building material, and as a constituent of various metal alloys, such as sterling silver used in jewelry, cupronickel used to make marine hardware and coins, and constantan used in strain gauges and thermocouples for temperature measurement. As with many natural resources, the total amount of copper on earth is vast, however, only a tiny fraction of these reserves is economically viable with present-day prices and technologies.

(2) Types of copper ore

There are two main types of copper ore: copper oxide ores and copper sulfide ores. Both ore types can be economically mined, though they are processed in different ways. Copper sulfide ores are typically the most profitable types of copper ores to mine. That is because they have high copper content, and the copper in them can more easily be separated from unwanted minerals. However, copper oxide ores are more abundant. Most copper ores contain only a small percentage of copper metal bound up within valuable ore minerals, with the remainder of the ore being unwanted rock or gangue minerals, typically silicate minerals or oxide minerals for which there is often no value. The average grade of copper ores in the 21st century is below 0.6 % copper, with a proportion of economic ore minerals (including copper) being less than 2 % of the total volume of the ore rock. A key objective in the metallurgical treatment of any ore is the separation of ore minerals from gangue minerals within the rock.

Copper extraction refers to the methods used to obtain copper from its ores. The conversion of copper consists of a series of physical and electrochemical processes. Methods have evolved and vary with country depending on the ore source, local environmental regulations, and other factors. If the ore is primarily sulfide copper minerals (such as chalcopyrite $CuFeS_2$), the ore is crushed and ground to liberate the valuable minerals from the waste (gangue) minerals. It is then concentrated using mineral flotation. As for the oxidized copper ores hydrometallurgical liberation process is used to concentrate copper, which uses the soluble nature of the ore minerals.

2. Leaching of oxide ores

Oxidized copper ore bodies may be treated via several processes, with hydrometallurgical processes used to treat oxide ores dominated by copper carbonate minerals such as azurite ($Cu_3(CO_3)_2(OH)_2$) and malachite ($Cu_2(OH)_2CO_3$), and other soluble minerals such as silicates like chrysocolla ($(Cu, Al)_2H_2Si_2O_5(OH)_4 \cdot nH_2O$), or chlorides such as atacamite ($Cu_2(OH)_3Cl$) and so on. Such oxide ores are usually leached by sulfuric acid, usually in a heap leaching or dump leaching process to liberate the copper minerals into a solution of sulfuric acid laden with copper sulfate in solution.

(1) The process of oxide copper ore heap leaching

The mined ore is usually crushed into small chunks and heaped on an impermeable plastic or clay lined leach pad where it can be irrigated with a leach solution (sulfuric acid) to dissolve the valuable metals. While sprinklers are occasionally used for irrigation, more often operations use drip irrigation to minimize evaporation, provide more uniform distribution of the leach solution, and avoid damaging the exposed mineral. The solution then percolates through the heap and leaches both the target and other minerals. This process is called the "leach cycle". The leach solution laden with copper sulfate is then collected, treated in a process plant via a solvent extraction and electrowinning, with the denuded sulfuric acid recycled back on to the heaps and reused on the leach pad. Alternatively, the copper can be precipitated out of the pregnant solution by contacting it with scrap iron; a process called cementation. Cement copper is normally less pure than solvent extraction and electrowinning copper. Commonly sulfuric acid is used as a leachant for copper oxide, although it is possible to use water, particularly for ores rich in ultra-soluble sulfate minerals. The whole process of copper ore heap leaching is shown in Figure 8.1.

(2) Chemistry of copper dissolution

The dissolution of copper minerals is basically simple decomposition in the case of oxidized minerals, and oxidization-reduction for sulfide minerals. The resultant products are water soluble. The reaction of the common copper minerals with sulfuric acid and ferric sulfate are briefly outlined below.

Azurite $Cu_3(CO_3)_2(OH)_2$

$$Cu_3(CO_3)_2(OH)_2+3H_2SO_4 \longrightarrow 3CuSO_4+2CO_2+4H_2O$$

Brochanite $Cu_4(OH)_6SO_4$

$$Cu_4(OH)_6SO_4+3H_2SO_4 \longrightarrow 4CuSO_4+6H_2O$$

Malachite $Cu_2(OH)_2CO_3$

$$Cu_2(OH)_2CO_3+2H_2SO_4 \longrightarrow 2CuSO_4+CO_2+3H_2O$$

Tenorite CuO

$$CuO+H_2SO_4 \longrightarrow CuSO_4+H_2O$$

Cuprite Cu_2S

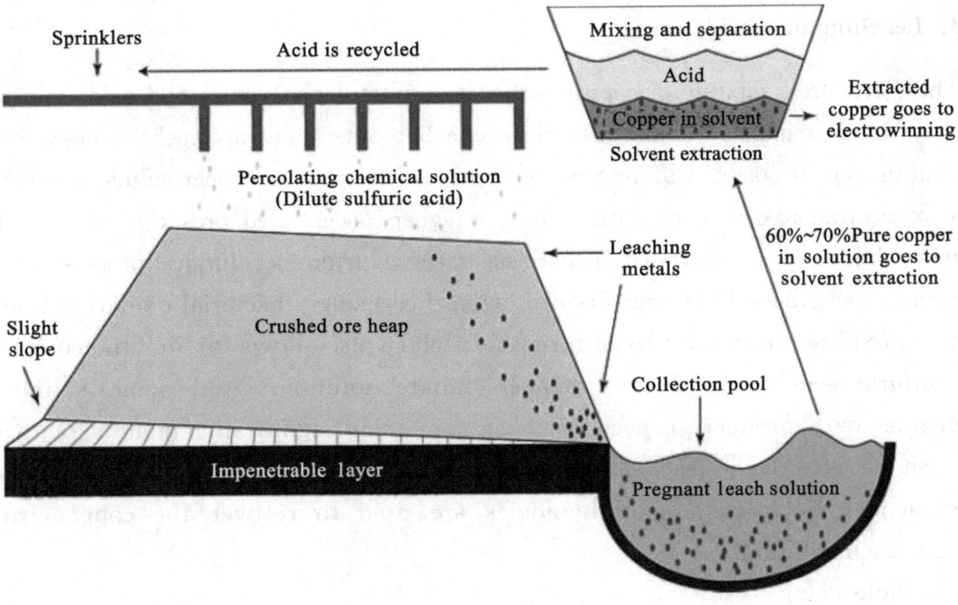

Figure 8.1 Heap leaching of copper ore

$$Cu_2S + 4H_2SO_4 \longrightarrow 2CuSO_4 + S\downarrow + 2SO_2\uparrow + 4H_2O$$

$$Cu + Fe_2(SO_4)_3 \longrightarrow CuSO_4 + 2FeSO_4$$

The dissolution of the various oxides, carbonates, silicates and sulfides occur at different rate which can greatly affect recovery. Each ore offers its own problems and the associated gangue may undergo reactions to form soluble salts to result in excessive consumption of solvent to produce an adverse effect on copper bearing solutions. These conditions can only be determined thorough adequate laboratory test and pilot plant test work to establish all factors essential to process development.

Although heap leaching is a low cost-process, it normally has recovery rates of 60%～70%. It is normally most profitable with low-grade ores. Higher-grade ores are usually put through more complex milling processes where higher recoveries justify the extra cost. The process chosen depends on the properties of the ore. The final product of copper ore heap leaching is cathode copper. Until recently, conventional acid-leaching of copper was utilized as a means of recovering copper from low-grade materials such as waste rock and flotation tailings and was conducted in dumps that were already in place. However, there are more and more restrictions that have impeded the acceptance of leaching as a major commercial process: ① The very slow rate of leaching. ② The inability to process sulfide mineral concentrates. ③ The inability to process chalcopyrite—the most abundant and, thus, the most economically important copper mineral. ④ The inability to recover precious metals.

3. Leaching of sulfide ores

These ores are a mixture of copper carbonate, sulfate, phosphate, and oxide minerals and secondary sulfide minerals, dominantly chalcocite but other minerals such as digenite can be important in some deposits. Different methods are used to process copper sulfide ores depending on the concentration of copper within the ore-higher-concentrated ores can be separated via smelting, while lower-concentrated ores are separated via hydrometallurgical processes.

Some supergene sulfide deposits can be leached using a bacterial oxidation heap leach process to oxidize the sulfides to sulfuric acid, which also allows for simultaneous leaching with sulfuric acid to produce a copper sulfate solution. And some sulfide ores, concentrates or byproduct, especially those containing significant amounts of carbonate rocks can be effectively leached by ammonia solutions. As with oxide ores, solvent extraction and electrowinning technologies are used to recover the copper from the pregnant leach solution.

(1) Bioleaching

Bioleaching is the extraction of metals from their ores through the use of living organisms. Bioleaching can involve numerous ferrous iron and sulfur oxidizing bacteria, including Acidithiobacillus ferrooxidans (Thiobacillus ferrooxidans) and Acidithiobacillus thiooxidans (Thiobacillus thiooxidans). As a general principle, Fe^{3+} ions are used to oxidize the ore. This step is entirely independent of microbes. The role of the bacteria is the further oxidation of the ore, but also the regeneration of the chemical oxidant Fe^{3+} from Fe^{2+}. For example, bacteria catalyse the breakdown of the mineral pyrite (FeS_2) by oxidising the sulfur and metal (in this case ferrous iron, (Fe^{2+})) using oxygen. This yields soluble products that can be further purified and refined to yield the desired metal.

Pyrite (FeS_2) *leaching*

In the first step, disulfide is spontaneously oxidized to thiosulfate by ferric ion (Fe^{3+}), which in turn is reduced to give ferrous ion (Fe^{2+}):

① $FeS_2 + 6Fe^{3+} + 3H_2O = 7Fe^{2+} + S_2O_3^{2-} + 6H^+$ (spontaneous).

The ferrous ion is then oxidized by bacteria using oxygen:

② $4Fe^{2+} + O_2 + 4H^+ = 4Fe^{3+} + 2H_2O$ (iron oxidizers).

Thiosulfate is also oxidized by bacteria to give sulfate:

③ $S_2O_3^{2-} + 2O_2 + H_2O = 2SO_4^{2-} + 2H^+$ (sulfur oxidizers).

The ferric ion produced in reaction ② oxidized more sulfide as in reaction ①, closing the cycle and given the net reaction:

④ $2FeS_2 + 7O_2 + 2H_2O = 2Fe^{2+} + 4SO_4^{2-} + 4H^+$

The net products of the reaction are soluble ferrous sulfate and sulfuric acid.

The microbial oxidation process occurs at the cell membrane of the bacteria. The electrons pass into the cells and are used in biochemical processes to produce energy for the bacteria while reducing oxygen to water. The critical reaction is the oxidation of sulfide by ferric iron. The main role of the bacterial step is the regeneration of this reactant.

The process for copper is very similar, but the efficiency and kinetics depend on the copper mineralogy. The most efficient minerals are supergene minerals such as chalcocite Cu_2S and covellite CuS. The main copper mineral chalcopyrite ($CuFeS_2$) is not leached very efficiently, which is why the dominant copper-producing technology remains flotation, followed by smelting and refining. The leaching of $CuFeS_2$ follows the two stages of being dissolved and then further oxidised, with Cu^{2+} ions being left in solution.

Chalcopyrite leaching

① $CuFeS_2 + 4Fe^{3+} = Cu^{2+} + 5Fe^{2+} + 2S^0$ (spontaneous)

② $4Fe^{2+} + O_2 + 4H^+ = 4Fe^{3+} + 2H_2O$ (iron oxidizers)

③ $2S^0 + 3O_2 + 2H_2O = 2SO_4^{2-} + 4H^+$ (sulfur oxidizers)

net reaction:

④ $CuFeS_2 + 4O_2 = Cu^{2+} + Fe^{2+} + 2SO_4^{2-}$

In general, sulfides are first oxidized to elemental sulfur, whereas disulfides are oxidized to give thiosulfate, and the processes above can be applied to other sulfide ores. Bioleaching of non-sulfide ores such as pitchblende also uses ferric iron as an oxidant (e. g., $UO_2 + 2Fe^{3+} = UO_2^{2+} + 2Fe^{2+}$). In this case, the sole purpose of the bacterial step is the regeneration of Fe^{3+}. Sulfide iron ores can be added to speed up the process and provide a source of iron. Bioleaching of non-sulfide ores by layering of waste sulfides and elemental sulfur, colonized by Acidithiobacillus spp., has been accomplished, which provides a strategy for accelerated leaching of materials that do not contain sulfide minerals.

Copper ore bioleaching

Covellite, chalcocite and chalcopyrite are generally used as copper ores for bioleaching processes. Apart from containing copper, the ores also contain other elements like iron, zinc and sulfur.

Bioleaching involves the use of microorganisms to catalyze the oxidation of iron sulfides to create ferric sulfate and sulfuric acid (Figure 8. 2). Ferric sulfate, which is a powerful oxidizing agent, then oxidizes the copper sulfide minerals and the copper contained is then leached by the sulfuric acid formed. For example, during the oxidation of chalcopyrite the following reaction occurs:

$$4FeSO_4 + O_2 + 2H_2SO_4 \longrightarrow 2Fe_2(SO_4)_3 + 2H_2O \tag{8.1}$$

$$4CuFeS_2 + 2H_2SO_4 + 17O_2 \longrightarrow 4CuSO_4 + 2Fe_2(SO_4)_3 + 2H_2O \tag{8.2}$$

$$CuFeS_2 + 2Fe_2(SO_4)_3 \longrightarrow CuSO_4 + 5FeSO_4 + 2S \tag{8.3}$$

$$2S + 2H_2O + 3O_2 \longrightarrow 2H_2SO_4 \tag{8.4}$$

Similarly, covellite and chalcocite is oxidized to copper sulphate:

$$CuS + Fe_2(SO_4)_3 \longrightarrow CuSO_4 + 2FeSO_4 + S \tag{8.5}$$

$$Cu_2S + Fe_2(SO_4)_3 \longrightarrow CuSO_4 + 2FeSO_4 + CuS \tag{8.6}$$

Bioleach processing differs depending on the type of resources to be processed. It is

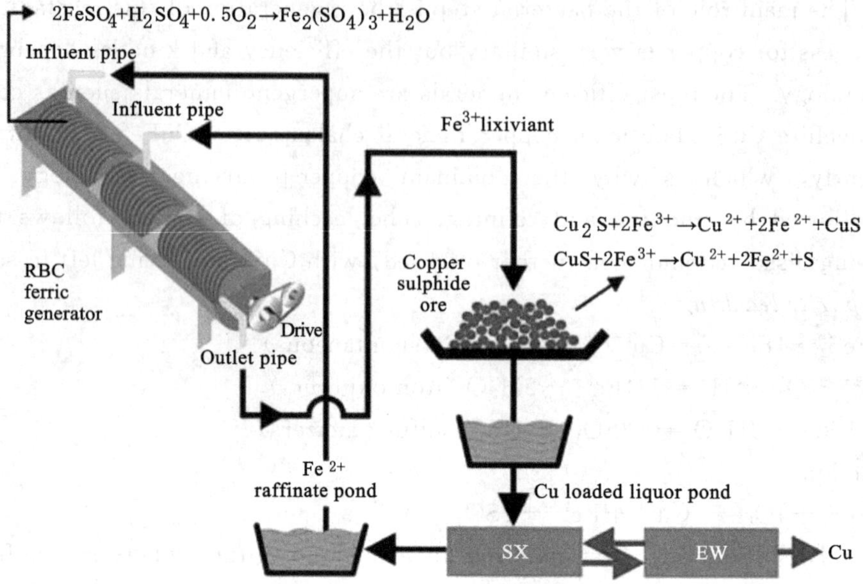

Figure 8. 2 The process of copper sulfide ore heap bioleaching (RBC—rotating biological contactor)

used today in three different regimes dependent upon the raw material types to be processed:

① Dump leaching—waste rock, low grade ore or concentrator tailings (low grade, oxides and secondary sulfides) are leached where placed for disposal. (Mature technology, widely used)

② Heap leaching—newly mined run-of-the-mine (ROM) material (intermediate grade, oxides and secondary sulfides) is deposited in a heap on an impervious natural surface or a pad and leached. ROM may be leached as mined or may be partially crushed and mixed with acid prior to depositing on heap. (Mature technology, increasing use); Primary sulfides; i. e. , chalcopyrite, leaching (Experimental only).

③ Agitated leaching—intermediate—to high-grade, chalcopyrite concentrates are deposited in a tank and leached using mechanical agitation.

There are two dominant views on the mechanisms involved in bioleaching. The first is that the overall leaching process occurs by the microbial oxidation of ferrous to ferric ions followed by the chemical oxidation of the sulfide mineral by the ferric ion. This is known as the indirect mechanism. The second view envisages the microbial catalysis of the overall dissolution of the mineral. It has been proposed that the microorganisms interact with the mineral directly, enhancing the rate of oxidation, over and above that achieved by chemical oxidation. This is known as the direct mechanism. Regardless of the mechanism, it is the enhanced oxidation of the mineral created by the microorganism that creates the acid-solubility required for leaching to take place.

(2) Ammonia leaching

Ammonia leaching is frequently used in hydrometallurgy for processing of oxide and sulfide ores, concentrates or byproduct, especially those containing significant amounts of carbonate rocks, which during the acidic leaching contribute to increased acid consumption. Ammonia and its salts (carbonate or sulfate: $(NH_4)_2CO_3$ or $(NH_4)_2SO_4$) are very effective and selective leaching agents. This method leaches the copper-containing materials with ammonia solution, so that the copper is transferred into the solution in the form of copper-ammonia coordination ions. The leaching method used by the Arbiter plant of Anaconda in the United States to treat copper sulfide concentrate is a typical example of the ammonia leaching method. Copper sulfide ore is leached in an oxygen-ammonia-ammonium sulfate system at a temperature of 373 K and close to normal pressure in a sealed stirred tank. The filtered leaching residue is subjected to flotation to recover the remaining copper and precious metals. The copper in the leaching solution is extracted with Lix-65N extractant, and then the cathode copper is obtained by electrowinning. This leaching method is suitable for the leaching of nickel-copper concentrates, copper scum from blast furnace of zinc smelting and other copper-containing wastes.

Notes

1. Oxidized copper ore bodies may be treated via several processes, with hydrometallurgical processes used to treat oxide ores dominated by copper carbonate minerals such as azurite ($Cu_3(CO_3)_2(OH)_2$) and malachite ($Cu_2(OH)_2CO_3$), and other soluble minerals such as silicates like chrysocolla ($(Cu,Al)_2H_2Si_2O_5(OH)_4 \cdot nH_2O$), or chlorides such as atacamite($Cu_2(OH)_3Cl$) and so on.

氧化铜矿可以通过多种工艺进行处理,湿法冶金工艺可用于处理以碳酸铜矿物(如蓝铜矿和孔雀石)以及其他可溶性矿物如硅酸盐(硅孔雀石)或氯化物(氯铜矿)等为主的氧化物矿。

2. This process is called the "leach cycle". The leach solution laden with copper sulfate is then collected, treated in a process plant via a solvent extraction and electrowinning, with the denuded sulfuric acid recycled back on to the heaps and reused on the leach pad.

此过程称为"浸出循环"。将载有硫酸铜的浸出溶液收集起来,在加工厂中通过溶剂萃取和电解沉积(萃取-电沉积法)处理,然后将余下的硫酸循环回矿石堆并在浸出垫上重复使用。

3. The dissolution of copper minerals is basically simple decomposition in the case of oxidized minerals, and oxidization-reduction for sulfide minerals.

铜矿物的溶解对于氧化矿基本上是简单分解,而对于硫化矿则发生氧化还原反应。

4. These ores are a mixture of copper carbonate, sulfate, phosphate, and oxide minerals and secondary sulfide minerals, dominantly chalcocite but other minerals such as digenite can be important in some deposits.

这些矿石(硫化矿)是碳酸铜、硫酸盐、磷酸盐以及氧化矿与次生硫化矿的混合物,以辉铜矿为主,但在某些矿床中蓝辉铜矿等矿物也很重要。

5. Covellite, chalcocite and chalcopyrite are generally used as copper ores for bioleaching processes.

铜矿石的生物浸出工艺通常应用于铜蓝、辉铜矿和黄铜矿这些矿石。

6. There are two dominant views on the mechanisms involved in bioleaching.

关于生物浸出的机制有两种主要观点。

7. The first is that the overall leaching process occurs by the microbial oxidation of ferrous to ferric ions followed by the chemical oxidation of the sulfide mineral by the ferric ion. This is known as the indirect mechanism.

第一个观点认为，整个浸出过程首先是亚铁离子被微生物氧化为三价铁离子，然后是硫化矿物被三价铁离子化学氧化。这种观点被称为间接氧化机制。

8. The second view envisages the microbial catalysis of the overall dissolution of the mineral.

第二种观点认为是微生物对矿物质整体溶解的催化作用。

Exercise

1. Answer the following questions according to the text.

(1) What are the two main types of copper ore of interest?

(2) What is copper extraction?

(3) What is the heap leaching process of oxide copper ores?

(4) What are sulfide ores?

(5) What are the mechanisms involved in bioleaching?

2. Translate the following phrases into English.

(1) 铜矿石　(2) 浸出循环　(3) 浸出垫　(4) 浸出溶剂

(5) 生物浸出　(6) 铜的硫化矿　(7) 铜的氧化矿　(8) 微生物

3. Translate the following paragraphs into Chinese.

Paragraphs 1 and 2 in section 1 (2) "There are two main types of copper ore ...".

4. Translate the following sentences into English.

(1) 铜矿有两种主要类型：氧化铜矿和硫化铜矿。

(2) 铜是一种柔软的、可塑的、有延展性的金属，具有非常高的导热性和导电性。

(3) 生物浸出是通过使用生物从矿石中提取金属的过程。

(4) 而一些硫化物矿石、精矿或副产品，特别是含有大量碳酸盐岩的矿石，可以用氨水溶液有效地浸出。

8.3 Gold Cyanidation(黄金氰化)

1. Gold cyanidation leaching

(1) Chemical properties of gold

Gold is a chemical element in the periodic table with the symbol Au and atomic number 79. Gold does not react with most chemicals but is attacked by chlorine, fluorine, aqua regia and cyanide. In particular, gold is insoluble in nitric acid, which will dissolve most other metals. Nitric acid has long been used to confirm the presence of gold in items.

Gold readily forms alloys with many other metals. These alloys can be produced to increase the hardness or to create exotic colors. Adding copper yields a redder metal, iron blue, aluminium purple, platinum metals white, and natural bismuth together with silver alloys produce black.

Heat, moisture, oxygen, and most corrosive agents have very little chemical effect on gold, making it well-suited for use in coins and jewelry; conversely, halogens will chemically alter gold, and aqua regia dissolves it.

Common oxidation states of gold include +1 (gold(Ⅰ) or aurous compounds) and +3 (gold(Ⅲ) or auric compounds). Gold ions in solution are readily reduced and precipitated out as gold metal by the addition of virtually any other metal as the reducing agent. The added metal is oxidized and dissolves allowing the gold to be displaced from solution and be recovered as a solid precipitate.

(2) Principles of cyanide leaching

Gold typically occurs at very low concentrations in ores, less than 10 g/t or 0.001% (mass basis). At these concentrations the use of aqueous chemical (hydrometallurgical) extraction processes is the only economically viable method of extracting the gold from the ore. Typical hydrometallurgical gold recovery involves a leaching step during which the gold is dissolved in an aqueous medium, followed by the separation of the gold bearing solution from the residues, or adsorption of the gold onto activated carbon. After elution from the activated carbon the gold is further concentrated by precipitation or electrodeposition.

A complexant, such as cyanide, which stabilizes the gold species in solution, and an oxidant such as oxygen are required to dissolve gold. The amount of cyanide in solution required for dissolution may be as low as 350 mg/L or 0.035% (as 100% NaCN).

Gold is dissolved as an aurocyanide complex in oxidizing alkaline cyanide solutions. The reaction is expressed in its simplest form by the Elsner equation:

$$4Au+8NaCN+O_2+2H_2O = 4Na[Au(CN)_2]+4NaOH \qquad (8.7)$$

The reaction of silver sulphide, commonly associated with gold ores, is not quite so straight-forward, but it is usually expressed in the following manner:

$$Ag_2S + 4NaCN \longrightarrow 2NaAg(CN)_2 + Na_2S \quad (8.8)$$

The reaction of silver sulphide with sodium cyanide is reversible. And it does not proceed far unless sodium sulphide is removed from the system or an excess of free sodium cyanide is available to drive the reaction to the right.

Fortunately, sodium sulphide is sensitive to oxidation, discomposing via side reactions to sulphocyanate and thiosulphate. The point is that oxygen is an indispensable ingredient for successful cyanide leaches of gold as well as the sulphide compounds of silver. Nevertheless, silver halides are attacked by cyanide without the necessity of oxygen.

Cyanide will also form complexes with other metals from the ore, including copper, iron and zinc. The formation of strongly bound complexes such as those with iron and copper will tie up cyanide that would otherwise be available to dissolve gold.

Copper cyanides are moderately stable. Their formation can cause both operational and environmental concerns, as wastewater or tailings from such operations may have significantly higher cyanide concentrations. High copper concentrations in the ore increase costs and lower recovery efficiencies by requiring higher cyanide application rates to compensate for reagent that reacts with copper rather than gold.

When gold is leached in an aqueous cyanide solution it forms a gold-cyanide complex by oxidizing with an oxidant such as dissolved oxygen. This complex is very stable and the cyanide required is only slightly in excess of the stoichiometric requirement. However, in practice the amount of cyanide used in leach solutions is dictated by the presence of other cyanide consumers, and the need to increase the rate of leaching to acceptable levels. Typical cyanide concentrations used in practice range from 300 mg/L to 500 mg/L (0.03% to 0.05% as NaCN) depending on the mineralogy of the ore.

(3) Cyanide leaching methods

Cyanide leaching can be applied in several forms: agitated leaching, heap or dump leaching, vat leaching, and intensive leaching.

Agitated leaching systems are used for the treatment of ground slurries or reclaimed tailings. The product from agitation leaching must either be subjected to one or more stages of solid-liquid separation to allow gold recovery from the solution or may be treated in pulp with carbon or resin for gold recovery. These in pulp processes, which include carbon in leach and resin in leach, can be also incorporated into the leaching circuit for treatment of mildly carbonaceous ores.

Heap or dump leaching can be applied to ores where gold occurs in a form that can be at least partially liberated without grinding. The process is performed on run-of-mine or crushed ore, which is most suitable for treatment of permeable ore types.

Vat leaching is essentially a flooded heap leach with the solution and the ore contained within a vessel or other suitable impermeable impoundment. Its application is limited to the leaching of unusual materials that do not require grinding for gold liberation, for example low grade oxide/free milling ores with most the gold present as coarse particles. This is a rarely used process option because of the generally superior economics of heap and agitated leaching systems.

Intensive cyanidation leaching has been used commercially for the treatment of gravity concentrates containing coarse gold. The leaching kinetics are increased by increasing cyanide and oxygen concentrations and, where necessary, by elevating temperature and pressure.

(4) Agitated leaching process

In a conventional milling and agitated leaching circuit, the ore is milled in semi-autogenous, ball or rod mills, and the milled ore (pulp) is conveyed to a series of leach tanks. The slurry is agitated in the leach tanks, either mechanically or by means of air injection, to increase the contact of cyanide and oxygen with the gold and enhance the efficiency of the leach process. The cyanide then dissolves gold from the ore and forms a stable gold-cyanide complex.

Oxygen is one of the reagents consumed during cyanidation, and a deficiency in dissolved oxygen slows leaching rate. The use of oxygen or peroxygen compounds instead of air as an oxidant increases the leach rate and decreases cyanide consumption, due to the inactivation of some of the cyanide consuming species present in the slurry. Oxygen can also be added by dosing the pulp with hydrogen peroxide solution.

At the head of the leach circuit, the pH of the slurry is raised to pH 10～11 using lime, which ensures that toxic hydrogen cyanide gas is not generated when cyanide is added and the cyanide is kept in solution to dissolve the gold. The slurry may also be subject to other preconditioning such as pre-oxidation at the head of the circuit before cyanide is added.

Activated carbon is used in the dissolved gold recovery process, either by introducing it directly into the CIL (carbon-in-leach) tanks or into separate CIP (carbon-in-pulp) tanks after leaching. The activated carbon adsorbs the dissolved gold from the leach slurry thereby concentrating it onto a smaller mass of solids. The carbon is then separated from the slurry by screening and subjected to further treatment to recover the adsorbed gold.

When carbon is not used to absorb the dissolved gold in the above-mentioned leach slurry, the gold bearing solution must be separated from the solids components utilizing filtration or thickening units. The resultant solution, referred to as pregnant solution, is subjected to further treatment to recover the dissolved gold. Gold is recovered from the solution using either cementation on zinc powder, or concentrating the gold using absorption on activated carbon followed by elution and concluding with either cementation with zinc or electrowinning. For efficient cementation, a clear solution prepared by filtration or counter current decantation is required.

2. Carbon in pulp process and carbon in leach process

Activated carbon is employed in many cyanidation plants to recover gold and silver from cyanide leach solutions from several processes: carbon in pulp (CIP), carbon in leach (CIL), and carbon in column (CIC). The activated carbon used in mining is called absorbent activated carbon and it's a solid with high porosity and superficial area of more than 1,000 m^2/g. Each pore is about 10～20 Angstrom.

(1) The carbon-in-pulp process

Normally, leaching must be intense in order to facilitate the processes of mass transference and have an optimum concentration of free cyanide and oxygen for a good rate of gold leaching. However, absorption of gold leached on activated carbon requires no intense agitation for avoiding attrition of carbon, and a lower concentration of cyanide for favoring absorption.

As consequence of these conditions, the carbon-in-pulp (CIP) process considers two circuits working independently, cyanidation and absorption (Figure 8.3). The CIP process usually comprises the following operations: absorption, elution, electrowinning, and thermal reactivation. The absorption process is perhaps the most important part of the process since the optimization of extraction determines not only the amount of soluble gold lost in the residues from the plant, but also the amount of gold locked up in the plant, and the quantity of carbon reporting for elution and reactivation. Unlike most hydrometallurgical processes, the rate of extraction of gold from a pulp is slow, and the kinetics is very important in order to determine optimum conditions for the operation.

(2) The carbon-in-leach process

The alternative of employing a process including cyanidation and absorption simultaneously is an interesting option since it reduces the cost of investment. This type of process is known as carbon-in-leaching (CIL) and was implemented for ores that have some organic material that can absorb the gold leached (Figure 8.4). These materials are carbonaceous species and can produce losses because they will go the tails.

CIL is a good alternative, but there are some disadvantages in this type of agitated cyanidation processes. Activated carbon losses by attrition during agitation and fine particles can go to the final tails. Usually, these losses can be 8% to 12% activated carbon employed in cyanidation.

(3) Absorption of the dissolved gold

The most cost-effective process is to create absorption of the dissolved gold onto activated carbon, resulting in an easier solid-solid separation based on size. To achieve this, the ore particles must typically be smaller than 100 μm while the carbon particles must be larger than 14 mesh (about 1.4 mm). Absorption is achieved by contacting the activated carbon with the agitated pulp, which can typically recover more than 99.5% of the gold in the solution in 8 to 24 hours, depending on the reactivity of the carbon, the

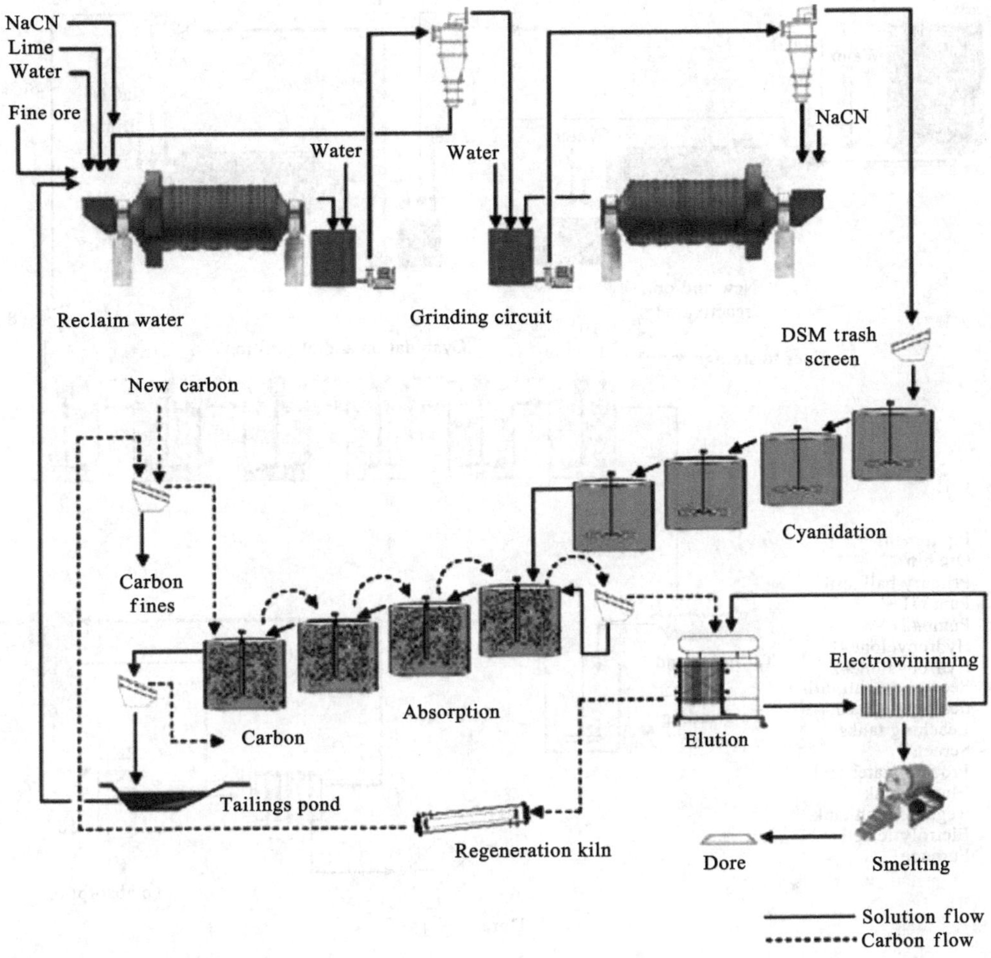

Figure 8.3 Grinding and carbon in pulp process

amount of carbon used and the mixer's efficiency.

The loaded carbon is then separated from the pulp by screens that are air or hydrodynamically swept, thus preventing blinding by the near sized carbon particles. The pulp residue is then either thickened to separate the cyanide containing solution for recovery/destruction of the cyanide, or sent directly to the tailings storage facility from which the cyanide containing solution is recycled to the leach plant.

(4) Elution of gold loaded carbon

The gold adsorbed on the activated carbon is recovered from the carbon by elution, typically with a hot caustic aqueous cyanide solution. The carbon is then regenerated and returned to the adsorption circuit while the gold is recovered from the eluate using either zinc cementation or electrowinning. The gold concentrate is then either calcined or directly smelted and refined to gold bullion that typically contains about 70%~90% gold. The bullion is then further refined to either 99.99% or 99.999% fineness using chlorination and electro-refining.

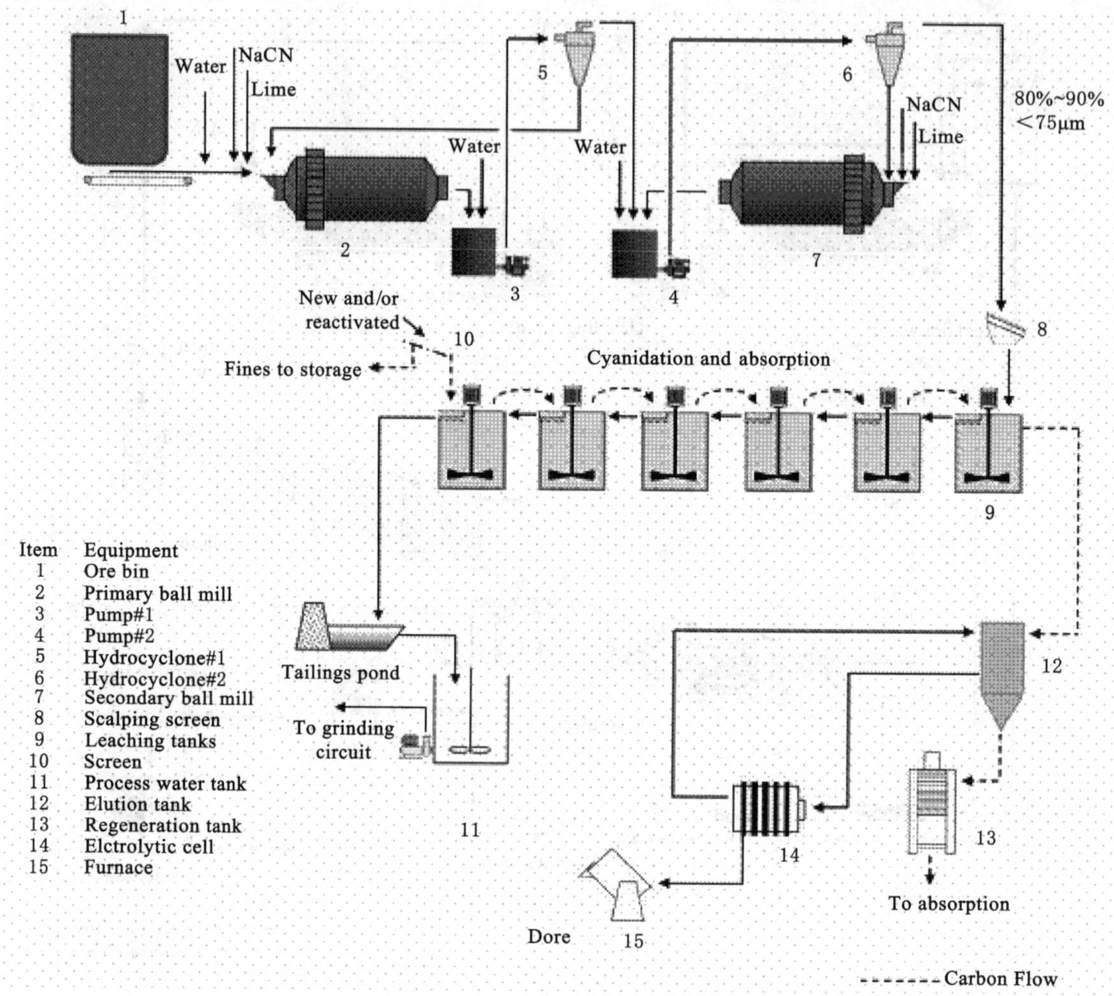

Figure 8.4 The carbon-in-leaching process

(5) Regeneration of activated carbon

Activated carbon can be used 8 to 10 times. The best way to assure a good period of life for activated carbon is not only avoiding losses by attrition during absorption else effecting thermic regeneration. Once carbon has been stripped of its mineral it has one more process to go through before it is ready to be put back into service. This is regeneration.

Carbon is not selective. It adsorbs other substances very readily. As the cyanide solution percolates through the ore mass it will pick up other elements that it is compatible with. These are often organic in nature. The stripping process regularly leaves these substances behind in the carbon. After a time they restrict the carbon's ability to adsorb. When this occurs, the carbon must be refreshed. To do this the carbon is heated to above 700 ℃. The problem is that the carbon will burn away at that temperature. To prevent this from happening, heat is applied in a negative oxygen atmosphere.

The thermal regeneration process generally follows three steps: drying at approximately 105 ℃, high temperature desorption and decomposition (500 ~ 900 ℃) under an inert atmosphere, and residual organic gasification by a non-oxidising gas (steam or carbon dioxide) at elevated temperatures (800 ℃). Per adsorption-thermal regeneration cycle between 5%~15% of the carbon bed is burnt off resulting in a loss of adsorptive capacity.

Notes

1. Gold does not react with most chemicals but is attacked by chlorine, fluorine, aqua regia and cyanide.

金不与大多数化学试剂发生反应,但它可以被氯、氟、王水和氰化物浸蚀。

2. The point is that oxygen is an indispensable ingredient for successful cyanide leaches of gold as well as the sulphide compounds of silver. Nevertheless, silver halides are attacked by cyanide without the necessity of oxygen.

氧气是成功氰化浸出金和含银硫化物必不可少的条件,但是,氰化物可在无氧的条件下溶解卤化银。

3. Cyanide leaching can be applied in several forms: agitated leaching, heap or dump leaching, vat leaching, and intensive leaching.

黄金的氰化浸出包括搅拌浸出、堆浸、槽浸和强化浸出。

4. Intensive cyanidation leaching has been used commercially for the treatment of gravity concentrates containing coarse gold. The leaching kinetics are increased by increasing cyanide and oxygen concentrations and, where necessary, by elevating temperature and pressure.

强化氰化工艺已应用于含粗粒金重选精矿的浸出。浸出动力学的增强是通过提高氰化物和氧气的浓度来实现的,如果需要,还可以提高氰化温度和压力。

5. The slurry is agitated in the leach tanks, either mechanically or by means of air injection, to increase the contact of cyanide and oxygen with the gold and enhance the efficiency of the leach process.

矿浆在浸出槽中可通过机械搅拌或充气搅拌增加金与氰化物、氧气的接触碰撞,从而提高浸出效率。

6. At the head of the leach circuit, the pH of the slurry is raised to pH 10~11 using lime, which ensures that toxic hydrogen cyanide gas is not generated when cyanide is added and the cyanide is kept in solution to dissolve the gold.

在浸出回路的开始阶段,用石灰将矿浆的 pH 值调整至 10~11,可确保氰化物加入不会产生有毒的氰化氢气体而是用于在矿浆中溶解金。

7. The pulp residue is then either thickened to separate the cyanide containing solution for recovery/destruction of the cyanide, or sent directly to the tailings storage facility from which the cyanide containing solution is recycled to the leach plant.

氰化后的尾浆或者浓缩脱除含氰溶液用于回用或分解氰化物,或者直接将尾浆排入尾矿库,含氰溶液再从尾矿库返回氰化厂。

8. The carbon is then regenerated and returned to the adsorption circuit while the gold is recovered from the eluate using either zinc cementation or electrowinning.

解吸后的活性炭再生后返回吸附系统，而洗脱液中的金可用锌粉置换法或电积法回收。

9. Activated carbon can be used 8 to 10 times. The best way to assure a good period of life for activated carbon is not only avoiding losses by attrition during absorption else effecting thermic regeneration.

活性炭一般可循环使用 8 到 10 次。确保活性炭使用寿命长的最好方法是，不仅要避免吸附过程中的炭磨损，还要确保热再生的效果。

Exercises

1. Answer the following questions according to the text.

(1) What are the chemical properties of gold?

(2) What operations does the CIP process usually comprise?

(3) What are the typical hydrometallurgical gold recovery steps?

(4) What is the carbon-in-leach process and what is its application?

(5) What is the best way to assure a good period of life for activated carbon?

2. Translate the following phrases into English.

(1) 金的氰化浸出　(2) 氧化还原反应　(3) 炭浆法　(4) 炭浸法

(5) 活性炭再生　(6) 搅拌浸出　(7) 络合物　(8) 王水

(9) 氰化物

3. Translate the following paragraphs into Chinese.

Paragraphs 1, 2 and 3 in section 1 (1) "Gold is a chemical element in the periodic table with ...".

4. Translate the following sentences into English.

(1) 溶液中的金离子很容易被还原，通过添加几乎任何其他金属作为还原剂而沉淀出金金属。

(2) 硝酸可以溶解大多数金属，但不能溶解黄金。

(3) 氰化浸出法是目前回收金的主要方法。

(4) 活性炭可以从氰化矿浆中吸附已溶解的金。

(5) 活性炭广泛应用于从炭浆法(CIP)、炭浸法(CIL)和炭柱法(CIC)氰化工艺的氰化浸出液中回收金和银。

8.4 Gold Smelting and Refining(黄金冶炼与精炼)

1. Gold smelting

The purpose of smelting is to remove base metals and other impurities from high grade concentrates to produce a gold-silver bullion containing typically more than 95% precious metals. The smelted product, called dore bullion, is suitable for direct sale and or further refining.

(1) Pyrometallurgy of gold

Gold melts at 1,063 ℃ and boils at 2,808 ℃ under atmospheric pressure. The melting and boiling point of other metals and compounds are listed in Table 8.1.

Table 8.1 The melting and boiling point of common metals and compounds

Metals/compounds	Melting Point/℃	Boiling Point/℃	Metals/compounds	Melting Point/℃	Boiling Point/℃
Silver	961	2,210	Copper	1,083	2,595
Platinum	1,769	4,530	SiO_2	1,723	2,230
Mercury	−38.9	357	Fe_2O_3	1,565	—
Zinc	420	907	FeS_2	1,171	—
Lead	327	1,744	Al_2O_3	2,072	2,980

At temperatures above the melting point, gold volatilizes as a red vapour, with the rate of volatilization increasing with increasing temperature. For example, the volatilization rate at 3,000 ℃ is sufficient to completely remove a gold film on porcelain within minutes. Even though the volatility of pure gold is negligible below approximately 1,050 ℃, the volatility increases with the presence of impurity metals, particularly Tellurium, e. g., an alloy containing 5% Tellurium loses between 2% and 4% of the contained gold in one hour at 1,245 ℃. Alloys containing 5% Hg lose approximately 0.2% of the gold under similar conditions.

Gold forms alloys with most metals. Those most commonly encountered are mercury, silver, lead, and copper.

Gold and mercury form alloys over the complete range of proportions. Mercury can be separated from gold by distillation, leaving a gold product containing 0.1%~1% Hg.

Gold and silver form alloys in all the proportions. Alloys containing more than 66% Ag can be separated by acid parting. Alloys with less silver must be inquarted with additional silver to increase the silver content to allow parting.

Gold and lead readily form a wide range of alloy compositions. Lead has a high affinity for gold and can be used as a collector for gold in pyrometallurgical processes. The lead can be subsequently separated by volatilization.

Gold and copper are miscible in all proportions when molten and form solid alloys of sensible uniform composition. Copper can be separate from gold by fusion with lead, by electrolysis, or by smelting to produce a copper matte. Copper imparts a red color to Au-Cu alloys, property that is used in the jewellery industry.

(2) Roasting of gold-bearing materials

Roasting is used to convert base metals, such as zinc, iron, lead and copper, to their respective oxides so that they can be easily removed into a slag during subsequent smelting process. Any minor amount of sulphides in the roasted material is also oxidized.

Oxidation is most effectively performed in the range 600～700 ℃ to enable conversion to the oxide within practical time scales, typically 12～18 hours, although temperatures well in excess of 700 ℃ have been used in practice to reduce roasting time. Above 600 ℃ most metals of importance have negative free energy of formation values, implying that the oxide should be formed under these conditions.

Slow roasting conditions are generally more efficient for converting these metals to their oxides. The reaction is usually limited by the mass transport of oxygen to the metal surfaces and an adequate flow of oxygen, as air, must be provided to all regions of the material. Thin layers of material (40～80 mm thick) in roaster trays should be used and care should be taken not to pack the material.

(3) Smelting fluxes

The smelting process is achieved by heating the material in the presence of slag-forming fluxes at temperatures in excess of the melting point of all the components of the charge, typically between 1,200～1,400 ℃. This maximum temperature is maintained for approximately 1.5 hours to ensure complete separation of impurities into the slag. The molten gold and silver form an alloy that is heavier than the slag and sinks to the bottom of the smelting vessel.

Silica has a high melting point (1,723 ℃) and tends to form a highly viscous slag which may entrain precious metals. The addition of sodium and boron oxides (sodium borate or borax) reduces both the melting point and viscosity. For example, a 1 : 1 molar ratio of sodium borate and silica has a melting point of approximately 800 ℃, which forms a much less viscous, but still strongly acid slag. The reaction of sodium borate with metal oxides is analogous to silica. Consequently, silica-sodium borate mixtures are commonly used as fluxes.

The amount of silica, sodium borate flux added with the charge depends on the quantity of base metals and other impurities in the charge. As smelting proceeds the acidity of the slag is gradually reduced as the base metals react with flux materials. Once the reaction is complete the slag should be neutral, or slightly acidic, to protect the furnace lining or crucible from basic corrosion.

Sodium carbonate improves slag clarity and decrease viscosity, thereby reducing precious metals entrainment.

Oxidizing agents such as sodium nitrate and manganese dioxide assist in the oxidation of unoxidized materials (base metals).

The addition of calcium fluoride and sodium carbonate may cause forming and/or increase precious metals volatilization losses, depending on the charge composition and smelting temperature. Consequently, care must be taken when applying such modified fluxes in smelting systems.

Although smelting is most efficient when treating thoroughly oxidized materials, there are occasions when direct smelting of unoxidized products is the most cost effective method. In these cases, oxidation must be achieved during the smelting stage itself. Manganese dioxide can be used when only a small proportion of the feed must be oxidized. Sodium nitrate is preferred when more severe oxidation is required. The use of oxidizing agents during smelting requires care since silver can be oxidized and lost into the slag, and gold losses can be increased.

Flux compositions are selected to optimize slag quality and to maximize crucible or furnace liner life. Some slag compositions are more corrosive than others, e.g., strongly oxidizing fluxes, or flux that react violently with the material to be smelted.

(4) Smelting with lead

This is a large scale procedure of the well-known fire assay procedure which uses lead to collect and separate precious metals from a molten charge. The precious metals are the recovered by vaporization, or less commonly by acid dissolution of the lead.

Fluxes, lead oxide (litharge) and, if necessary, carbon are mixed with the charge or material to be smelted. Smelting is performed at similar temperatures to those required for simple flux smelting. The lead drains through the molten charge as fine droplets and collect the contained gold and silver. The collection process is highly efficient provided that the ratio of lead to precious metals is kept high (>10 : 1), the lead precious metals alloy is tapped out of the crucible and cooled.

This alloy is then treated by cupellation in a strong stream of air to oxidize the lead back to lead oxide, which is removed as a vapor and can be recovered as a solid on cooling to allow recycling to the fusion stage. The gold and silver remain and can be poured or remelted into bars.

The process is slow, primarily because it requires two stages. It is labour intensive and lead fumes are produced, which present a health hazard. Despite these disadvantages the process is effective for treatment of low-grade refinery products.

2. Gold refining

(1) Miller process

The Miller process is one of the mostly used in the gold refining. In this process the charge is melted in the induction furnace. Chlorine gas is introduced into the molten raw gold through a rotating lance in a very precise and controlled way, so that no break through of chlorine gas occurs. By-metals like silver, copper, nickel etc. as well as amphoteric elements like S, Se, Te, As, Bi are removed into the slag as molten chlorides which float on top of the gold melt.

The slag is first manually poured off into a de-golding crucible, which is then slowly cooled to allow any trapped gold to sink to the bottom of the crucible to solidify. The molten chloride is then carefully poured off into separation moulds leaving the gold. The remaining gold particles can be removed manually from the crucible and separation mould when they have cooled. After the slag removal the gold melt is poured through a ceramic filter directly into the granulation equipment or anode moulds. The granulation water is treated in a wash water treatment system before discharge. The gold granules are then moved to the dissolving reactor for wet chemical refining, the anodes go to electrolysis. The slag from the de-golding process is returned to the slimes smelter. The main advantages are summarized as follows:

① Short process time—quick refining;

② Ceramic rotating chlorine lance for optimum diffusion minimizes chlorine consumption, metal losses and scrubber neutralization costs;

③ Induction furnaces incorporate high efficiency coil with advanced solid state power supply systems to minimize power consumption;

④ Ceramic crucible with special spout for long service life;

⑤ Computerized control panel to monitor all functions like furnace temperature, flow control of nitrogen/chlorine gas, scrubber functions, exhaust measurement, chlorine leakage detection, including alarm signals;

⑥ A specially designed in-line 2 stage integrated Prior scrubber is placed next to furnace to provide maximum extraction and gas scrubbing efficiency.

(2) Aqua regia process

Treatment of zinc precipitates

A zinc precipitate can have a chemical composition like this: 12% gold, 5% silver, 12% CaO, 20% Cu, 11% Zn. The process involves the following stages (Figure 8.5).

① Treatment with nitric acid

Zinc precipitated is leached with hot nitric acid so that silver and common metals can be dissolved. Then, the slurry is filtered and the residue is washed with hot water. The result depends mainly of nitric acid dosage and temperature.

② Silver precipitation

The obtained silver nitrate solution is treated with sodium chloride in order to precipitate silver as silver chloride. The reaction is shown below.

$$NaCl + AgNO_3 = AgCl \downarrow + NaNO_3 \qquad (8.9)$$

The amount of silver in the solution from acid leaching can be calculated by simple calculation. Now, it is necessary to know the addition of sodium chloride. The rule that

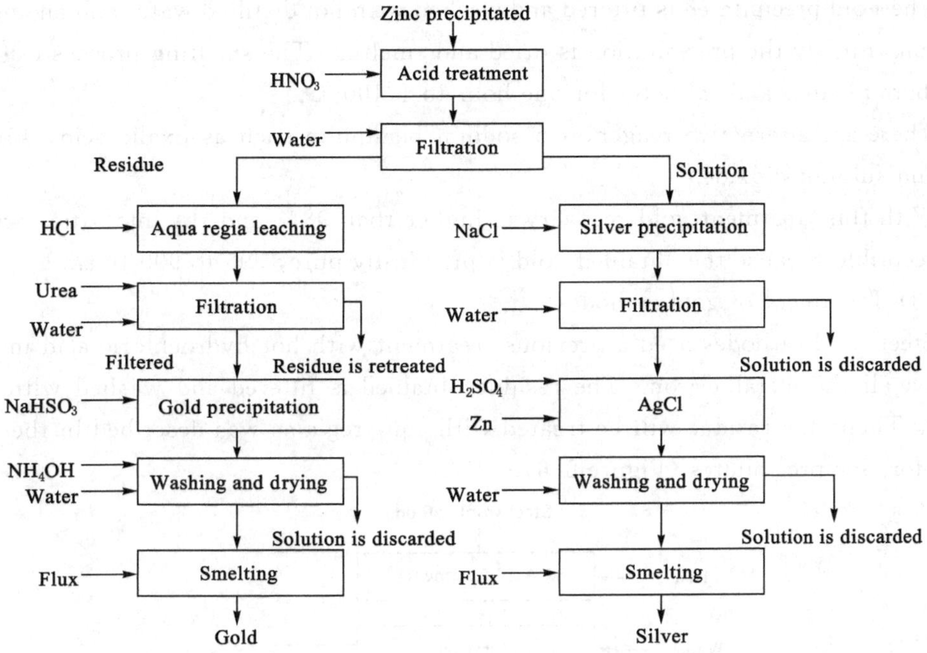

Figure 8. 5 Refining of zinc precipitates

each gram of silver in solution requires 0. 55 grams NaCl must be considered.

The silver chloride precipitated is filtered and washed with distilled water. Then it must be transferred to a beaker for its treatment with zinc dust, which has been previously acidified lightly with sulphuric acid. The metallic product obtained is filtered and washed with hot distilled water. The final product is dried and finally smelted in order to get metallic silver.

③ Gold dissolution with aqua regia

The precipitated obtained is leached with hot aqua regia (85～90 ℃). The amount of aqua regia to be employed must be in excess (at least 100 times) to the required theoretically. The reaction is shown below.

$$Au+3HNO_3+4HCl \longrightarrow HAuCl_4+3NO_2+3H_2O \tag{8. 10}$$

After attacked by aqua regia the slurry will be added urea, which decreases the residual HNO_3.

$$6HNO_3+5CO(NH_2)_2 \longrightarrow 8N_2+5CO_2+13H_2O \tag{8. 11}$$

The slurry is then filtered and the residue is washed with hot distilled water until a yellowish coloration in the filtered solution disappears.

④ Gold precipitation

Metallic gold is precipitated from the solution by adding sodium bisulphite. In theory, each gram of gold needs 1. 89 grams of sodium bisulphite. However, sodium bisulphite must be added more, approximately 1. 5 more times.

$$2HAuCl_4+3NaHSO_3+3H_2O \longrightarrow 3NaHSO_4+2Au\downarrow+8HCl \tag{8. 12}$$

The gold precipitated is filtered and washed with hot distilled water and an ammonium solution. Finally the precipitation is dried and smelted. The smelting process can be done with borax, silica and saltpeter for one hour to 1,100 ℃.

There are alternative reagents to sodium bisulphite such as oxalic acid, sulphurous acid and sulphur dioxide.

With this treatment gold recovery is higher than 98% and the impurities are almost imperceptible because the obtained gold is practically pure, 995 to 999 fine.

(3) Treatment of gold cathodes

Steel wool cathodes need a previous treatment with hot hydrochloric acid in order to dissolve all the metallic iron. The residue obtained is filtered and washed with distilled water. Then, the residue will be treated with aqua regia as was described in the previous steps for zinc precipitates (Figure 8.6).

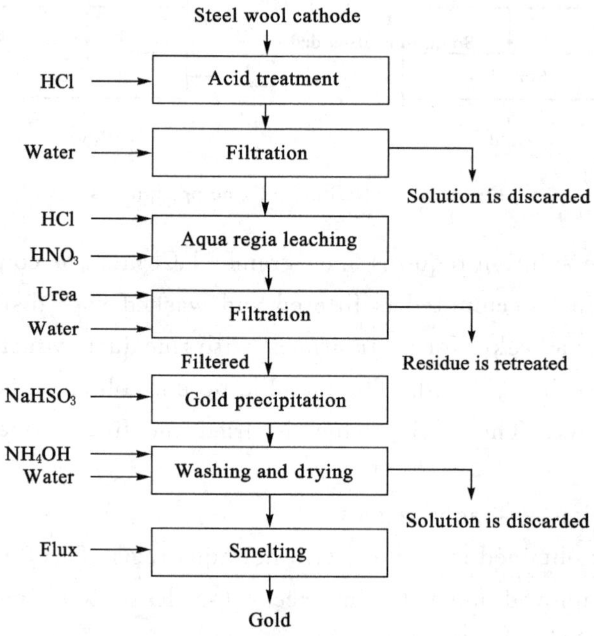

Figure 8.6 Refining of steel wool cathode

Notes

1. The purpose of smelting is to remove base metals and other impurities from high grade concentrates to produce a gold-silver bullion containing typically more than 95% precious metals.

金冶炼的目的是从高品位的精矿中去除贱金属和其他杂质，产出贵金属含量大于95%的金银锭。

2. Roasting is used to convert base metals, such as zinc, iron, lead and copper, to their respective oxides so that they can be easily removed into a slag during subsequent smelting process.

焙烧用于将贱金属锌、铁、铅和铜转化为各自的氧化物，以便在随后的冶炼过程中容易将其去除，变成熔渣。

3. For example, a 1:1 molar ratio of sodium borate and silica has a melting point of approximately 800 ℃, which forms a much less viscous, but still strongly acid slag.

例如，物质的量比 1∶1 的硼酸钠和硅石的熔点大约是 800 ℃，它们形成的熔渣具有更小的黏性，但仍是强酸性渣。

4. This alloy is then treated by cupellation in a strong stream of air to oxidize the lead back to lead oxide, which is removed as a vapor and can be recovered as a solid on cooling to allow recycling to the fusion stage.

然后将该合金(铅金合金)在强空气流中采用灰吹法将铅氧化成氧化铅，氧化铅以蒸气的形式脱除，随后可以冷却回收并返回熔炼阶段使用。

5. The process is slow, primarily because it requires two stages, it is labour intensive and lead fumes are produced, which present a health hazard.

该工艺速度慢，主要是由于它需要两个步骤。它的劳动强度大，同时产生铅烟，存在健康危害。

6. Chlorine gas is introduced into the molten raw gold through a rotating lance in a very precise and controlled way, so that no break through of chlorine gas occurs.

氯气沿着特定的路线通过旋转喷枪非常精确地通入熔化的金原料中，以防止氯气泄漏事故的发生。

7. The slag is first manually poured off into a de-golding crucible, which is then slowly cooled to allow any trapped gold to sink to the bottom of the crucible to solidify.

熔炼渣首先人工(手动)倒入沉金坩埚，然后将坩埚缓慢冷却，金沉淀至坩埚底部凝固。

8. Zinc precipitated is leached with hot nitric acid so that silver and common metals can be dissolved.

锌置换金泥用热硝酸浸出，可以溶解银和其他贱金属。

9. The silver chloride precipitated is filtered and washed with distilled water. Then it must be transferred to a beaker for its treatment with zinc dust, which has been previously acidified lightly with sulphuric acid.

将氯化银沉淀过滤并用蒸馏水洗涤后，转移至烧杯中用硫酸预先轻微酸化过的锌粉置换银。

10. With this treatment gold recovery is higher than 98% and the impurities are almost imperceptible because the obtained gold is practically pure, 995 to 999 fine.

通过精炼处理后金的回收率高达 98%以上，其中的杂质含量微乎其微，因为精炼后的金几乎是纯的，纯度为 995 到 999。

Exercises

1. Answer the following questions according to the text.

(1) What is the purpose of smelting?

(2) What are the common gold alloys?

(3) What is the function of roasting of gold-bearing materials?

(4) What is the process of smelting?

(5) How does the Miller process work?

2. Translate the following phrases into English.

(1) 金的熔炼	(2) 贱金属	(3) 贵金属	(4) 造渣溶剂
(5) 金的精炼	(6) 氯气	(7) 硫酸	(8) 焙烧
(9) 冶炼溶剂	(10) 熔点		

3. Translate the following paragraphs into Chinese.

Paragraphs 1, 2 and 3 in section 1 (2) "Roasting is used to convert base metals ...".

4. Translate the following sentences into English.

(1) 在高于金熔点的温度下,金会以红色蒸气的形式挥发,挥发的速度随着温度的升高而增加。

(2) 铅灰吹法是从低品位含金物料中提炼金的有效方法。

(3) 冶炼渣一般呈中性或偏酸性,有利于保护炉衬和坩埚。

(4) 黄金精炼的目的是进一步除去银、铜等其他杂质。

9 PRODUCT DEWATERING AND TAILINGS DISPOSAL (产品脱水与尾矿处理)

9.1 Sedimentation(沉淀)

With few exceptions, most mineral separation processes involve the use of substantial quantities of water. The products are dewatered to minimize shipping costs, improve handling, and increase market values. In general, dewatering methods can be classified into three groups: sedimentation, filtration and thermal drying.

Dewatering in mineral processing is normally a combination of the above methods. The bulk of water is first removed by sedimentation which produces a thickened pulp of 55%～65% solid by weight and up to 80% of water can be separated at this stage. Sedimentation methods include gravity sedimentation and centrifugal sedimentation.

Sedimentation is the first stage of dewatering process, which is most efficient when there is a large density difference between liquid and solid. The efficiency of sedimentation dewatering by gravity decreases as the particle size. For very fine particles, if only a few microns in diameter, settling extremely slowly by gravity alone, centrifugal sedimentation may have to be performed. Alternatively, the particle may be agglomerated, or flocculated, into relatively large lumps, called flocs, so that increases the efficiency of very fine particle's sedimentation dewatering.

1. Coagulation and flocculation

(1) Coagulation

Coagulation causes extremely fine colloidal particles to adhere directly to each other. All particles exert mutual attraction forces, known as London—Van der Waal's forces, which are effective only at very close range. Normally, the adhesion due to these forces is prevented by the presence of an electrically charged atmosphere around each particle, which generates repulsion forces between particles approaching each other. There is, therefore, in any given system, a balance between the attractive forces and the electrical repulsion present at the solid-liquid interface.

In any given system the electrical charges on the particle surfaces will be of the same sign, in aqueous suspensions of pH 4.0 and above generally being negative. Positively charged surfaces occur mainly in strong acid solutions.

The repulsion forces not only prevent coagulation of the particles, but also retard their settlement by keeping them in constant motion, this effect being more pronounced the smaller the particle.

Coagulants are electrolytes having an opposite charge to the particles, thus causing charge neutralization when dispersed in the system, allowing the particles to come into contact and adhere as a result of molecular forces. Inorganic salts have long been used for this purpose, and as counter-ions in aqueous system are most frequently positively charged, salts containing highly charged cations, such as Al^{3+}, Fe^{3+}, and Ca^{2+} are mainly used. Lime or sulphuric acid, depending on the surface charge of the particles, can also be used to cause coagulation. Most pronounced coagulation occurs when the particles have zero charge in relation to the suspending medium, this occurring when zeta potential is zero. The nature of the zeta potential can be explained from the model of the electrical double layer at the surface of a particle (Figure 9.1). The surface shown has a negative charge, such that positive ions from solution will be attracted to it, forming a bound layer of positive ions, known as the Stern layer and a diffuse layer of counter ions decaying in concentration with increasing distance until the solution equilibrium concentration is attained. These layers of ions close to the surface constitute the electrical double layer. When a particle moves in the liquid, shear takes place between the bound layer, which moves with the particle, and the diffuse layer, the potential at the plane of shear being known as the zeta potential. The magnitude of the zeta potential depends on the surface potential, and the concentration and charge of the counter-ions. In general, the greater the counter-ion charge and counter-ion concentration, the lower is the zeta potential, although ions of high charge may cause complete charge reversal, therefore, optimum doses of electrolyte are critical.

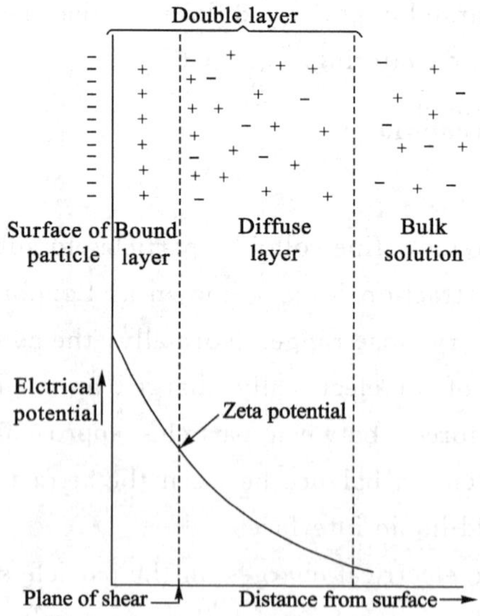

Figure 9.1 The electrical double layer

(2) Flocculation

Flocculation involves the formation of much more open agglomerates than those resulting from coagulation and relies upon molecules of reagent acting as bridges between separate suspended particles. The reagents used to form the "bridges" are long-chain organic polymers, which were formerly natural materials, such as starch, glue, gelatine, and guar gum, but which are now increasing synthetic materials, loosely termed poly-electrolytes. The majority of these are anionic in character but some of them are non-ionic, and some cationic, but these form a minor proportion of the commercially available products of today's flocculant market. Inorganic salts are not able to perform this bridging function, but they are sometimes used in conjuction with an organic reagent as a cheaper means of charge neutralization, although an ionic polyelectolyte can and often does perform both functions.

The polyacrylamides, called No. 3 flocculants in China, which vary widely in molecular weight and charge density, are extensively used as flocculants. The charge density refers to the percentage of acrylic monomer segments which carry a charge.

The mode of action of the anionic polyacrylamide depends on a segment of the very long molecule being adsorbed on the surface of a particle, leaving a large proportion of the molecule free to be adsorbed on another particle, so forming an actual molecular linkage or bridge, between particles.

The factors influencing the degree of flocculation are the efficiency or strength of adsorption of the polymer on the surface, the degree of agitation during flocculation and the subsequent agitation, which can result in breakdown of flocs.

The maximum effect of a flocculant is achieved at an optimum dosage rate, excess polymer can cause dispersion of the particles due to floc breakdown. Polyelectrolytes are normally made up to stock solutions of about 0.5%～1%, which are diluted to about 0.01% before adding to the slurry. The diluted solution must be added at enough points in the stream to ensure its contact with every portion of the system. A shower pipe is frequently used for this purpose (Figure 9.2).

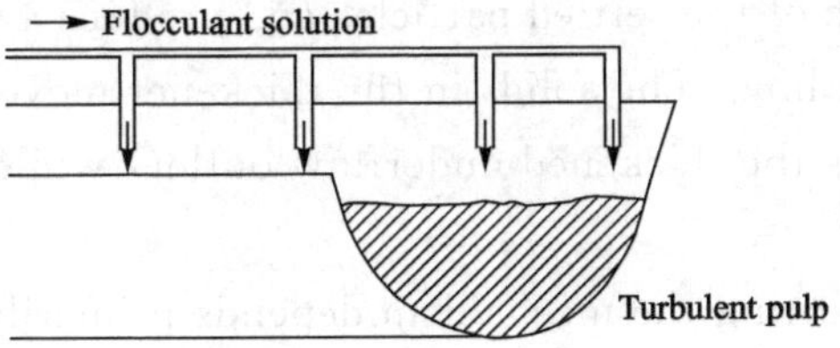

Figure 9.2 Typical method of flocculant addition

Mild agitation is essential at the addition points, and shortly thereafter, to assist in flocculant dispersion in the process stream. Care should be taken to avoid severe agitation after the flocs have been formed.

2. Gravity sedimentation

Gravity sedimentation, or thickening, is the most widely applied dewatering technique in mineral processing, as it is a relatively cheap, high capacity process, which involves very low shear forces, thus providing good conditions for flocculation of fine particles.

The thickener is used to increase the concentration of the suspension by sedimentation, accompanied by the formation of a clear liquid. In most cases the concentration of suspensions is high and hindered settling takes place. The design considerations of thickener are based on the settling rates of the slowest settling particles and conditions for minimum disturbance of medium (water) through which the solid particles are allowed to settle. To achieve these objectives cylindrical tanks with conical or flat bottoms are used and the velocity of the feed slurry is minimized to reduce turbulence in the settling tank. The clarifier is similar in design, but is less robust, handling suspensions of much lower solid content than the thickener. The basic principle of the thickener is shown in Figure 9. 3.

The continuous thickener consists of a cylindrical tank, the diameter ranging from about 2 m to 200 m in diameter, and of depth 1～7 m. The feed slurry is carried to the center via a feed-well placed up to 1 m below the surface of the tank, which is designed to break the fall of the slurry and dissipate the energy. The clarified liquid overflows a peripheral launder, while the solids which settle over the entire bottom of the tank are withdrawn as a thickened pulp from an outlet at the center. The settling rates of particles in a fluid are governed by Stock's or Newton's laws, depending on the particle size. Very fine particles settle extremely slowly by gravity alone and flocculants may have to be employed. Within the tank are one or more rotating radial arms, from each of which are suspended a series of blades, shaped so as to rake the settled solids towards the central outlet (Figure 9. 4). On most modern thickeners these arms rise automatically if the torque exceeds a certain value, thus preventing damage due to overloading. The blades also assist the compaction of the settled particles and produce a thicker underflow than can be achieved by simple settling. The solids in the thickener move continuously downwards, and then inwards towards the thickened underflow outlet, while the liquid moves upwards and radially outwards.

The method of supporting the mechanism depends primarily on the tank diameter. In relatively small thickeners, of diameter less than about 45 m, the drive head is usually supported on a superstructure spanning the tank, with the rakes being attached to the drive shaft. Such machines are referred to as bridge or beam thickeners (Figure 9. 5). The underflow is usually drawn from the apex of a cone located at the centre of the sloping bottom.

A common arrangement for larger thickeners, of up to about 180 m in diameter, is to support the drive mechanism on a stationary steel or concrete centre column. In most

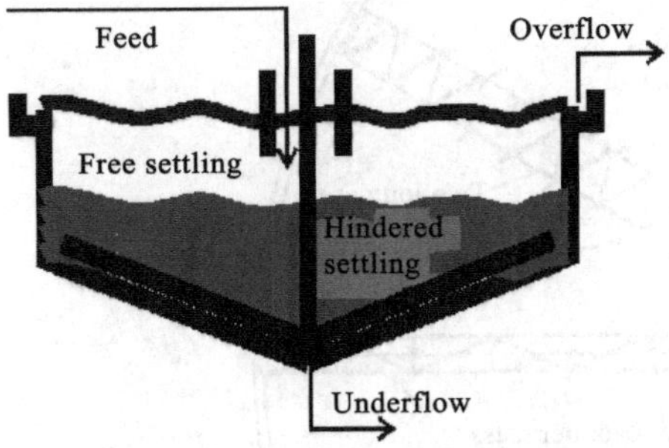

Figure 9.3 Principle of the thickener

Figure 9.4 Structure of the thickener

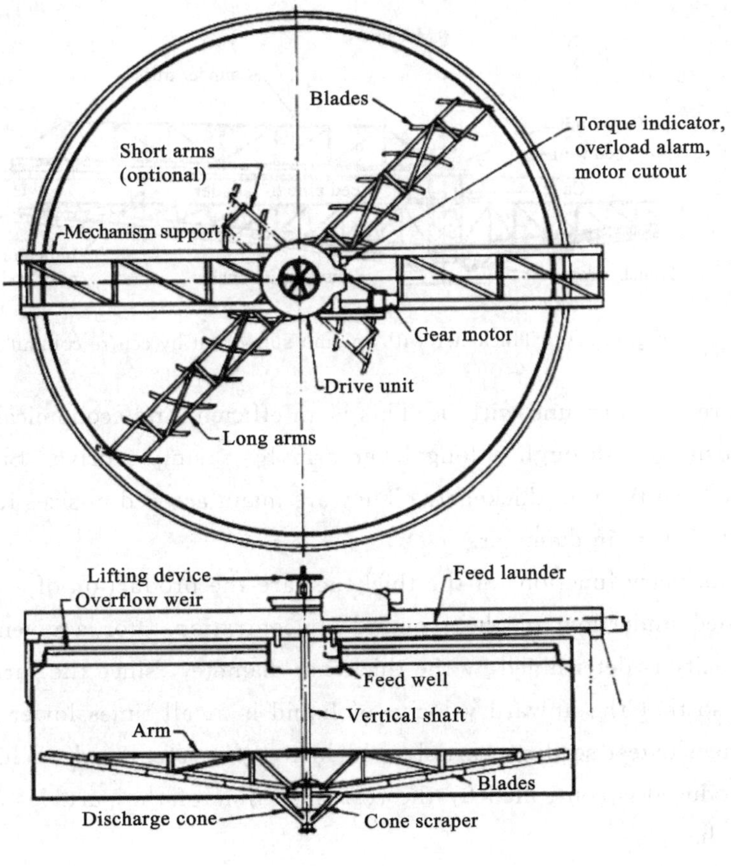

Figure 9.5 Thickener with mechanism supported by superstructure

cases, the rake arms are attached to a drive cage, surrounding the central column, which is connected to the drive mechanism. The thickened solids are discharged through an annular trench encircling the centre column (Figure 9.6).

In larger thickeners, a single long rake is mounted with one end on the central support column while to the other are fixed traction wheels that run on a rail on top of the tank wall. The wheels are driven by motors which are mounted on the end of the rake and

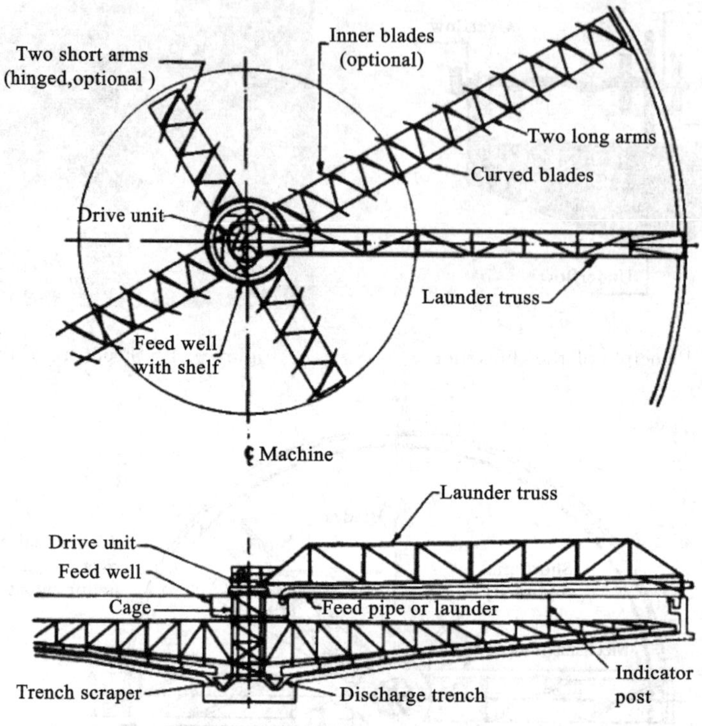

Figure 9.6　Thickener with column supported by centre column

which therefore travel around with it. This is an efficient and economical design since the torque is transmitted through a long lever arm by a simple drive. Such machines are normally called the traction thickeners. They are manufactured in sizes ranging from 60 m to approximate 120 m in diameter.

The two primary functions of the thickener are the production of a clarified overflow and a thickened underflow of the required concentration. For a given throughput the clarifying capacity is determined by the thickener diameter, since the surface area must be large enough so that the upward velocity of liquid is at all times lower than the settling velocity of the slowest-settling particle which is to be recovered. While the degree of thickening produced is controlled by the residence time of the particles and hence by the thickener depth.

3. Centrifugal sedimentation

Centrifugal separation can be regarded as the extension of gravity separation, as the settling rates of particles are increased under the influence of centrifugal force.

Centrifugal separation can be performed either by hydrocyclone or centrifuges. By comparison, centrifuges are much more costly and complex, but have a much greater clarifying power and are generally more flexible. Various types of centrifuge are used industrially, the solid bowl scroll centrifuge having widest used in the minerals industry due to its ability to discharge the solids continuously. The basic principles of a typical machine are shown in Figure 9. 7.

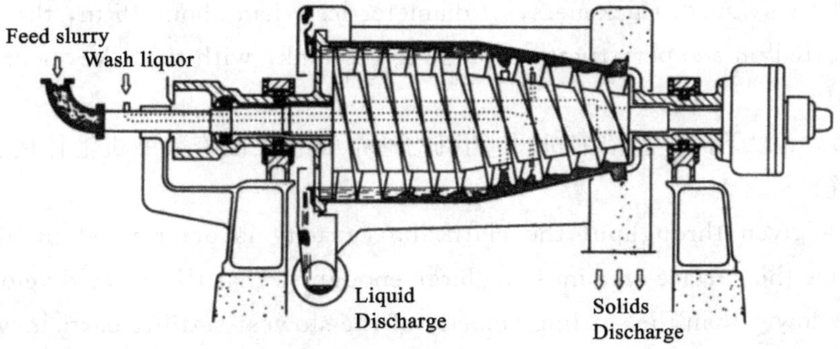

Figure 9. 7 Continuous solid bowl scroll centrifuge

It consists essentially of a horizontal revolving shell or bowl, cylindroconical in form, inside which a screw conveyor of similar section rotates in the same direction at a slightly higher or lower speed. The feed pulp is admitted to the bowl through the center tube of the revolving-screw conveyor. On leaving the feed pipe the slurry is immediately subjected to a high centrifugal force causing the solids to settle on the inner surface of the bowl at a rate which depends on the rotational speed employed, this normally being between 1,600 r/min and 8,500 r/min. The separated solids are conveyed by the scroll out of the liquid and discharged through outlets at the smaller end of the bowl. The solids are continuously dewatered by centrifugal force as they proceed from the liquid zone to discharge.

Centrifuges are manufactured with bowl diameters ranging from 15 cm to 150 cm, the length generally being about twice the diameter. Throughputs vary from about 0. 5 m^3/h to 50 m^3/h of liquid and from about 0. 25 t/h to 100 t/h of solids depending on the feed concentration, which may vary widely from 0. 5% to 70% solids, and on the particle size, which may range from about 12 mm to as fine as 2 μm, or even less when flocculation is used.

The actual size and geometry of centrifuges varies according to the throughput required and the application. The length of the cylindrical section largely determines the clarifying power and is thus made a maximum where overflow clarity is of prime importance. The length of the conical section decides the residual moisture content of the solids, so that a long shallow cone is used where maximum dryness is required.

Notes

1. Gravity sedimentation, or thickening, is the most widely applied dewatering technique in mineral processing, as it is a relatively cheap, high capacity process, which involves very low shear forces, thus providing good conditions for flocculation of fine particles.

重力沉降，或称浓缩，由于成本相对较低以及处理能力大而成为选矿过程中应用最广泛的脱水技术。该过程产生的剪切力很小，从而为细粒物料的絮凝提供了良好的条件。

2. In relatively small thickeners, of diameter less than about 45 m, the drive head is usually supported on a superstructure spanning the tank, with the rakes being attached to the drive shaft.

直径在 45 m 以下的小型浓缩机，传动机构通常安装在横跨浓缩机池体的桁架上，耙臂与传动轴相连。

3. For a given throughput the clarifying capacity is determined by the thickener diameter, since the surface area must be large enough so that the upward velocity of liquid is at all times lower than the settling velocity of the slowest-settling particle which is to be recovered.

对于既定处理量，浓缩机的澄清能力取决于它的直径，因为其表面积必须足够大，以使流体的上向速度总是低于回收颗粒中沉降最慢的颗粒沉降速度。

4. Centrifugal separation can be performed either by hydrocyclone or centrifuges.

离心分离可以通过水力旋流器或离心机进行。

5. It consists essentially of a horizontal revolving shell or bowl, cylindroconical in form, inside which a screw conveyor of similar section rotates in the same direction at a slightly higher or lower speed.

它主要由一个水平旋转的、圆锥形的壳体组成，在圆锥形壳体的内部，有一个类似截面的螺旋输送机，以相同方向、稍高或稍低的速度旋转。

6. On leaving the feed pipe the slurry is immediately subjected to a high centrifugal force causing the solids to settle on the inner surface of the bowl at a rate which depends on the rotational speed employed, this normally being between 1,600 r/min and 8,500 r/min.

矿浆离开给料管，立即受到一个很高的离心力的作用，使固体颗粒以一定速度沉淀到圆筒的内表面，该速度由所采用的旋转速度决定，旋转速度一般在 1 600～8 500 r/min 之间。

Exercises

1. Answer the following questions according to the text.

(1) What are the purposes to dewater the products?

(2) How to classify the dewatering methods?

(3) What is flocculation?

(4) Why is gravity sedimentation the most widely applied dewatering technique in mineral processing?

(5) What are the two primary functions of the thickeners?

(6) Which equipment can perform centrifugal separation?

2. Translate the following phrases into English.

(1) 重力沉降　(2) 离心沉降　(3) 浓密机　(4) 絮凝

(5) 凝聚　(6) 澄清能力　(7) 静电斥力　(8) 有机高分子

3. Translate the following paragraphs into Chinese.

Paragraphs 1 and 2 in section 2 "Gravity sedimentation, or thickening ...".

4. Translate the following sentences into English.

(1) 沉降方法包括重力沉降和离心沉降。

(2) 浓密机的两个主要功能是产生澄清溢流和一定浓度的浓缩底流。

(3) 浓密机运动机构的支撑方式主要取决于池体的直径。

(4) 在离心力作用下颗粒的沉降速度增大,因此,离心分离被认为是重力分离的延伸。

(5) 沉淀设备的处理能力与沉淀池面积成正比,并由悬浮颗粒的自由沉降速度决定。

9.2 Filtration and Drying(过滤与干燥)

Sedimentation is only an initial dewatering process. The obtained thick pulp is then handled by filtration to produce moist filter cake of between 80%~90% solids, which may require thermal drying to produce a final product of about 95% solids by weight.

1. Principles of filtration

(1) Factors affecting the rate of filtration

Filtration is the process of separating solids from liquid by means of a porous medium which retains the solid but allows the liquid to pass.

Filters can be operated in two basic modes. Constant pressure filtration maintains a constant pressure, so that the flow rate falls slowly from a maximum at the start of the cycle. Most continuous filters can be considered to operate on this principle, using a vacuum to provide the pressure difference. Constant rate filtration requires gradually increasing pressure as the cake builds up and increases the resistance to flow. A common approach is to use a constant flow rate until the pressure builds up to a certain level, and then use constant pressure filtration for the remainder of the time. This cycle can conveniently be achieved using centrifugal pumping and has the advantage of forming a rather loosely knit initial cake, which minimizes the quantity of solids forced into and through the medium, typically it is used for pressure filtration, where pressures greater than 1 atm are possible.

The conditions under which filtration are carried out are many and varied, and the choice of the most suitable type of equipment will depend on a large number of factors. Whatever type of equipment is used, a filter cake gradually builds up on the medium and the resistance to flow progressively increases throughout the operation. Factors affecting the rate of filtration include:

① The pressure drops from the feed to the far side of the filter medium. This is achieved in pressure filters by applying a positive pressure at the feed end and in vacuum filters by applying a vacuum to the far side of the medium, the feed side being at atmospheric pressure.

② The area of the filtering surface.

③ The viscosity of the filtrate.

④ The resistance of the filter cake.

⑤ The resistance of the filter medium and initial layers of cake.

Filtration in mineral processing applications normally follows thickening. The thickened pulp may be fed to storage agitators from where it is drawn off at uniform rate to the filters. Flocculants are sometimes added to the agitators in order to aid filtration. Slimes have an adverse effect on filtration, as they tend to "blind" the filter medium;

flocculation reduces this and increases the voidage between particles, making filtrate flow easier. The lower molecular weight flocculants tend to be used in filtration, as the flocs formed by high molecular weight products are relatively large, and entrain water within the structure, increasing the moisture content of the cake, even with the lower molecular weight flocculants, which have a higher shear-resistance, and the resultant filter cake is a uniform porous structure which allows rapid dewatering, yet prevents migration of the finer particles through the cake to the medium. Other filter aids are used to reduce the liquid surface tension, thus assisting flow through the medium.

(2) The filter medium

The choice of the filter medium is often the most important consideration in assuring efficient operation of a filter. Its function is generally to act as a support for the filter cake, while the initial layers of cake provide the true filter. The filter medium should be selected primarily for its ability to retain solids without blinding. It should be mechanically strong, corrosion resistant, and offer as little resistance to flow of filtrate as possible. Relatively coarse materials are normally used and clear filtrate is not obtained until the initial layers of cake are formed, the initial cloudy filtrate being recycled.

The media generally used in industrial practice are woven fabric, woven synthetic material and non-woven synthetic material. As all cotton material is attacked by acidic and alkaline slurries, they are best used under neutral conditions. Cotton fabrics shrink or stretch under load and sometimes swell. As a result, the structure of pores and pore sizes are affected which can adversely affect the size of particles in the filtrate. The woven synthetic materials are usually polypropylenes, polyesters and various polyamides.

2. Types of filter

Cake filters are the type most widely used in mineral processing, where the recovery of large amounts of solids from fairly concentrated slurries is the main requirement. Those where the main requirement is the removal of small amounts of solid from relatively dilute suspensions are known as screening or clarification filters. Cake filters may be pressure, vacuum, batch, or continuous types.

(1) Pressure filters

Because of the virtual incompressibility of solids, filtration under pressure has certain advantages over vacuum. Higher flow rates and better washing and drying may result from the higher pressures that can be used. However, the continuous removal of solids from the pressure-filter chamber can be extremely difficult and consequently, although continuous pressure filters do exist, the vast majority operate as batch units.

Filter presses are the most frequently used type of pressure filter. They are made in two forms—the plate and frame press and the recessed plate or chamber press.

The plate and frame press (Figure 9. 8) consists of plates and frames arranged alternately. The hollow frame is separated from the plate by the filter cloth. The filter press is closed by means of a screw or hydraulic piston device and compression of the filter cloth between plates and frames helps to prevent leakages. A tight chamber is therefore formed between each pair of plates. The slurry is introduced to the empty frames of the press through a continuous channel formed by the holes in the corners of the plates and frames. The filtrate passes through the cloth and runs down the grooved surfaces of the plates and is removed through a continuous channel. The cake remains in the frame and, when the frame is full, the filter cake can be washed, after which the pressure is released and the plates and frames separated one by one. The filter cake in the frames can then be discharged, the filter press closed again and the cycle repeated.

The chamber press (Figure 9.9) is similar to the plate and frame type except for the fact that the filter elements consist solely of the recessed filter plates. The individual filter chambers are therefore formed between successive plates. All the chambers are connected by means of a comparatively large hole in the centre of each plate. The filter cloth with a central hole covers the plate and slurry is led through the inlet channel. The clear filtrate passing through the cloth is removed by means of smaller holes in the plate, the cake gradually depositing in the chambers.

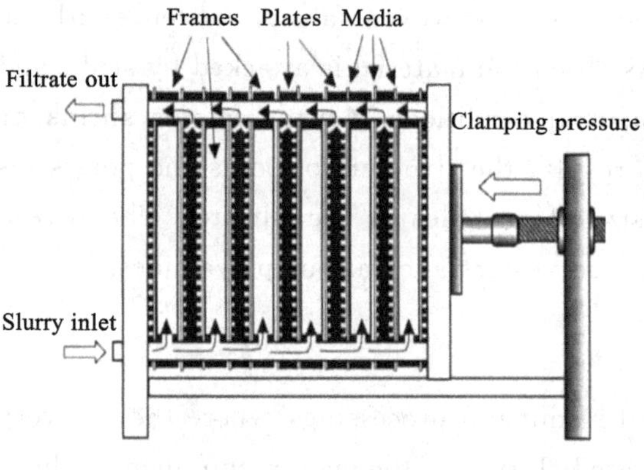

Figure 9.8 Plate and frame filter press

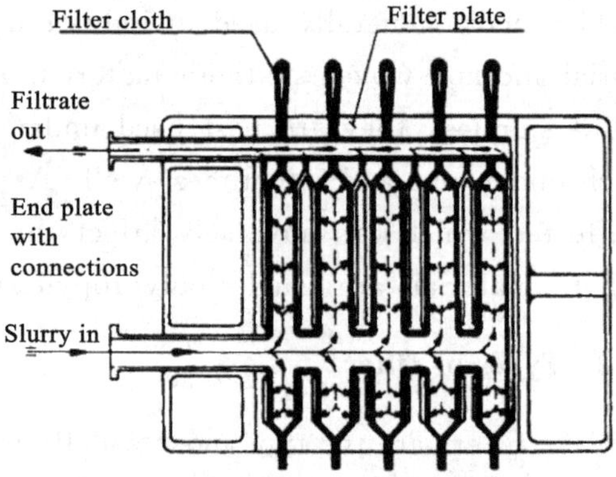

Figure 9.9 Chamber filter press

(2) Vacuum filters

There are many different types of vacuum filter, but they all incorporate filter media suitably supported on a drainage system, beneath which the pressure is reduced by connection to a vacuum system. Vacuum filters may be batch or continuous. Continuous vacuum filters are the most widely used filters in mineral processing applications and fall into three classes—drums, discs, and horizontal filters.

The rotary-drum filter

It is the most widely used type in industry (Figure 9.10), finding application both where cake washing is required and where it is unnecessary. The drum is mounted horizontally and is partially submerged in the filter trough, into which the feed slurry is fed and maintained in suspension by agitators. The periphery of the drum is divided into compartments, each of which is provided with a number of drain lines, which pass through the inside of the drum, terminating at one end as a ring of ports which are covered by a rotary valve to which vacuum is applied. The filter medium is wrapped tightly around the drum surface which is rotated at low speed, usually in the range 0.1～0.3 r/min, but up to 3 r/min for very free-filtering materials.

Figure 9.10 Rotary-drum filter with belt discharge

As the drum rotates, each compartment goes through the same cycle of operations, the duration of each being determined by the drum speed, the depth of submergence of the drum, and the arrangement of the valve. The normal cycle of operations consists of filtration, drying, and discharge, but it is possible to introduce other operations into the basic cycle, such as cake washing and cloth cleaning.

Various methods are used for discharging the solids from the drum, depending on the material being filtered. The most common form makes use of a reversed blast of air, which lifts the cake so that it can be removed by a knife, without the latter actually contacting the medium. Another method is belt discharge, as shown in Figure 9.10, where the filter medium itself leaves the filter and passes over the external roller, before returning to the drum. This has a number of advantages in that very much thinner cakes can be handled, with consequently increased filtration and draining rates and hence better washing and drying products. At the same time, the cloth can be washed on both sides by means of sprays before it returns to the drum, thus minimising the extent of blinding.

Disc filters

The principle of operation of disc filters (Figure 9.11), is similar to that of rotary drum filters. The solids cake is formed on both sides of the circular discs, which are connected to the horizontal shaft of the machine. The discs rotate and lift the cake above the level of the slurry in the trough, whereupon the cake is suction-dried and is then removed by a pulsating air blow with the assistance of a scraper. The discs can be located along the shaft at about 30 cm spacing and consequently a large filtration area can be accommodated in a small floor space. Cost per unit area is thus lower than for drum filters, but cake washing is virtually impossible and the disc filter is not as adaptable as a drum filter.

Figure 9.11　Rotary-disc filters

The horizontal belt filter

The horizontal belt filter (Figure 9.12), consists of an endless perforated rubber drainage deck supporting a separate belt made from a suitable filter cloth.

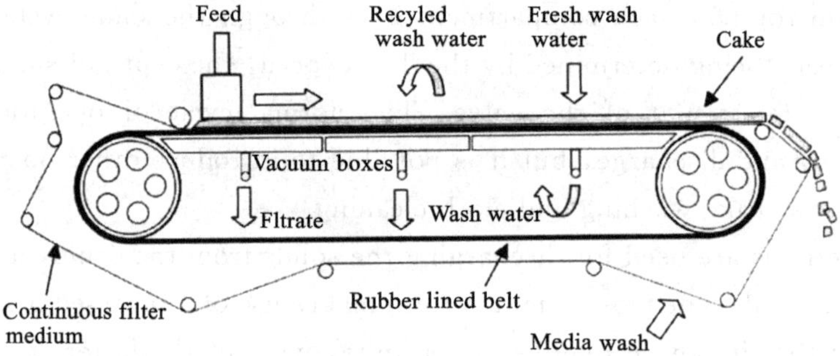

Figure 9.12　Horizontal belt filter

The feed slurry is added onto the belt that has vacuum boxes installed on its bottom. The vacuum box has compartments that are adjustable along the length. Filtration takes place in the first compartment under vacuum and the filtrate is withdrawn from the bottom. In subsequent compartments the cake is washed under vacuum by recycled or fresh water. The washed cake is dewatered, dried by drawing air through it, and then discharged as the belt reverses. Because steel belt is employed, hot air can be used to further reduce product moisture. In addition to their low installed capital cost when compared with discs and drum filters, relatively low operating cost mean that these filters offer a particularly cost-effective and reliable solution to filtration problems, especially with low-value material such as mine tailings.

3. Drying

The drying of concentrates prior to shipping is the last operation performed in the mineral-processing plant. It reduces the cost of transport and is usually aimed at reducing the moisture content to about 5% by weight. Dust losses are often a problem if the moisture content is lower.

Rotary thermal dryers are often used. These consist of a relatively long cylindrical shell mounted on rollers and driven at a speed of up to 25 r/min. The shell is at a slight slope, so that material moves from the feed to discharge end under gravity. Hot gases, or air, are fed in either at the feed end to give parallel flow or at the discharge to give counter-current flow.

The method of heating may be either direct, in which case the hot gases pass through the material in the dryer, or indirect, where the material is in an inner shell, heated externally by hot gases. The direct-fired is the dryer most commonly used in the minerals industry, the indirect-fired type being used when the material must not contact the hot combustion gases. Parallel flow dryers (Figure 9.13) are used in the majority of current operations because they are more fuel efficient and have greater capacity than counterflow types. Since heat is applied at the feed end, build-up of wet feed is avoided, and in general these units are designed to dry material to not less than 1% moisture. Since counter-flow dryers apply heat at the discharge end, a completely dry product can be achieved, but its use with heat-sensitive materials is limited because the dried material comes into direct contact with the heating medium at its highest temperature.

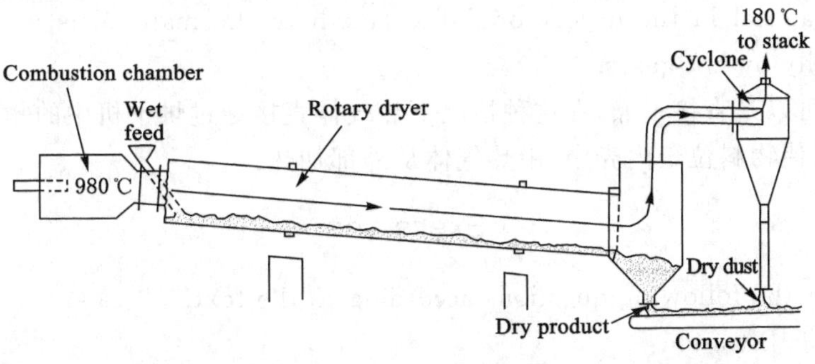

Figure 9.13 Direct fired, parallel flow rotary dryer (after Kram (1980))

Notes

1. Filtration is the process of separating solids from liquid by means of a porous medium which retains the solid but allows the liquid to pass.

过滤是利用能够保留固体但允许液体通过的多孔介质将固体颗粒从液体中分离的过程。

2. This cycle can conveniently be achieved using centrifugal pumping and has the advantage of forming a rather loosely knit initial cake, which minimizes the quantity of solids forced into and through the medium, typically it is used for pressure filtration, where pressures greater than 1 atm are possible.

使用离心泵很容易实现这种操作循环,其优点是形成相当疏松的初期滤饼,这样可以将压入并穿过介质的固体数量减少到最小值，该方法一般用于压力大于 1 atm 的压滤作业。

3. Relatively coarse materials are normally used and clear filtrate is not obtained until the initial layers of cake are formed, the initial cloudy filtrate being recycle.

通常使用相对较粗的物料，所以到初始滤饼层形成时才会得到澄清的滤液，而初始的浑浊滤液需返回处理。

4. Cake filters are the type most widely used in mineral processing, where the recovery of large amounts of solids from fairly concentrated slurries is the main requirement.

滤饼式过滤机是矿物加工中使用最广泛的类型，其主要作用是从较浓的矿浆中回收大量固体。

5. The filter press is closed by means of a screw or hydraulic piston device and compression of the filter cloth between plates and frames helps to prevent leakages.

压滤机利用螺旋系统或液压柱塞装置进行封闭，压紧板和框之间的滤布，有助于防止渗漏。

6. The slurry is introduced to the empty frames of the press through a continuous channel formed by the holes in the corners of the plates and frames.

矿浆通过板框角落上小孔组成的连续通道进入压滤机的空框架中。

7. The method of heating may be either direct, in which case the hot gases pass through the material in the dryer, or indirect, where the material is in an inner shell, heated externally by hot gases.

干燥方法可以是直接干燥，在这种情况下热气体直接穿过烘干机中的物料；也可以是间接干燥，该方法使物料位于内壳中，由热气体从外部加热。

Exercises

1. Answer the following questions according to the text.

(1) What is filtration?

(2) What are the two basic modes of filter operation?

(3) What are the advantages of filtration under pressure over vacuum?

(4) Which type of filters are the most widely used in mineral processing

(5) What are the methods of heating?

2. Translate the following phrases into English.

(1) 多孔介质　(2) 真空过滤机　(3) 恒压过滤　(4) 盘式过滤机

(5) 恒速过滤　(6) 滤液　(7) 压滤机　(8) 滤饼

3. Translate the following paragraphs into Chinese.

Paragraphs 1 and 2 in section 1 (1) "Filtration is the process of separating solids from ...".

4. Translate the following sentences into English.

(1) 过滤是一种利用多孔介质从液体中分离固体颗粒的方法。

(2) 过滤包括恒压过滤和恒速过滤两种基本模式。

(3) 压滤过程中,滤饼在过滤介质表面逐渐积聚,致使液体流动阻力不断增大。

(4) 筒式过滤机和盘式过滤机是矿物加工中最常见的过滤设备,两者的工作原理相同。

(5) 过滤介质应有高的机械强度、耐腐蚀且对滤液的阻力尽可能小。

9.3 Tailings Disposal(尾矿处理)

1. Tailings disposal methods

The disposal of mill tailings is a major environmental problem, which is becoming more serious with the increasing exploration for metals and the working of lower-grade deposits. Apart from the visual effect on the landscape of tailings disposal, the major ecological effect is usually water pollution, arising from the discharge of water contaminated with solids, heavy metals, mill reagents, sulphur compounds, etc. Waste must therefore be disposed of in both an environmentally acceptable and, if possible, economically viable manner.

Mill tailing usually transported and disposed as a slurry with high moisture content. Early disposal methods included discharge of tailings into rivers and streams, which is still practiced at some mines, and the dumping of coarse dewatered tailings on to land. Due to the damage caused by such methods, and the much finer grinding necessary on most modern ores, other techniques have been developed. The most satisfactory way of dealing with tailings is to make positive use of them, such as reprocessing in order to recover additional values, or to use them as a useful product in their own right, e. g. , the use of coarse (20～30 mm) DMS float as railway ballast and aggregate.

In underground mines, it is common practice that returning the coarser fraction of the mill tailings underground to fill the mined-out areas. This method has been used since the beginning of the century in South Africa's gold mines. Back-filling worked-out stopes reduces the volume of tailings which must be impounded on the surface, but not all tailings are suited as back-fill materials. It is invariably necessary to deslime the tailings, the resultant slimes, which may account for up to 50% of the total weight requiring surface disposal. Some tailings swell or shrink after the fill has been placed, and some have the useful property of being self-cementing, which removes the necessity of adding cement to the back-fill, which is common practice prior to placement underground.

Tailings do not have to be stored in ponds or sent as slurries into oceans, rivers or streams. There is a growing use of the practice of dewatering tailings using vacuum or pressure filters so the tailings can then be stacked. Advances in the disposal of tailings using semidry or dry techniques offer a number of advantages over the wet disposal methods. Dry disposal techniques require that tailings be thickened or dewatered prior to disposal. The dried tailings can then be disposed by dry stacking, thickened tailings disposal or paste fill for back-filling underground mines. These are all schemes that improve water and reagent recovery and decrease tailings volumes and footprint, which greatly assists site rehabilitation. Although semi-dry or dry disposal of tailings has benefits, these techniques are not as capital cost-effective as the more traditional wet disposal of tailings and require a detailed understanding of the rheology and transport of the dried tailings.

For operations that are close to the sea, submarine tailings disposal is an alternative to conventional tailings disposal provided the governmental regulations permit disposal in such a manner. The basic submarine tailings disposal design comprises a tailings line to a de-aeration/mixing chamber, with a seawater intake line, and discharge to location and depth allowing gravity flow of a coherent density to the final sedimentation area. Such systems can place mine tailings at locations and depths constraining environmental impact to restricted areas of the seabed and deep water turbidity. This form of tailings disposal attracts considerable attention from environmental groups as the final disposal of the tailings is not in a controlled impoundment but is released directly into the lower levels of the ocean and can therefore affect the deep sea ecosystem.

Up till now, the most widely used method is to contain the tailings within a purpose-built dam.

2. Tailings dams

The design, construction, and operation of tailings dams is rapidly becoming a major consideration for most new mining developments, as well as for many existing operations.

The disposal of tailings adds to the production costs, so it is essential to make disposal as cheap as possible. It is economically advantageous to site the impoundment close to the mine, but this imposes limits on site selection. The ground underlying the dam must be structurally sound and able to bear the weight of the impoundment. If such a site cannot be found close to the mine, it may be necessary to pump the tailings, at a high pulp density, to a suitable location.

Tailings dams may be built across river valleys, or as curved, or multisided dam walls on valley sides, this latter design facilitating drainage. On flat or gently sloping ground, lagoons are built with walls on all sides of the impoundment.

There are three mainly types of tailings embankment which is generally determined by the local seismic activity, water clarification, tailings natures, geological and hydrological conditions and the environmental conditions, etc.

(1) Upstream tailings dam

The upstream method has been employed for many tailings dam because of the low cost. In this method, a small starter dam is placed at the extreme downstream point (Figure 9.14) and the dam wall is progressively raised on the upstream side. The tailings are discharged by spigoting off the top of the dyke and, when the initial pond is nearly filled, the dyke is raised and the cycle repeated. Various methods are used to raise the dam material: material may be taken from the dried surface of the previously deposited tailings and the cycle repeat, or more commonly the wall may be built from the coarse fraction of the tailings, separated by cyclones, or spigots, the fines being directly into the pond.

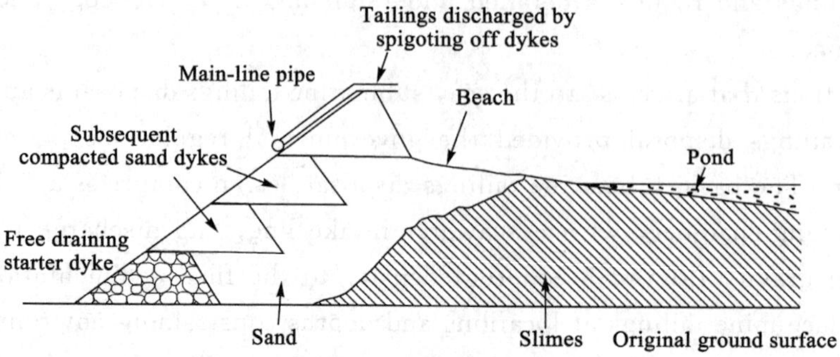

Figure 9. 14 Upstream tailings dam

The method suffers from the disadvantage that the dam wall is built on the top of previously deposited unconsolidated slimes retained behind the wall. There is a limiting height to which this type of dam can be built before failure occurs and the tailings flow out, often with disastrous results.

(2) Downstream tailings dam

The downstream method is a relatively new development which has evolved as a result of efforts to devise methods for constructing larger and safer tailings dams (Figure 9. 15). It is essentially the reverse of the upstream method, in that as the dam wall is raised, the centerline shifts downstream and the dam remain founded on coarse tailings. Downstream dam building is the only method that permits design and construction of tailings dams to acceptable engineering standards.

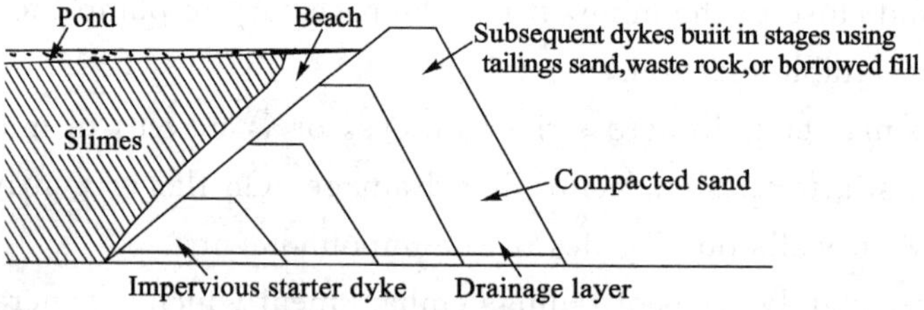

Figure 9. 15 Downstream tailings dam

All tailings dams in seismic areas, and all major dams, regardless of their location, should be constructed using some form of the downstream method. The major disadvantage of the technique is the large amount of sand required to raise the dam wall. It may not be possible, especially in the early stages of operation, to produce sufficient sand volumes to maintain the crest of the tailings dam above the rising pond levels. In such cases, either a higher starter dam is required or the sand supply must be augmented with borrowed fill, such procedures increasing the cost of tailings disposal.

(3) Centre-line tailings dam

The center-line method (Figure 9. 16) is a variation of that used to construct the downstream dam and the crest remains in the same horizontal position as the dam wall is raised. It has the advantage of requiring smaller volumes of sand-fill to raise the crest to any given height. The dam can thus be raised more quickly and early and there is less trouble keeping it ahead of the tailing pond during the early stages of construction. Care, however, must be exercised in raising the upstream face of the dam to ensure that unstable slopes do not develop temporarily.

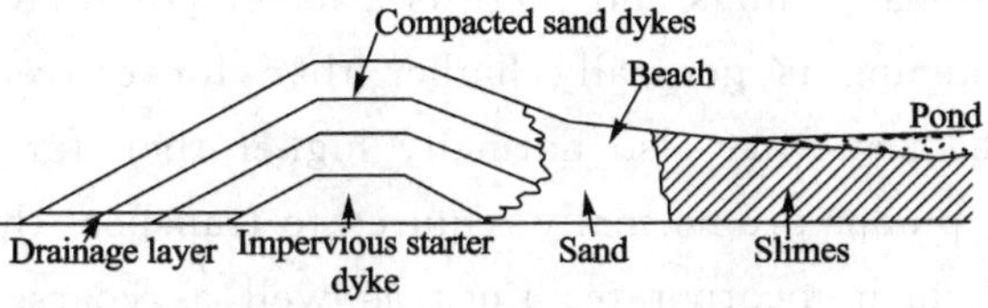

Figure 9.16 Centre-line tailings dam

Erosion of dams due to wind and rain can affect the stability and produce environmental problems. Many methods are used to combat this, such as vegetation of the dam banks and chemical stabilization to form an air and water-resistant crust.

The most serious problem associated with the disposal of tailings is the release of polluted water. Recycling of water in tailings disposal process is becoming more important due to the environmental pressures. As much water as possible must be reclaimed from the tailings pond for re-use in the mill and the volume of fresh make-up water used be kept to a minimum. The main disadvantage of water reclamation is the recirculation of pollutants to the mill, which can interfere with processes such as flotation. Water treatment can usually overcome this, at little, or no extra cost, because similar treatment would be required for the effluent discharge in any case. A number of wastewater treatment techniques are available, e. g. , active carbon adsorption, biological oxidation of organics, removal of ionic species by ion exchange resins, and relatively new techniques such as reverse osmosis.

3. Backfilling mined out areas

(1) Paste tailings and paste backfill

Disposal into exhausted open pits is generally a straightforward operation, while disposal into underground voids is more complex. A common modern approach is to mix a certain quantity of tailings with waste aggregate and cement, creating a product that can be used to backfill underground voids and stopes. A common term for this is HDPF—high density paste fill. HDPF is a more expensive method of tailings disposal than pond storage, however it has many other benefits-not just environmental but it can significantly increase the stability of underground excavations.

Paste tailings is a modification to the conventional methods of disposal of tailings (pond storage). A conventional tailings slurry has a low solids and a relatively high water content (normally ranging from 20% to 60% solids) and when deposited into the tailings pond the solids and liquids separate. In paste tailings the solids of the tailings slurry is increased through the use of paste thickeners to produce a product where the minimal separation of water and solids occurs and the material is deposited into a storage area as a paste (with a consistency somewhat like toothpaste). Paste tailings has the advantage that more water is recycled in the processing plant and therefore the process is more water efficient than conventional tailings and there is a lower potential for seepage. However, the cost of the thickening is generally higher than for conventional tailings and the pumping costs for the paste are also normally higher than for conventional tailings as positive displacement pumps are normally required to transport the tailings.

Paste backfill, which incorporates fines as well as coarse particles, such as the complete particle size distribution of mill tailings, is a more recent development. Paste is produced by filtration or thickening. If backfill strength is required, the paste is mixed with a binder, such as Portland cement and water is added to adjust the slump as necessary for pumping and pipeline flow.

The slump required in the mine is a function of the mining plan and cycle time for accessing stopes. Typically, a slump of 6～10 inches is required, and if the tailings are too fine to produce the required strength with binder addition the tailings may be mixed with alluvial sand or crushed rock to produce a wider particle size distribution and higher strengths.

A paste backfill can be engineered to meet the mine requirements for strength by working with the parameters of paste solids content and the amount of binder addition. Some of the advantages of paste backfill systems over hydraulic sandfill systems are identified as follows:

① Greater strengths are generated with less cement.

② Decanting water from stopes is unnecessary.

③ Generally total tailings can be used.

④ Lower mining cycle time is possible because strength is achieved earlier.

⑤ Barricades may be eliminated or simplified with low slump paste.

(2) Preparation of paste backfill with a paste thickener

Thickeners, including paste thickeners, are fundamentally continuous processes. In many cases the paste is intermittently required in the mine, depending on the mining plan. Development of procedures and mechanical design to meet a requirement for discontinuous operation has been a significant development for a paste thickener in mine backfill. In these cases the paste thickener is operated as a semi-batch process where the paste may be stored for several hours or even days depending on the mining cycle time. Figure 9.17 shows a typical flowsheet for preparing paste backfill with addition of a binder.

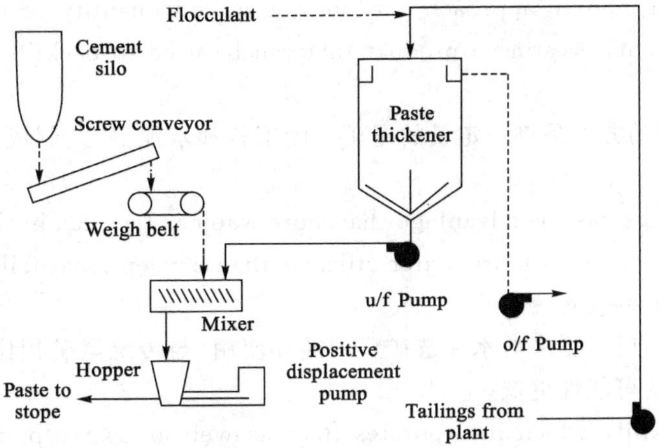

Figure 9.17 Preparing backfill with a paste thickener

Notes

1. Early disposal methods included discharge of tailings into rivers and streams, which is still practiced at some mines, and the dumping of coarse dewatered tailings on to land.

早期的处置方法包括将尾矿排入河流和溪流(某些矿山仍在使用这种做法),以及将脱水后的粗粒尾矿堆存在地表。

2. In underground mines, it is common practice that returning the coarser fraction of the mill tailings underground to fill the mined-out areas.

在地下矿山,常见的做法是将选矿尾矿中的粗粒级返回到地下以充填采空区。

3. Dry disposal techniques require that tailings be thickened or dewatered prior to disposal.

干法处置技术要求尾矿在处理前进行浓缩或脱水。

4. The design, construction, and operation of tailings dams is rapidly becoming a major consideration for most new mining developments, as well as for many existing operations.

对大多数新建矿山和许多现有矿山,尾矿坝的设计、建造和运行都是首要考虑的问题。

5. There are three mainly types of tailings embankment which is generally determined by the local seismic activity, water clarification, tailings natures, geological and hydrological conditions and the environmental conditions, etc.

尾矿坝主要有三种类型,一般由当地地震活动、水的澄清度、尾矿的性质、地质和水文条件以及环境条件等决定。

6. It is essentially the reverse of the upstream method, in that as the dam wall is raised, the centerline shifts downstream and the dam remains founded on coarse tailings.

下游式尾矿坝本质上与上游式筑法相反,随着坝堤的提高,中心线向下游方向移动,而坝体依然是建立在粗粒尾矿之上。

7. A common modern approach is to mix a certain quantity of tailings with waste aggregate and cement, creating a product that can be used to backfill underground voids and stopes.

现在一种常见的方法是将一定量的尾矿与废填料和水泥混合,制成可用于回填地下空隙和采场的产品。

8. Paste tailings has the advantage that more water is recycled in the processing plant and therefore the process is more water efficient than conventional tailings and there is a lower potential for seepage.

膏状尾矿的优点是,更多的水在选矿厂中循环使用,与传统尾矿相比,该工艺的用水效率更高,并且渗漏的可能性更低。

9. Paste backfill, which incorporates fines as well as coarse particles, such as the complete particle size distribution of mill tailings, is a more recent development.

采用选矿厂尾矿的全粒级那样既有细颗粒又有粗颗粒的膏体充填是最近发展的一种技术。

Exercises

1. Answer the following questions according to the text.

(1) Why is the disposal of mill tailings a major environmental problem?

(2) What are the methods of tailings disposal?

(3) What are the three mainly types of tailings embankment?

(4) What is HDPF?

(5) What is tailings paste?

(6) What is the disadvantage of upstream dam?

2. Translate the following phrases into English.

(1) 尾矿处理　(2) 上游式尾矿坝　(3) 中线式尾矿坝　(4) 下游式尾矿坝

(5) 回填采空区　(6) 膏体充填　(7) 初期坝　(8) 膏体浓密机

3. Translate the following paragraphs into Chinese.

Paragraphs 1, 2 and 3 in section 1 "The disposal of mill tailings is a major environmental problem ...".

4. Translate the following sentences into English.

(1) 选矿厂尾矿的处理是一个主要的环境问题,随着金属勘探的增加和低品位矿床的开采,这个问题变得越来越严重。

(2) 在地下采矿中,通常采用粗粒级尾矿充填采空区。

(3) 下游式尾矿坝是唯一一种尾矿坝设计和建造均符合工程标准的方法。

(4) 风雨对于尾矿坝的腐蚀能够影响坝的稳定性,并可能造成环境污染问题。

(5) 尾矿处理过程中,水的循环利用非常重要。

9.4 Paste Thickening Technology for Tailings Disposal (膏体浓缩技术在尾矿处理中的应用)

1. Introduction

Thickening process plays an important role in mineral, mining industries and waste water treatment plants. It is the most widely used technique for solid-liquid separation in mineral processing plants. Thickening is used to increase solids content by removing water for the purpose of recycling. This reduces the plant water requirement and the need to find a new waste disposal site.

Paste thickeners were introduced by British National Coal Board in the 1970s. They have the ability to deliver higher product concentration than conventional or high rate thickeners, which translates to significant improvements in water savings. This is done by utilizing a deeper sediment layer, which forms a significant level of compressible bed to augment the thickening capacity. Thickening by providing high volume of bed depth (which is the case for paste thickeners) is considered more economical than those require greater area with a rather swallow bed level (which is the case for conventional and high rate thickeners). Therefore, paste thickeners have attracted considerable attentions from the mineral processing industry.

Paste thickening is a relatively new, rapidly growing technology within the minerals industry for thickening solids to a higher concentration than achievable with conventional thickeners. The concentrated solids produced by a paste thickener have the physical attributes commonly associated with the word "paste": uniform properties throughout and flows slowly or not at all under gravity. An example of coal fines paste is shown in Figure 9. 18, and paste sample of a fine iron ore tailings is shown in Figure 9. 19.

Figure 9. 18 Coal fines paste

Figure 9. 19 Paste of fine iron ore tailings

Tailings from metal mining operations were conventionally disposed of as a slurry in a surface impoundment near a mineral processing mill. Such disposal can lead to a variety of environmental and other hazards, such as dust generation, dam failure, and leakage. Additional disadvantages may result from the requirement for large amounts of land, as well as aesthetic impacts.

The use of a thickener, without filtration, to produce a non-settling paste (also called thickened tailings) with non-Newtonian rheology has been practiced in Australia for many years for red mud disposal. The application of paste thickening technology outside of alumina is relatively new. Interest and installations in paste thickening are rapidly growing. Water shortages and environmental pressures make consideration of thickening tailings to solids concentrations within the paste range necessary in most new tailings projects. The paste thickeners at the DeBeers CTP plant in Kimberley, South Africa are an example of paste tailings disposal on the surface, also called "stacking". Nowadays, paste thickening has become an alternative to conventional thickening and filtration for water recovery and tailings disposal in minerals industry.

2. Nature of paste suspensions

The commonly used term "paste" is a suspension of solids characterized by relatively non-settling, non-segregating particles at high concentration as compared to settling slurries. Paste is generally described as a Bingham plastic, characterized by a yield stress, which is measured in units of pressure and is related to the force required to make a paste flow. The shape exhibited by a paste that is not flowing is a result of the presence of a yield stress. Yield stress can be measured directly with a laboratory viscometer. The yield stress curve is correlation of yield stress with solids concentration, as shown in Figure 9.20.

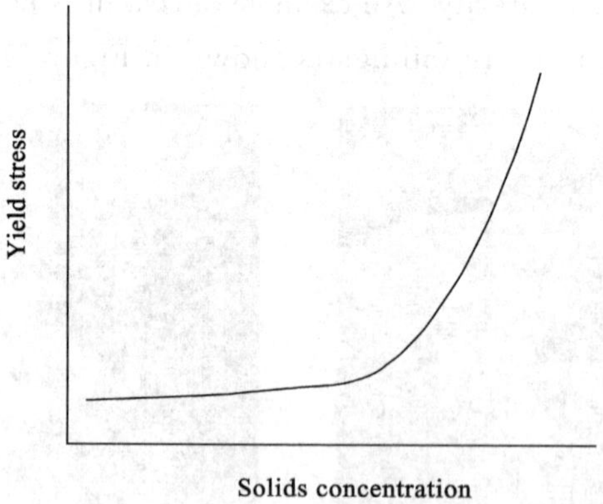

Figure 9.20 Yield stress curve

Slump, measured in units of distance, is an indirect measurement of yield stress. As shown in Figure 9.21, a cylinder is filled with sample, the cylinder is manually lifted, and the distance the paste sample "slumps" is measured. Slump and yield stress have an inverse relationship.

The yield stress and non-settling nature of paste is produced from naturally occurring clays, silts, and fine sands. The solids content for a given yield stress is determined by the specific gravity and particle size distribution of the solids. Rheological and transport characteristics of paste are mainly dependent on the content of fine particles (less than 20 micron) as shown in Figure 9. 22. The rheological character of paste is a result of the interaction of solid particles of different diameters and volumes making percent volume an appropriate way to characterize solid concentrations. However, industry convention is to use weight percent. In these terms paste concentrations can be as high as 75%~85% for base metal tailings depending on particle size and solids specific gravity.

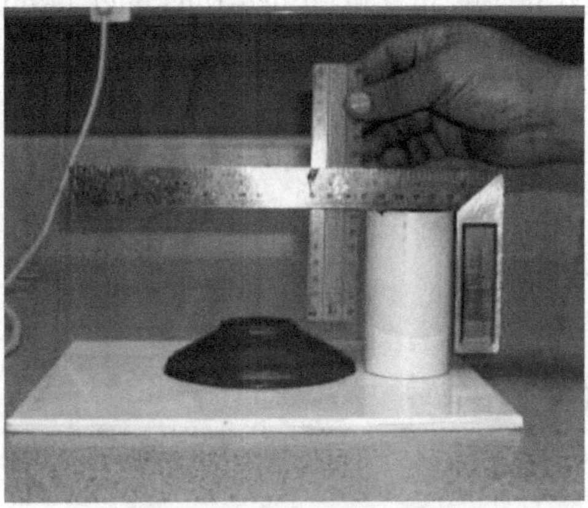

Figure 9. 21 Slump measurement

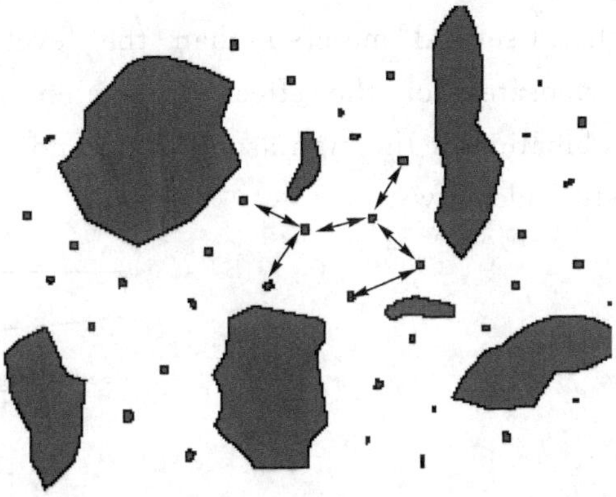

Figure 9. 22 Fine particles interact producing paste properties

The high viscosity of pastes makes pipeline design critical. Pastes are generally pseudoplastic where the viscosity decreases with shearing such as in pumping and pipe line flow. The yield stress of paste produced by a thickener, for example, may be higher than the yield stress at the end of the pipeline delivering the paste to the stack.

Tailings paste is defined as a dense, viscous mixture of tailings and water which, unlike slurries, does not segregate when not being transported. Paste has a working consistency similar to wet concrete. One of the most distinguishing properties of paste is the grain size distribution of the solids. Based on a large volume of empirical data and operational experience, it has been determined that a paste must contain at least 15 % by weight passing 20 micron to exhibit the typical paste flow properties and retain sufficient colloidal water to create a non-segregating mixture. According to Robinsky (1999), one of the pioneers of paste technology, virtually all mineral processing methodologies generate tailings amenable to paste production. When being transported either by gravity or through pumping, paste produces a plug flow, with the fine particles creating an outer annulus, thereby reducing friction. The coarse particles are forced into the center of the conduit with the finer fraction acting as the carrier. This allows for conveyance of very coarse fragments, the size of which is only limited by the pipe diameter.

3. Paste thickener design

The concept of deep cone thickening was developed in the 1960 and 70s in the British coal industry. Over the years the deep cone thickening idea has been combined with modern day flocculation techniques, feedwell design, and improvements on the design of the tank, rake and underflow discharge system to produce the modern deep cone type paste thickener. This thickener is characterized by a height to diameter ratio typically greater than one to one with unit areas [$m^2/(t/h)$] several times smaller than for conventional and high-rate thickeners (Figure 9.23). The mud bed height in a deep bed thickener is much higher (several meters) than the level in a high rate thickener which more than compensates for the effect of smaller area on underflow solids concentration. The combination of the high aspect ratio and steep cone angle also facilitates the discharge of paste underflows.

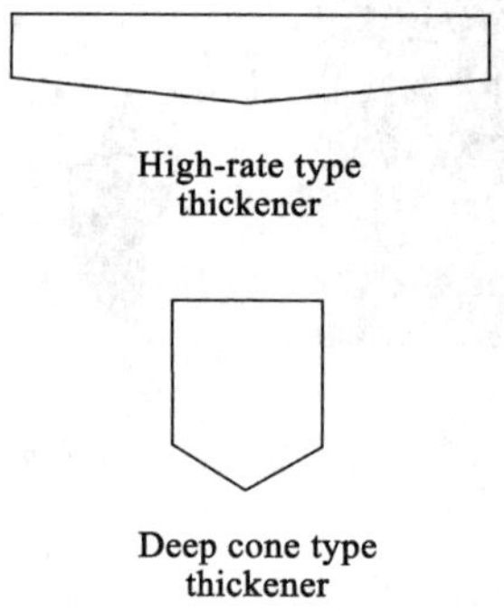

Figure 9.23 Comparison of high-rate and deep-cone thickeners

Understanding yield stress is a key parameter in paste thickener design. Under gravity a tailings paste will flow to a point dictated by the yield stress and stop. The paste must move through the thickener and be delivered to the pumping system for transportation to the application point. The operating range of a paste thickener can be located on the yield stress curve revealing the maximum underflow paste solids concentration and the minimum concentration below which the suspension will be settling slurry.

Bed depth distinguishes paste thickeners from other types of thickeners. Bed depths in a deep cone type paste thickener may be 10 m or higher, creating compression forces which drive up the underflow solids concentration. Because of these depths, the concept of unit area, has little meaning for a deep cone style paste thickener. For a given diameter and tonnage throughput, a paste thickener may have a wide range of bed depths, producing a similarly wide range of underflow densities. Because of the understanding required for relating paste rheology, bed depth, and thickener design to paste production, the number of producers of this style of thickener is limited and currently includes GL&V, Outokumpu, and WesTech.

The WesTech Deep Bed™ Paste Thickener (Figure 9.24) can thicken a tailings stream to non-settling paste at solids concentrations approaching that of a filter cake. Design characteristics of the Deep Bed™ result in maximum water recovery, maximum underflow density, and paste with minimal slump. The main features of WesTech Deep Bed™ Paste Thickener include: height-to-diameter ratios typically as high as 2 : 1, floor slope up to 45 degrees, self-diluting feedwell, diameters to over 20 m, and rake mechanism with full-length dewatering pickets. An 8 m diameter paste thickener, at the Associated Manganese (ASSMANG) ferrochrome plant in Machadodorp, South Africa, treats approximately 15 t/h of ultra-fine tailings with a D_{80} of less than 30 micron at a flow rate of 160 m^3/h, resulting in an underflow density of 1.6～1.7 t/m^3 (greater than 60% solids) with a yield stress of greater than 250 Pa.

Figure 9.24 WesTech Deep Bed™ paste thickener

4. Benefits from use of paste

Of all potential advantages associated with disposal of tailings in paste form, the environmental benefits are the most promising. Use of paste technology may provide an avenue for minimizing or even eliminating various environmental issues.

(1) Surface stacking of paste

Surface stacking is the disposal of tailings as a paste (commonly referred to as "thickened tailings") on the surface of the ground, as a non-settling, non-segregating suspension of solids with minimal water release. There are various methods of accomplishing the stacking depending on the topography of the disposal site and the rheology of the paste. For slopes up to 5%, paste thickeners can be used.

Figure 9.25 shows how a paste thickener is used to produce the disposal stack. The paste thickener's underflow concentration must meet the requirements of the disposal site (stack slope) and still allow the paste to be pumped out of the thickener. Depending on the distance from the thickener to the disposal site, centrifugal or positive displacement pumps are used for transporting the tailings.

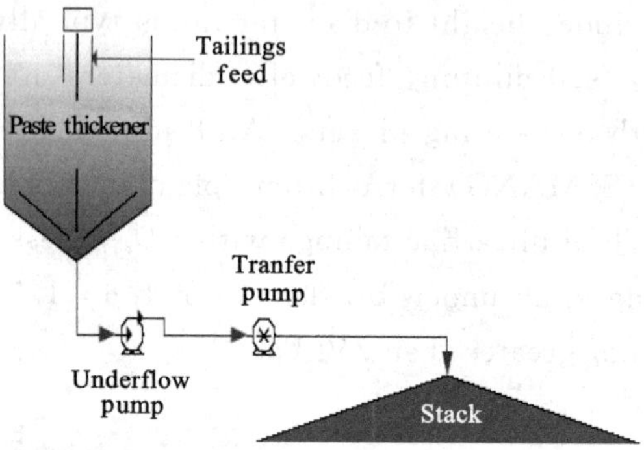

Figure 9.25 Paste thickener flowsheet for stacking

Stacking as a method of tailings disposal, offers significant advantages over ponding. The most obvious benefit is the smaller impoundment area. In addition, there is less water in the pond, improved water or chemical recovery, less groundwater contamination and easier final reclamation of the site.

(2) Paste backfill

Backfill is employed to reduce surface environmental impacts by storing tailings underground. With the use of binders, such as cement, paste backfill also becomes a process for mine productivity improvement by mining methods such as undercut-and-fill.

In the conventional mine paste backfill process a filter, proceeded by a high-rate slurry thickener to reduce the hydraulic loading, is typically used to produce paste backfill. The filter cake is discharged to a batch mixing hopper or continuous mixer where a binder is added to produce the required strength underground. The cemented paste is then pumped below ground.

There is a range of paste solids content required for any backfill application. A thickener/filter combination can be used for the entire range, by diluting the filter cake as necessary to achieve the paste properties required in the mine such as solids concentration and rheology (measured by slump). A paste thickener is an option for mine backfill applications. For mine paste backfill the paste thickener produces a paste at a required solids concentration and slump. The underflow is pumped with centrifugal pumps to the mixer as shown in Figure 9.26.

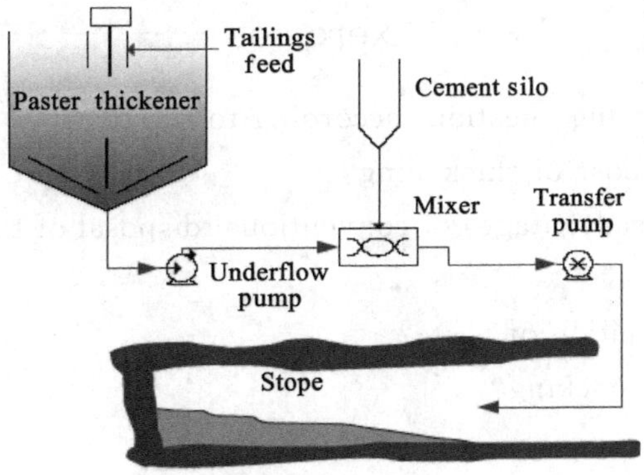

Figure 9. 26 A Paste thickener producing mine backfill

Notes

1. Water shortages and environmental pressures make consideration of thickening tailings to solids concentrations within the paste range necessary in most new tailings projects.

缺水和环境压力使大多数新的尾矿项目必须考虑将尾矿浓缩到膏体范围的固体浓度。

2. The commonly used term "paste" is a suspension of solids characterized by relatively non-settling, non-segregating particles at high concentration as compared to settling slurries.

通常使用的术语"膏体"是一种固体悬浮液,与沉降的矿浆相比,其特点是在高浓度下颗粒相对不沉降、不离析。

3. The rheological character of paste is a result of the interaction of solid particles of different diameters and volumes making percent volume an appropriate way to characterize solid concentrations.

膏体的流变特性是不同直径和体积的固体颗粒相互作用的结果,因此体积百分比是表征固体浓度的适当指标。

4. Tailings paste is defined as a dense, viscous mixture of tailings and water which, unlike slurries, does not segregate when not being transported.

尾矿膏体是一种尾矿和水的浓的、黏性混合物,与矿浆不同,在不输送时也不会离析。

5. The concept of deep cone thickening was developed in the 1960 and 70s in the British coal industry.

深锥浓密机的概念是 20 世纪 60 年代和 70 年代在英国煤炭业发展起来的。

6. With the use of binders, such as cement, paste backfill also becomes a process for mine productivity improvement by mining methods such as undercut-and-fill.

随着水泥等黏合剂的使用,膏体充填也成为采矿方法(如暗挖回填)中提高矿山生产力的一种工艺。

Exercises

1. Answer the following questions according to the text.

(1) What is the purpose of thickening?

(2) What are the disadvantages of conventional disposal of tailings from metal mining operations?

(3) What is the definition of paste?

(4) What is surface stacking?

(5) What is backfill?

2. Translate the following phrases into English.

(1) 膏体浓缩　(2) 溃坝　(3) 屈服应力　(4) 膏体浓密机

(5) 深锥浓密机　(6) 黏性混合物　(7) 假塑性　(8) 坍落度

3. Translate the following paragraphs into Chinese.

Paragraphs 1 and 2 in section 1: "Thickening process plays an important ... ".

4. Translate the following sentences into English.

(1) 浓缩技术在矿物、采矿业和废水处理厂中发挥着重要作用。

(2) "膏体"是一种固体悬浮液，其特点是在高浓度下颗粒不沉降、不离析。

(3) 尾矿膏体是一种尾矿和水的浓的、黏性混合物。

(4) 采用回填法将尾矿储存在地下，以减少对地表环境的影响。

10 ORE SORTING AND E-WASTE RECYCLING (矿石分拣与电子垃圾回收)

10.1 Ore Sorting(矿石分拣)

1. Introduction

Ore sorting is the original concentration process, having probably been used several thousand years ago. It involves the appraisal of individual ore particles and the rejection of those particles that do not warrant further treatment.

Hand sorting has declined in importance due to the need to treat large quantities of low-grade ore which requires extremely fine grinding. Hand sorting of some kind, however, is still practiced at some mines, even though it may only be the removal of large pieces of timber, tramp iron, etc. from the run-of-mine ore.

Electronic ore-sorting equipment was first produced in the late 1940s, and although its application is fairly limited, it is an important technique for the processing of certain minerals.

Sorting can be applied to pre-concentration, in which barren waste is eliminated to reduce the tonnage reporting to the downstream concentration processes, such as in uranium or gold ore sorting, or to the production of a final product, such as in diamond sorting. The ore must be sufficiently liberated at a coarse size (greater than 5～10 mm) to allow barren waste to be discarded without significant loss of value. Preconcentration by sorting is seen as a method of improving the sustainability of mineral processing operations by reducing the consumption of energy and water in grinding and concentration, and achieving more benign tailings disposal.

2. Electronic sorting principles

Many rock properties have been used as the basis of electronic sorting, including reflectance and colour in visible light (magnesite, limestone, base metal and gold ores, phosphates, talc, coal), ultraviolet (scheelite), natural gamma radiation (uranium ore), magnetism (iron ore), conductivity (sulphides), and X-ray luminescence (diamonds). Infrared, Raman, microwave attenuation, and other properties have also been tested.

Electronic sorters inspect the particles to determine the value of some property (e. g., light reflectance) and then eject those particles which meet some criterion (e. g., light vs dark rocks). Either valuables or waste may be selected for ejection. It is essential, therefore, that a distinct difference in the required physical property is apparent between the valuable minerals and the gangue.

The particle surfaces must be thoroughly washed before sorting, so that blurring of the signal does not occur and, as it is not practical to attempt to feed very wide rock size ranges to a single machine, the feed must undergo preliminary sizing. The ore must be fed in a monolayer, as display of individual particles to the sorting device must be affected.

(1) Photometric sorting

Photometric sorting is the mechanised form of hand-sorting, in which the ore is divided into components of differing value by visual examination. The basis of the photometric sorter (Figure 10. 1) is a laser light source and sensitive photomultiplier, used in a scanning system to detect light reflected from the surfaces of rocks passing through the sorting zone. Electronic circuitry analyses the photomultiplier signal, which changes with the intensity of the reflected light and produces control signals to actuate the appropriate valves of an air-blast rejection device to remove certain particles selected by means of the analysing process. The sorter is fully automatic and can be attended by one operator on a part-time basis. Typical throughput per machine ranges from 25 t/h for a −25+5 mm feed to 300 t/h for a −300+80 mm material.

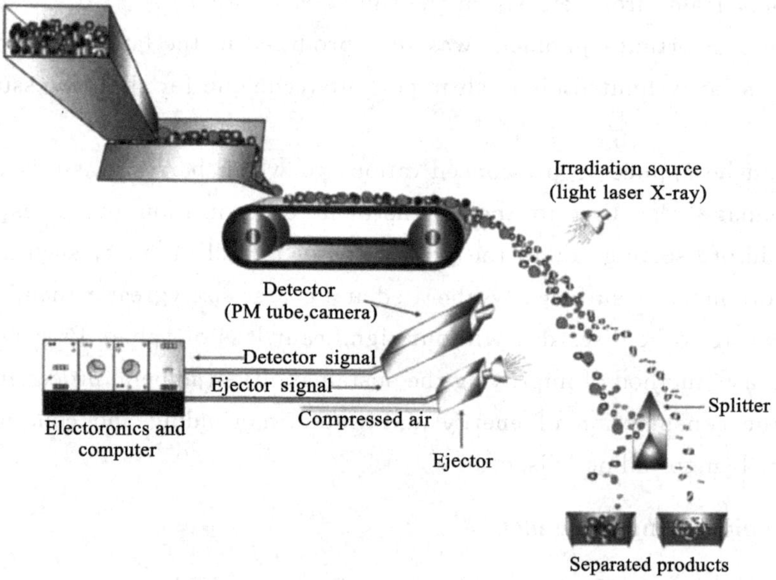

Figure 10. 1 Principles of photometric sorting

The modem photometric sorter is typified by the UltraSort UFS 120, which is used in the processing of magnesite, feldspar, limestone, and talc. Ore passes from a vibrating feeder to high pressure water sprays and counterweight feeder where water is removed and the rocks are accelerated to form a monolayer. The rock layer, 0. 8～1. 2 m wide, is scanned by a laser beam at up to 4,000 times per second, and the reflection analyzed in less than 0.25 μs by photomultiplier tubes and high-speed parallel processors operating in excess of 80 MB/s. One or more of 120 air ejectors are fired to divert the value or waste past a cutter and into the accept/reject bins. As the position of the rock is accurately identified, and the ejector firing duration is less than 1 ms, the sorter can operate very selectively. A range of lasers of different wavelengths can be used, from ultraviolet to infrared, to achieve optimum discrimination.

(2) X-ray sorting

Electronic sorting has been employed in diamond recovery since the 1960s, initially using simple optical sorters and more recently machines based on the fact that diamonds luminesce when irradiated by X-rays. X-ray sorters are used in almost all diamond operations for the final stages of recovery after the ore has been concentrated by DMS. They replace grease separation which is now used only in rare cases where the diamonds luminesce weakly. Luminescence is a more consistent diamond property than oleophilicity, and sorters are more secure than grease belts or tables. In an early dry X-ray sorter, the DMS concentrates are exposed to a beam of X-rays in free fall from a conveyor belt, the luminescence detected by photomultiplier tubes and the diamonds ejected by air ejectors. Both dry and wet X-ray machines are now available, and the process is usually multistage to ensure efficient rejection of waste with very high diamond recoveries.

(3) Radiometric sorting

Radiometric sorting has been used to preconcentrate uranium ore in South Africa, Namibia, Australia, and Canada. A sorter installed at the Rossing Mine in Namibia detects gamma radiation from the higher grade ore pieces using scintillation counters comprising NaI crystals and photomultiplier tubes mounted under the belt (Figure 10.2). Lead shielding is used to achieve improved resolution of detection. A laser-based optical system similar to that used in photometric sorters is used to determine rock position and size for ejection, and can be adapted to determine additional optical characteristics of the rocks.

(4) Other sorting methods

Several other physical properties of ores and minerals have been exploited in a range of sorting machines.

Neutron absorption separation has been used for the sorting of boron minerals. The ore is delivered by a conveyor belt between a slow neutron source and a scintillation neutron detector. The neutron flux attenuation by the ore particles is detected and used as the means of sorting. The method is most applicable in the size range 25～150 mm. Boron minerals are easy to sort by neutron absorption since the neutron capture cross-section of the boron atom is very large compared with those of common associated elements and thus the neutron absorption is almost proportional to the boron content of the particles.

Photoneutron separation is recommended for the sorting of beryllium ores, since when a beryllium isotope in the mineral is exposed to gamma radiation of a certain energy, a

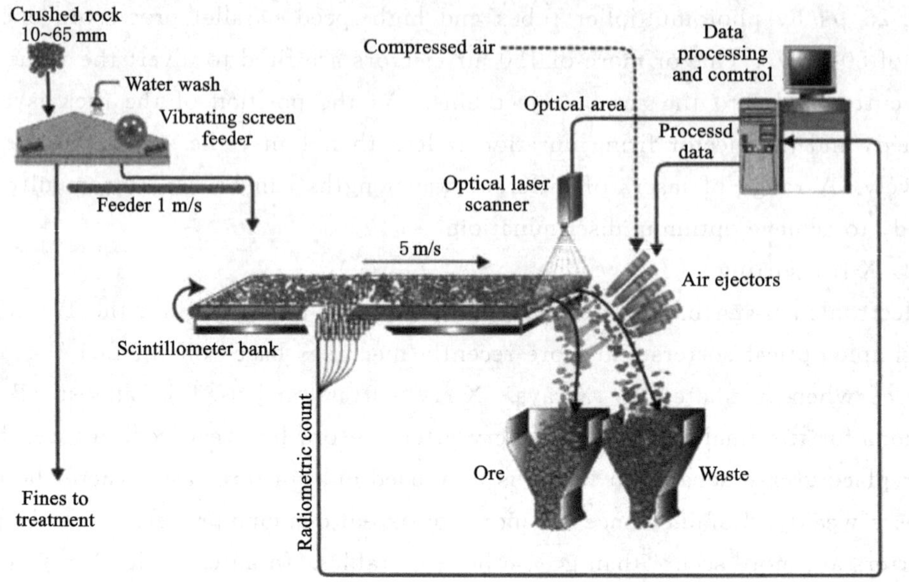

Figure 10.2 The Ultrasort radiometric sorter (Courtesy Ultrasort Pty Ltd)

photoneutron is released and this may be detected by scintillation or by a gas counter.

The RTZ Ore Sorters Model 19 sorter measured conductivity and magnetic properties and had application to a wide variety of ores including sulphides, oxides, and native metals. The machine treated 25～150 mm rocks at up to 120 t/h. Such systems employ a tuned coil under the belt which is influenced by the conductivity and/or magnetic susceptibility of the rocks in its proximity.

Outokumpu developed the "Precon" sorter and installed it at its Hammaslahti copper mine. It used gamma scattering analysis to evaluate the total metal content, and had a capacity of 7 t/h for 35 mm lumps rising to 40 t/h for 150 mm lumps. It preconcentrated primary crushed ore, rejecting about 25% as waste grading 0.2% copper, compared with an average feed grade of 1.2%.

Microwave attenuation has been used to sort diamond-bearing kimberlite from waste rock. The development was notable for the first use of high-speed pulsed water ejectors.

Equipment to sort asbestos ore has been developed. The detection technique was based on the low thermal conductivity of asbestos fibres and used sequential heating and infra-red scanning to detect the asbestos seams.

A similar machine was installed at King Island Scheelite in Tasmania, where the scheelite was sensed by its fluorescence under ultra-violet radiation.

3. Intelligent disposal of gangue from coal

To improve the quality of coal, save costs and simplify the follow-up process, effectively separating gangue from lump coal before its utilization is economically viable. Presently, manual selection based on the differences of color and texture between coal and gangue is widely used in coal preparation plant (Figure 10. 3). Such manually selecting method to remove bulk gangue is time-consuming and labor-intensive with poor working environment. Therefore, automatic separation of gangue from coal has been developed in industry production. In recent years, X-ray or γ-ray transmission method, three-dimensional laser scanning and machine image processing method have been applied to automatically separate gangue from coal.

Figure 10. 3 Manual selection of coal

(1) Sorting based on X-ray/γ-ray transmission

Traditional coal preparation methods include heavy media, jigging, and flotation. These methods have been applied for a long history. All of them, however, have a common feature is that high water consumption and needs disposal of sewage. While, the X-ray coal preparation method does not require water in the process of coal gangue separation and has the advantages of low cost and no pollution.

Intelligent dry type separator is a coal sorting equipment based on the advanced imaging technology and artificial intelligence (AI) technology. The coal and gangue are automatically analyzed and identified by X-ray imaging system, which can obtain the material characteristics; and then the high-pressure air is controlled to blow the target materials away. Then the full-automatic raw coal separation is realized. Based on X-ray and fiber optic sensor technology, the sorting system mainly composed of X-ray source module, detector module, fiber optic sensor module, pneumatic regulating valve sorting device, feeding device, conveying device, etc (Figure 10. 4). Besides, an intelligent dry sorting system typed GDRT using γ-ray as gangue detection sensor and high-pressure air gun as actuator have been successfully applied in some coal mining companies in China.

(2) Machine vision applied in coal and gangue separation

Although the X-ray or γ-ray transmission method is an advanced approach in this

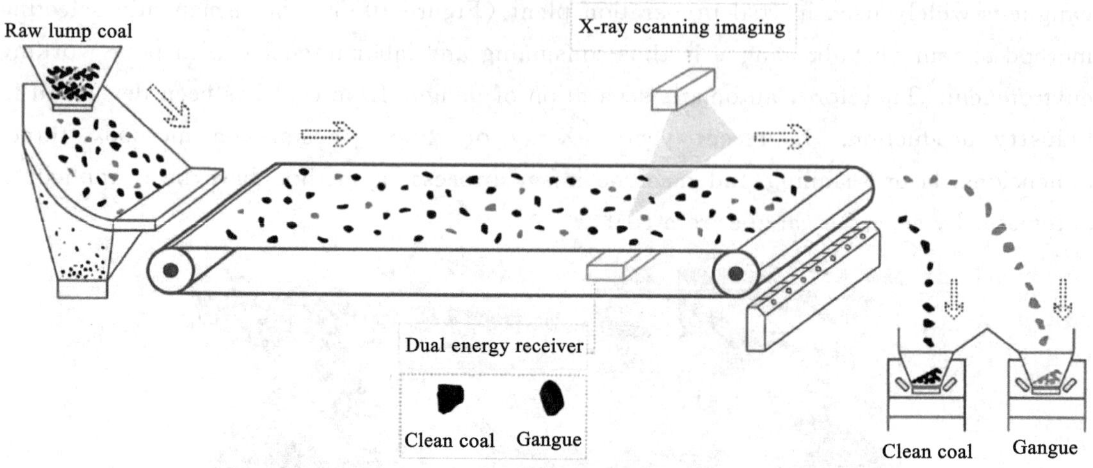

Figure 10.4 The working system of X-ray sorting

context for separating coal and gangue, its application is also limited because of the need for complicated equipment and an associated radiation hazard. In response to the above problems, researchers have begun to explore the application of machine vision for identifying gangue from coal, because it is a real-time and contactless inspection, offering low costs, simple operation, and strong environmental adaptability.

Machine vision uses cameras instead of human eyes for collecting information, computers instead of human brains for analysis, and controls on-site execution equipment to remove gangue. The camera converts the raw coal into an image signal, which is passed to the image recognition system, and the features (color features and/or texture features) are extracted according to the image information, and finally the recognition judgment is made. Based on machine vision technology, the intelligent recognition system mainly composed of image collection unit, data operation unit, sorting execution unit, sorting control unit, feeding device, conveying device and other structures, as shown in Figure 10.5.

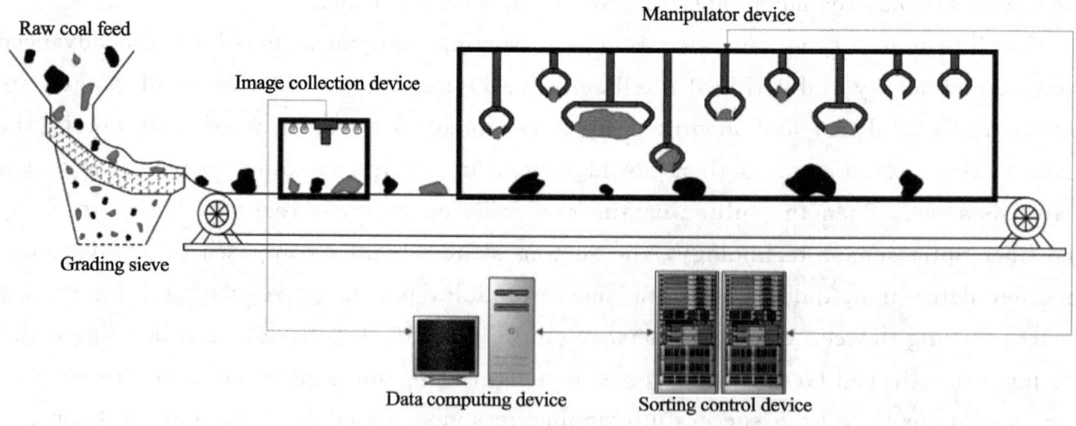

Figure 10.5 Coal gangue intelligent recognition system based on deep learning

Notes

1. Ore sorting is the original concentration process, having probably been used several thousand years ago.

矿石拣选是最原始的选矿工艺,可能在几千年前就已经开始使用了。

2. Sorting can be applied to pre-concentration, in which barren waste is eliminated to reduce the tonnage reporting to the downstream concentration processes, such as in uranium or gold ore sorting, or to the production of a final product, such as in diamond sorting.

拣选可应用于预选,去除低品位尾矿可以减少流向下游分选作业的处理量,例如铀或金矿石的拣选;或者用于产出最终产品,例如金刚石的拣选。

3. Many rock properties have been used as the basis of electronic sorting, including reflectance and colour in visible light (magnesite, limestone, base metal and gold ores, phosphates, talc, coal), ultraviolet (scheelite), natural gamma radiation (uranium ore), magnetism (iron ore), conductivity (sulphides), and X-ray luminescence (diamonds), Infrared, Raman, microwave attenuation, and other properties have also been tested.

岩石的许多性质已被用作电子拣选的基础,包括在可见光中的反射率和颜色(菱镁矿、石灰石、贱金属矿石和金矿石、磷酸盐、滑石、煤)、紫外辐射(白钨矿)、天然伽马射线辐射(铀矿石)、磁性(铁矿石)、导电性(硫化物)和 X 射线发光(金刚石)。红外、拉曼、微波衰减和其他特性也进行过试验。

4. Photometric sorting is the mechanised form of hand-sorting, in which the ore is divided into components of differing value by visual examination.

光度分选是人工拣选的机械化形式,该方法通过视觉检查将矿石分成价值不同的组分。

5. A laser-based optical system similar to that used in photometric sorters is used to determine rock position and size for ejection, and can be adapted to determine additional optical characteristics of the rocks.

一个类似于光度分选机中使用的激光光学系统被用来确定需要排除的岩石位置和大小,并可用于确定岩石的其他光学特性。

6. Therefore, automatic separation of gangue from coal has been developed in industry production. In recent years, X-ray or γ-ray transmission method, three-dimensional laser scanning and machine image processing method have been applied to automatically separate gangue from coal.

因此,在工业生产中发展了煤矸石的自动分选。近年来,应用 X 射线或 γ 射线透射法、三维激光扫描和机器图像处理等方法实现煤矸石的自动分选。

7. Machine vision uses cameras instead of human eyes for collecting information, computers instead of human brains for analysis, and controls on-site execution equipment to remove gangue.

机器视觉使用摄像头代替人眼收集信息,使用计算机代替人脑进行分析,并控制现场执行设备去除矸石。

Exercise

1. Answer the following questions according to the text.

(1) What are the applications of sorting?

(2) What properties of rocks have been used as the basis of electronic sorting?

(3) What are the requirements of particle and ore feed before sorting?

(4) What are the new techniques developed in manually selecting method to remove bulk gangue?

(5) What are the processes of coal gangue intelligent recognition system based on deep learning?

2. Translate the following phrases into English.

(1) 矿石拣选	(2) 手选	(3) 电子拣选	(4) 光度分选
(5) 放射性分选	(6) 自动排矸	(7) 煤矸分选	(8) 机器视觉

3. Translate the following paragraphs into Chinese.

Paragraphs 3 and 4 in section 1 "Electronic ore-sorting equipment was first produced ...".

4. Translate the following sentences into English.

(1) 由于需要处理大量的低品位矿石,需要极其细的研磨,手选的重要性已经下降。

(2) 矿石拣选涉及对单个矿石颗粒的评估,以及去除那些不值得进一步处理的颗粒。

(3) 几乎所有的金刚石分选作业都使用 X 射线分选机,用于矿石经 DMS 富集后的最后阶段的回收。

10.2 Recycling of Electronic Waste(电子垃圾的回收利用)

1. Background

(1) E-waste

E-waste comprises of waste electric and electronic equipment (WEEE) or goods which are not fit for their originally intended use. Such WEEEs may be TVs, telephones, radios, computers, printers, fax machines, DVDs, CDs, washing machines, refrigerators, dryers, vacuum cleaners, etc (Figure 10.6). The United Nations (UNs) estimate of the global E-waste production was 49 million tons in 2012 and more than 50 million tons today (Figure 10.7).

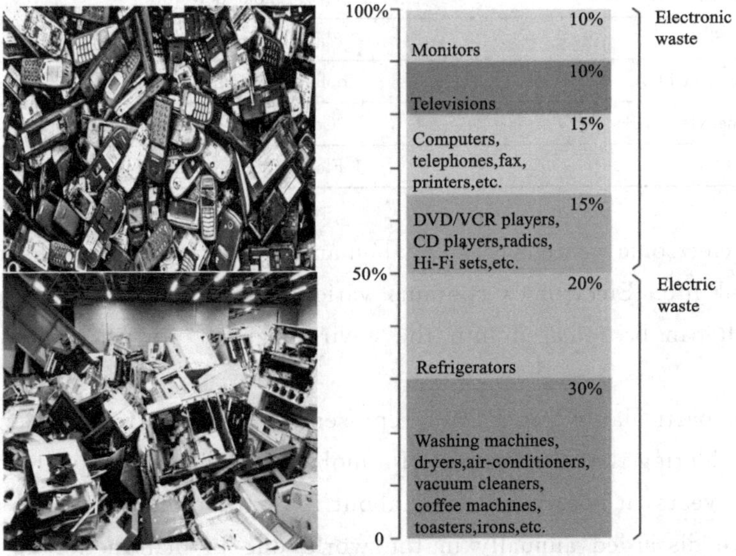

Figure 10.6 Types of E-waste

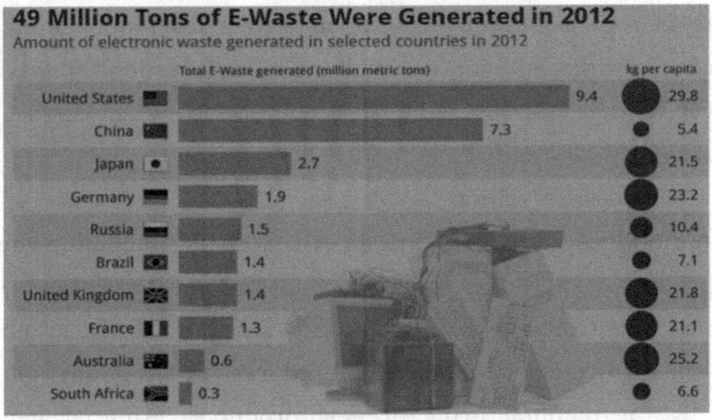

Figure 10.7 Quantity and distribution of E-waste

E-waste contains lots of valuable resources together with plenty of heavy metals and hazardous materials (Table 10.1), which are considered both an attractive polymetallic secondary source and an environmental contaminant.

Table 10.1 Hazardous substances in WEEE

Substances	Occurrence in WEEE
Lead (Pb)	CRT screens, batteries, PCBs
Mercury (Hg)	Fluorescent lambs, some alkaline, switches
Chromium VI (Cr^6)	Data tapes, floppy-discs
Barium (Ba)	Getters in CRT
Cadmium (Cd)	NiCd batteries, fluorescent layer (CRT screens), printer inks and toners
Arsenic (As)	Gallium arsenide in light emitting diodes (LED)
Americium (Am)	Smoke detectors
Antimony (Sb)	Flame retardants in plastics
Chlorofluoro carbon (CFC)	Cooling units, insulation foams
Polychlorinated biphenyls (PCB)	Condensers, transformers
PBDEs, PBBs	Flame retardants in plastics

E-waste or electronic waste is created when an electronic product is discarded after the end of its useful life. Electronics contains various toxic and hazardous chemicals and materials, which can be released into the environment if we do not dispose of them properly.

E-waste, in particular waste PCBs, represents a rapidly growing disposal problem worldwide. Considering that the lifetime of a mobile phone is approximately 1 year and of a computer 2～5 years, it is estimated that about 100 million mobile phones and 17 million of computers are discarded annually in the world due to malfunctioning equipment or because technologies become obsolete.

Unfortunately, the majority of these electronic products end up in landfills, and just 12.5% of e-waste is recycled. Many of the materials used in making these electronic devices can be recovered, reused, and recycled-including plastics, metals, and glass. In a report, Apple revealed that it recovered 2,204 pounds of gold—worth of $40 million—from recycled iPhones, Macs, and iPads in 2015.

(2) Benefits of e-waste recycling

Most electronic devices contain a variety of materials, including metals that can be recovered for future uses (Table 10.2). By dismantling and providing reuse possibilities, natural resources can be conserved and air and water pollution caused by hazardous disposal is avoided. Additionally, recycling reduces the amount of greenhouse gas emissions caused by the manufacturing of new products. Another benefit of recycling e-waste is that many of the materials can be recycled and re-used again.

Table 10.2 Various types of e-wastes and their metal content

E-waste	Fe /%	Cu /%	Al /%	Pb /%	Ni /%	Ag /%	Au /%	Pd /%
TV board scrap	28	10	10	1.0	0.3	280	20	10
PC board scrap	7	20	5	1.5	1	1,000	250	110
Mobile phone scrap	5	13	1	0.3	0.1	1,380	350	210
Portable audio scrap	23	21	1	0.14	0.03	150	10	4
DVD player scrap	62	5	2	0.3	0.05	115	15	4
Calculator scrap	4	3	5	0.1	0.5	260	50	5
PC mainboard scrap	4.5	14.3	2.8	2.2	1.1	639	566	124
Printed circuit boards scrap	12	10	7	1.2	0.85	280	110	—
TV scrap (CRT's removed)	—	.34	1.2	0.2	0.038	20	<10	<10
Electronic scrap	8.3	8.5	0.71	3.15	2.0	29	12	—
PC scrap	20	7	14	6	0.85	189	16	3
Typical electronic scrap	8	20	2	2	2	2,000	1,000	50
E-scrap sample 1	37.4	18.2	19	1.6	—	6	12	—
E-scrap sample 2	27.3	16.4	11.	1.1	—	210	15	20
Printed circuit boards	5.3	26.8	1.9	—	0.47	3,300	80	—
E-scrap (1972 sample)	26.2	18.6	—	—	—	1,800	220	30
E-waste mixture	36	4.1	4.9	0.29	1.0	—	—	—

Materials that can be recycled include ferrous (iron-based) and non-ferrous metals, glass, and various types of plastic. Non-ferrous metals, mainly aluminum and copper can all be re-smelted and re-manufactured. Ferrous metals such as steel and iron also can be re-used.

Improper management of e-waste is resulting in a significant loss of scarce and valuable raw materials, such as gold, platinum, cobalt and rare earth elements. As much as 7% of the world's gold may currently be contained in e-waste, with 100 times more gold in a tonne of e-waste than in a tonne of gold ore.

Recycling is an essential element of e-waste management. Properly carried out, it should greatly reduce the leakage of toxic materials into the environment and mitigate against the exhaustion of natural resources. However, it does need to be encouraged by local authorities and through community education. Less than 20% of e-waste is formally recycled, with 80% either ending up in landfill or being informally recycled in developing countries, exposing workers to hazardous and carcinogenic substances such as mercury, lead and cadmium.

(3) Recycling of the printed circuit boards

One of the major challenges is recycling the printed circuit boards (PCBs) from the electronic wastes. The circuit boards contain such precious metals as gold, silver, platinum, etc. and such base metals as copper, iron, aluminum, etc. The composition of metals, ceramic and plastics in PCBs could reach 40%, 30% and 30%, respectively (Figure 10. 8). Meanwhile, the concentrations of precious metals such as Au in waste PCBs are richer than in natural ores.

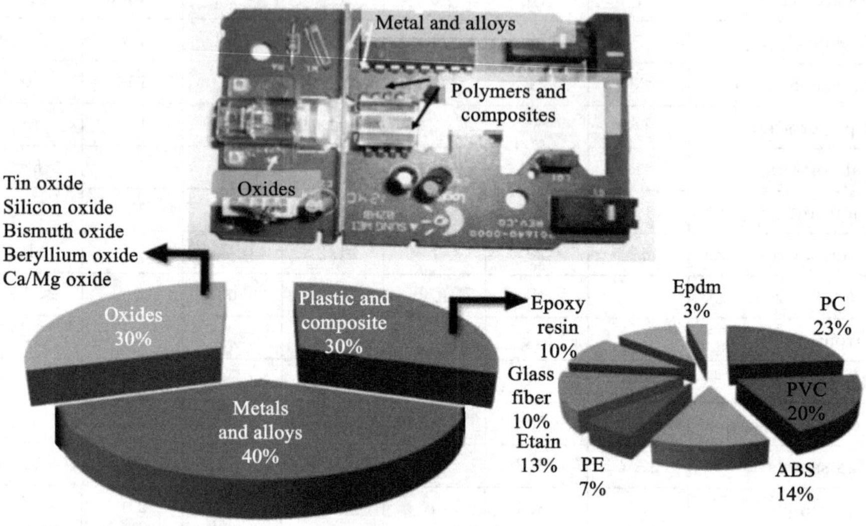

Figure 10. 8 The composition of metals, ceramic and plastics in PCBs

Besides all the hazards originating from e-waste, manufacturing mobile phones and PCs consumes considerable fractions of the Au, Ag and Pd mined annually worldwide. 43% of total production of gold in the world is used in electronics. A large fraction of the WEEE precious metals is found on the PCBs. Since PCBs as becoming more complex and smaller, the amount of materials is constantly changing. 1 tons of circuit boards can contain between 80 g and 1,500 g of Au and between 160 kg and 210 kg of Cu. These concentrations are 40～800 times the amount of gold in Au ore, and 30～40 times the concentration of copper in Cu ore mined in the USA. Globally, 267. 3 tons of Au and 7,275 tons of Ag are consumed annually by electronic industry. According to 2014 UNs' e-waste report, yearly about 300 tons of Au were recovered and the e-waste market size was 52 billion dollars. It has been proved that it is worthwhile to recycle electronic scrap in spite of the fact that the content of precious metals (Au, Ag and Pd) steadily decreases.

One way e-waste is processed is by melting circuit boards, burning cable sheathing to recover copper wire and open-pit acid leaching for separating metals of value. Conventional method employed is mechanical shredding and separation but the recycling efficiency is low. Alternative methods such as cryogenic decomposition have been studied for printed circuit board recycling, and some other methods are still under investigation. Properly disposing of or reusing electronics can help prevent health problems, reduce greenhouse-gas emissions, and create jobs.

Electronics recycling can be challenging because discarded electronic devices are sophisticated devices manufactured from varying proportions of glass, metals, and plastics. The process of recycling can vary, depending on the materials being recycled and the technologies employed.

PCB recycling process with the aim of highest recovery of metallic fraction usually includes three stages: ① pretreatment (selective disassembly of the reusable and toxic electronic parts); ② size reduction and separation including comminution and separation of materials using mechanical or physical processing (such as shredding, crushing, pulverizing, screening, and upgrading); ③ metallurgical or chemical refining (purification).

2. The recycling process

(1) Comminution

An effective liberation of various materials like metals and plastics is a crucial step towards mechanical separation. In addition, classification of electronic scrap is also important to be able to provide an appropriate feed material for the subsequent physical/metallurgical separation process. The purpose of shredding/crushing is to strip metals from PCBs. Crushing technology is intimately related to not only energy consumption, but also further selective efficiency. Waste PCBs are comprised of reinforced resin and metal parts such as wires and joints. Comminution of waste PCBs and high effective liberation of the metal composition from nonmetals is the prerequisite of the following sequence separation for better recovery of waste PCBs.

After removal of the hazardous components (i. e., capacitors, cathode ray tubes (CRTs), liquid crystal display (LCD) and batteries), different mineral processing unit operations such as shredding, crushing and grinding can be used to liberate metals from cladding materials such as resin, fiberglass and plastics. Various types of hammer crushers, rotary crushers, disc crushers, shredders, cutters equipped with a bottom sieve, are used for liberation. Ball milling and disc milling are also reported for pulverizing the PCBs after cutting into small sizes.

Low speed high torque shear shredders (−10 mm) are basically ideal for the primary crushing, while a number of mills have been reported for the finer comminution, with the swing hammer types appearing to have become the industry standard. It breaks mostly by impact, with shear, compression and abrasion; a combination of stress modes appropriate for the diverse material types, from brittle to ductile, present on PCBs. Discharge size below 500 μm is necessary for liberation. If we use froth flotation for separation, then −74 μm is enough for total liberation.

(2) Gravity separation

Gravity concentration methods separate materials of different specific gravity by their relative movement in response to gravity. Nonetheless, this separation is not only dependent on the density of the components, but also on their size.

By using different heavy liquids (such as tetrabromoethane), the metals can be separated from the plastics or ceramics. Water or airflow tables, heavy media separation and sifting are common gravity separators used in e-waste recycling.

Different metal particles can be further separated. For this purpose, the PCB material is processed on concentrating tables. The shaking tables exploit the difference in the specific gravity and particle size to achieve desired separation.

Density separation techniques which are well-known in the mineral processing industry have found their way into e-waste recycling based on the fact that electronic scrap consists essentially of plastics, with a density less than 2.0 g/cm^3, light metal, primarily Al and glass, with a density of 2.7 g/cm^3, and heavy metals, predominantly Cu and ferromagnetics, with a density more than 7 g/cm^3, In sink-float separation, both PC and PCB scraps ~50% (weight) of floats which is primarily plastics can be separated out at the specific density of 2.0 g/cm^3.

(3) Magnetic separation

Magnetic separation can be used to separate magnetic metallic particles (ferrous) from non-magnetic particles (i.e., plastics, glass, Al, etc.) Magnetic separators, in particular low-intensity drum separators are widely used for the recovery of ferromagnetic metals from nonferrous metals and other nonmagnetic wastes. Over the past decade, there have been many advances in the design and operation of high-intensity magnetic separators (HIMS), mainly as a result of the introduction of rare earth alloy permanent magnets capable of providing very high field strengths and gradients.

There are some problems associated with this method. One of the major issues is the agglomeration of the particles which results in the attraction of some nonferrous fraction attached to the ferrous fraction. This will lead to the low efficiency of this method. Through the process of magnetic separation, it is possible to obtain two fractions: magnetic fraction, which includes Fe, steel, Ni, etc. and non-magnetic fraction, which includes Cu.

It is reported that a magnetic field of 6,000~6,500 Gauss (G) can be used to separate the ferromagnetic elements, such as Fe and Ni. The chemical concentration of the magnetic fraction was 43% Fe and 15.2% Ni on average. However, there was a considerable amount of Cu impurity in the magnetic fraction as well. Some researchers used a two-stage magnetic separation to treat milled PCBs. In the first stage, a low magnetic field of 700 G was applied which led to the separation of 83% of Ni and Fe in the magnetic fraction and 92% of Cu in the non-magnetic fraction. The second magnetic separation stage was conducted at 3,000 G which resulted in a reduction in the grade of the Ni-Fe and an increase in the Cu grade.

(4) Electric conductivity based separation

There are three typical electric conductivity-based separation techniques: ① Corona electrostatic separation, ② Triboelectric separation, ③ Eddy current separation.

The corona-electrostatic method is perhaps the most effective separation technology for the metallic and non-metallic fractions at present. The method has the advantage of being environmentally friendly, producing no wastewater and no gaseous emissions. In the corona electrostatic separation, electrode system, rotor speed, moisture content and particle size have the greatest effect in determining the separation results.

Tribo-electricity or contact charging is useful for charging and separating materials that have similar conductivities.

Eddy current separation technique is used for separating metals from nonmetal component; for example, aluminum from glass, which cannot be done by gravity methods as the difference in the densities of aluminum and glass is very small. The structure and principle of an eddy current separator is shown in Figure 10. 9.

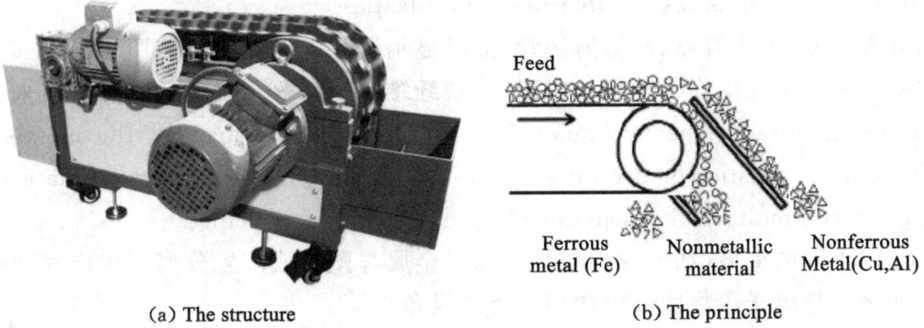

Figure 10. 9 The structure and principle of an eddy current separator

An alternating magnetic field is produced by a high-speed cylindrical assembly of permanent magnets rotating inside an outer drum over which the material is passed. The alternating magnetic field induces eddy currents in the metallic (nonferrous) particles. The eddy currents can produce an induced magnetic field that repels the particle from the alternating magnetic field. The materials leave the drum in various products non-ferrous metal component, which is thrown clear of the drum; non-metallics which drop from the drum by gravity; and ferrous metals, which are attracted to the drum and brushed off.

(5) Froth flotation

Application of flotation has been reported in waste processing. A major challenge to the physical processing alternative is the poor recovery of the base and precious metal values deported to the 74 μm fine fraction generated during the comminution operation. Assessing this as an applied minerals processing problem, froth flotation has been advanced as a promising beneficiation technique for this fine fraction. It is mainly used for drawing back nonferrous metal. The use of froth flotation in environment protection, plastic recycling and sorting were reviewed by Saitoh et al. Sarvar et al. used mechanical separation from computer PCBs with jigging (for +0. 59 mm to 1. 68 mm size) and flotation (for −0. 59 mm size). Collectorless flotation with gasoline was promising method for separating non-metals (i. e., polymers) of PCBs.

Notes

1. E-waste comprises of waste electric and electronic equipment (WEEE) or goods which are not fit for their originally intended use.

电子废弃物包括废弃的电气和电子设备(WEEE)或不适合其原定用途的商品。

2. An effective liberation of various materials like metals and plastics is a crucial step towards mechanical separation.

金属和塑料等各种物料的有效解离是实现机械分离的关键一步。

3. After removal of the hazardous components (i. e., capacitors, cathode ray tubes (CRTs), liquid crystal display (LCD) and batteries), different mineral processing unit operations such as shredding, crushing and grinding can be used to liberate metals from cladding materials such as resin, fiberglass and plastics.

去除有害成分(即电容器、阴极射线管、液晶显示器和电池)后,可使用切碎、压碎和磨矿等不同的矿物加工单元操作将金属从树脂、玻璃纤维和塑料等覆层材料中解离出来。

4. By using different heavy liquids (such as tetrabromoethane), the metals can be separated from the plastics or ceramics. Water or airflow tables, heavy media separation and sifting are common gravity separators used in e-waste recycling.

通过使用不同的重液(如四溴乙烷),可以将金属与塑料或陶瓷分离。水或气动摇床、重介质分选和筛分是电子废物回收中常用的重选设备。

5. Magnetic separators, in particular low-intensity drum separators are widely used for the recovery of ferromagnetic metals from nonferrous metals and other nonmagnetic wastes.

磁选机,特别是弱磁场筒式磁选机,广泛用于从有色金属和其他非磁性废弃物中回收铁磁性金属。

6. There are three typical electric conductivity-based separation techniques: ① Corona electrostatic separation, ② Triboelectric separation, ③ Eddy current separation.

有三种典型的基于导电性的分选技术:① 电晕静电分选,② 摩擦电选,③ 涡流分选。

Exercises

1. Answer the following questions according to the text.

(1) What is electronic waste?

(2) What are the recovery methods of metal in electronic waste?

(3) What harm does electronic waste do to the environment?

(4) What are the three stages of PCB recycling process?

2. Translate the following phrases into English.

(1) 电子废弃物　(2) 有色金属　(3) 黑色金属　(4) 电路板

(5) 颗粒团聚　(6) 摩擦电选　(7) 涡流分选　(8) 预处理

3. Translate the following paragraphs into Chinese.

Paragraph 1 in section 2 (1) "An effective liberation of various materials like metals ...".

4. Translate the following sentences into English.

(1) 电子垃圾含有大量有价值的资源以及大量的重金属和有害物质,它们被认为是具有吸引力的多金属二次资源,并且是环境污染物。

(2) 电子废物管理不当导致稀缺和有价值的原材料大量流失,例如金、铂、钴和稀土元素。

(3) 主要挑战之一是从电子废物中回收印刷电路板(PCB)。电路板含有金、银、铂等贵金属。

(4) 可以回收的物料包括黑色金属、有色金属、玻璃和各种类型的塑料。

(5) 磁选可用于将磁性金属颗粒(铁)与非磁性颗粒(即塑料、玻璃、铝等)分离。

APPENDIX NEW WORDS AND EXPRESSIONS (附录 生词和短语)

A

association *n.* 共生
abrasion *n.* 研磨
abrasive *n.* 磨料
acceleration *n.* 加速(度)
activator *n.* 活化剂
aerofloat *n.* 黑药
aerophilic *adj.* 亲气的
affinity *n.* 吸引力,亲和力
aggregate *n.*; *v.* 填料,聚合物;聚集
agitation *n.* 搅拌,搅动(作用)
air tables 空气摇床
alkyl sulfate 烷基硫酸盐(酯)
aluminium *n.* 铝
amalgamation process 混汞法
amine *n.* 胺
amplitude *n.* 振幅
amyl *n.* 戊(烷)基
andesite *n.* 安山岩
anionic collector 阴离子型捕收剂
anthracite *n.* 无烟煤
aperture *n.* 孔,口
apex *n.*(旋流器)沉砂口
appraisal *n.* 评价,估计,鉴定
aqua regia 王水
archaea *n.* 古细菌
arcing *n.* 电弧
array *n.* 排列
arsenopyrite *n.* 毒砂,含砷黄铁矿
asbestos *n.* 石棉
ash *n.* 灰分
ash fusion temperatures 灰熔融温度
as-mined ore 原矿石,采出矿石
associated gangue minerals 连生脉石矿物
ASTM (American Society of Testing Materials)美国材料试验标准
atacamite *n.* 石笋
attrition *n.* 磨损,研磨
auric *adj.* 金的,含金的,三价金的
aurocyanide *n.* 氰亚金酸盐
aurous *adj.* 亚金的,一价金的
aurum *n.* 金
autogenous mills 自磨机
automatic control 自动控制
azurite *n.* 蓝铜矿

B

back-filling 回填
baffle *n.* 导流板,栅板
banana screens 香蕉筛
bar *n.* 棒,杆,条
barite *n.* 重晶石
basalt *n.* 玄武岩
base metal 贱金属
batch flotation cell 单槽浮选机
bauxite *n.* 铝土矿
beach sand 海滨砂矿
beneficiation *n.* 选矿
benzene *n.* 苯
biodegradation *n.* 生物降解
biooxidation *n.* 生物氧化
bituminous coal 烟煤
blade *n.* 叶片
blasting *n.* 爆破
blinded particles 堵塞的颗粒

bolt *n.* 锚杆,螺栓,桁架
bomb calorimeter 弹式量热计
Bond work index 邦德功指数
borax *n.* 硼砂
bouyant force 浮力
bowl *n.* 转鼓,滚筒
brass *n.* 黄铜
bridge thickener 桥式浓缩机
brochanite *n.* 水胆矾
bromoform *n.* 溴仿,三溴甲烷
buffers *n.* 缓冲
bulk-oil flotation 全油浮选

C

cake filter 滤饼过滤机
calcination *n.* 煅烧
calcite *n.* 方解石
calcium chloride 氯化钙
calcium fluoride 氟化钙
carat(=karat) *n.* 克拉
carbinol *n.* 甲醇
carbon tetrachloride *n.* 四氯化碳
carbonaceous *adj.* 含碳的,碳质的
carbonate *n.* 碳酸盐
carboniferous time 石炭纪
carbon in pulp (CIP)炭浆工艺
carbon in leach (CIL)炭浸工艺
carbon in column (CIC)炭柱工艺
carboxyl *n.* 羧基
carboxylate *n.* 羧酸盐,羧化物
carnallite *n.* 光卤石,砂金卤石
carousel ring 转环
carrier *n.* 矿石搬运
cascade cell 喷流槽,泻落槽
cassiterite *n.* 锡石
cathode copper 电阴极铜
cationic collector 阳离子型捕收剂
caustic aqueous cyanide solution 碱性氰化水溶液
centre-line tailings dam 中线式尾矿坝
cement clinker *n.* 水泥熟料
cement *n.* 水泥,黏结剂
cementation *n.* 胶结
centrifugal force 离心力
centrifugal separator 离心分选(选矿)机
centrifuge *n.* 离心机,离心作用
cerussite *n.* 白铅矿
"cell-to-cell" flotation machines 隔槽式浮选机
chalcocite *n.* 辉铜矿
chalcopyrite *n.* 黄铜矿
chalk *n.* 滑石粉
charge *n.* 电荷
chelating agents 螯合剂
chemical bond 化学键
chemisorption *n.* 化学吸附
chlorauric acid 氯金酸
chlorination *n.* 氯化,用氯处理
choke *n.*; *v.* 堵塞
chrome-bearing *adj.* 含铬的
chrysocolla *n.* 硅孔雀石
chute *n.* 溜井(槽),装矿溜口
circulating load 循环负荷
clarain *n.* 亮煤
classifier *n.* 分级机
cleaner cell 精选槽
cleft across 交叉碎裂
Clerici solution 克列里奇液(重液)
coagulation *n.* 凝聚
coal and gangue separation 煤矸分选
coal assay 煤质分析
coal seam 煤层
cobalt slag 钴渣
coil *n.* 线圈
coke *n.* 焦炭
coalification *n.* 煤化
collection bin 料箱
collector *n.* 捕收剂

colloidal particle 胶体颗粒
columbite *n.* 铌铁矿
combustible carbon 可燃炭
comminute *v.* 使成粉末,粉碎
comminution *n.* 粉碎
complex *n.* 络合物,螯合物
complexant *n.* 配位剂
compression *n.* 压力,压缩
concentrate *n.*；*v.* 精矿;富集
concentration criteria 可选性准则
concentration *n.* 富集,浓缩
conditioning *n.* 调浆
conduit *n.* 管道,导管,水道
cone crushers 圆锥破碎机
cone separator 圆锥分选机
configuration *n.* 结构,外形,配置
conglomerate *n.* 砾岩
consolidation trickling 固结滴流
constantan *n.* 康铜(一种铜镍合金)
constriction *n.* 收(压)缩
contact angle 接触角
contaminant 污染物,杂质
contra-flow *n.* 逆流
conveyor *n.* 运输机
copper sulfate 硫酸铜
corona discharge 电晕放电
corona electrostatic separation 电晕型静电分选
covellite *n.* 铜蓝
crack tip 裂隙末端
cresol *n.* 甲酚,甲氧基
crest *n.* 顶部
critical size 临界粒度
cross-member 横构件
cross-section *n.* 横断面
crucible *n.* 坩埚
crucible swelling index 坩埚膨胀系数
crush *v.* 破碎,碎矿
crushing *n.* 破碎
crystal structure 晶体结构
crystallization *n.* 结晶化,结晶体
cumulative percentage 累积百分比
cupellation *n.* 灰吹法
cuprite *n.* 赤铜矿
cupronickel *n.* 白铜
cyanidation *n.* 氰化
cyanide *n.* 氰化物
cyclic high-gradient separator 周期式高梯度磁选机

D

Davcra cell 达夫克拉浮选机
dacite *n.* 英安岩
DC（direct current ）直流电
deck *n.* 摇床床面
decomposition *n.* 分解
deep cone thickening 深锥浓密机
deformation temperature 变形温度
degradation *n.* 粉碎,减少
degree of liberation 解离度
de-gritting *n.* 脱砂
dense medium 重介质
dense medium separation 重介质分选
deposit *n.* 矿床
depressant *n.* 抑制剂
de-sliming *n.* 脱泥
dewatering *n.* 脱水(作用)
dextrin *n.* 糊精
diacetone alcohol 二丙酮醇
diamagnetic *adj.* 逆磁性的
diaphaneity *n.* 透明度
dichromate *n.* 重铬酸盐
dielectric *n.* 电介质,绝缘体
differential motion 差速运动
digenite *n.* 重碳酸盐
dilute medium 稀介质
dilution *n.* 稀释,冲淡
dimensionless *n.* 无量纲

diorite *n.* 闪长岩
dipole *n.* 偶极,双极子
direct flotation 正浮选
discharge *n.* 放电
disk *n.* 盘
distillation *n.* 蒸馏
dithiocarbonate *n.* 二硫代碳酸盐,黑药
dixanthogen *n.* 双黄药
downstream tailings dam 下游式尾矿坝
drain *n.*; *v.* 排水沟;排空
draining screen 脱介筛
drier *n.* 干燥机,干燥剂
drip irrigation 滴灌
drum *n.* 鼓,滚筒
dump leaching 堆浸
durain *n.* 暗煤
dyke *n.* 堤坝
dynamics *n.* 动力学,动态特征

E

earthed cathode 接地电极
eccentric *n.* 偏心圆,偏心轮
ecosystem *n.* 生态系统
effective density 有效密度
elastic *n.* 弹性
electrical conductivity 电导率
electrical repulsion 静电斥力
electrode assembly 电极组
electrodeposition *n.* 电沉积
electrolysis *n.* 电解
electronic waste 电子废弃物
electro-refining 电解精炼
electrostatic *n.* 静电
electrostatic separation 静电分离,电选
electrowinning *n.* 电积
elution *n.* 解吸
elutriation sizing 淘析分级
elutriation technique 淘析技术
elutriator *n.* 淘析器
embankment *n.* 筑堤,堤防
empirical exponent 经验指数
end product 最终产品
energy output 发热量
enrichment ratio 富集比
entrapment *n.* 夹带
envelope of zero velocity 零速度包络面
equigranular *adj.* 等粒状的
equilibrium *n.* 平衡
equipotential *adj.* 等电位的,等电势的
equivalent diameter 等效直径
ester *n.* 酯
ethyl acetal 乙基二乙醇
ethyl *n.* 乙烷基
extraneous moisture 外在水分

F

Falcon concentrator 法尔肯分选机
fast-floating *adj.* 易浮的
fat coal 肥煤
feed rate 给矿速度
feldspar *n.* 长石
ferric sulfate 硫酸铁
ferromagnetic *adj.* 铁磁性的
ferromagnetic minerals 铁磁性矿物
ferrosilicate *n.* 铁硅酸盐
ferrosilicon *n.* 硅铁
ferrous metal 黑色金属
filter aids 助滤剂
filter cake 滤饼
filter *n.* 过滤机
filtrate *n.* 滤液
filtration *n.* 过滤
finely disseminated 微细粒嵌布的
flotation column 浮选柱
flake *n.* 片状
flame coal 长焰煤
flaw *n.* 裂隙
flight *n.* 刮板

float product *n.* 上浮物
floatability *n.* 可浮性,浮动性
flocculant *n.* 絮凝剂
flocculation *n.* 絮凝
flotation circuit 浮选流程,浮选回路
flotation reagents 浮选药剂
flowrate *n.* 流量
flowsheet *n.* 工艺流程图
fluidized bed 流化床
fluidized bed separators 流化床分选机
fluorite *n.* 萤石
forge coal 锻炉煤
frame *n.* 框架
free settling 自由沉降
free swelling index 自由膨胀系数
free-fall *n.* 自由落体,自由沉降
frother *n.* 起泡剂
frothing agent 起泡剂
froth crowders 泡沫聚集装置
"free-flow" machines 敞开式浮选机
froth flotation 泡沫浮选
fume *n.* 烟雾,烟气
fusain *n.* 丝炭

G

gabbro *n.* 辉长岩
galena *n.* 方铅矿
gangue *n.* 脉石
gape *n.* 裂缝
garnet *n.* 石榴石
gas coal 气煤
gas flame coal 长焰煤
gasification *n.* 气化
gasoline *n.* 汽油
gauze *n.* 纱网
gear *n.* 齿轮,传动装置
Gibbs free energy 吉布斯自由能
glancing impulse *n.* 瞬时冲力
gneiss *n.* 片麻岩
gold refining 黄金精炼
gold smelting 黄金冶炼
grade *n.* 品位
gradient *n.* 梯度
grading screen 分级筛
grain size 粒度
granite *n.* 花岗岩
granular *adj.*; *n.* 粒状的;粒状
graphite *n.* 石墨
gravitational force 重力
gravity separation/concentration 重力分选
grid *n.* 栅条,篦条,网,格
grind *v.* 研磨,磨矿
grinding circuit 磨矿流程
grinding discs 磨盘
grizzly *n.* 格筛
ground particles 磨碎的颗粒
group *n.* 基团,组
gypsum *n.* 石膏
gyratory crushers 旋回破碎机

H

halide *n.* 卤化物
halogens *n.* 卤素
hammer mill 锤式粉碎机
hand selection 手选
Hardgrove grindability index 哈氏可磨性指数
heap leaching 堆浸
heavy metal 重金属
heavy-duty machine *n.* 重型机械设备
hematite *n.* 赤铁矿
heteropolar *adj.* 异极的
high frequency screens 高频筛
high gradient magnetic separator 高梯度磁选机
high tension separator 高压电选机
high-intensity field 强磁场
high-intensity magnetic separators 强磁场磁

选机
high-pressure grinding rolls 高压辊磨机
high-tension separation 高压电选
hindered settling 干涉沉降
hollow cone 空心锥
hollow trunnions 空心耳轴
homogeneous magnetic field 均匀磁场
hopper *n.* 料斗,漏斗
horizontal current 水平流
horizontal screens 平面筛
hornblende schist *n.* 角闪片岩
hutch *n.* 筛下室
hydrated layer 水化层(膜)
hydraulic classifier 水力分级机
hydrazine *n.* 联胺
hydrocarbon *n.* 碳氢化合物,烃
hydrocyclone *n.* 水力旋流器
hydrodynamic force 水动力,液体动力
hydrogen peroxide 过氧化氢
hydrogenation *n.* 加氢
hydrometallurgy *n.* 湿法冶金
hydrophobic *adj.* 疏水的,不易被水沾湿的
hydrophobicity *n.* 疏水性
hydroscopic moisture 吸湿水分
hydrosizer *n.* 水力分级机
hysteresis *n.* 磁滞

I

igneous rock 火成岩
ilmenite *n.* 钛铁矿
image force 镜像力
impact crusher 冲击式破碎机
impeller *n.* 叶轮
impeller-stator system 叶轮定子系统
impermeable plastic 防渗塑料
inactivation *n.* 钝化(作用)
inclined screen 斜筛
incompressibility *n.* 不可压缩性
increment *n.* 增加,增量,增值
induction furnace 感应炉
inertial force 惯性力
in-flow *n.* 顺流,流入量,进水量
inherent moisture 内在水分
initial interfacial energies 初态(初始)界面能
inlet *n.* 入口
integrated gasification combined cycle (IGCC) 整体煤气化联合循环
inter-atomic bond 原子间键
intergrown *n.* 共生
interpolation *n.* 插值
intersection *n.* 交叉,交点,横断
interstice *n.* 间隙
interstitial trickling 钻隙滴流作用
ion exchange 离子交换
ion *n.* 离子
ionic collector 离子型捕收剂
ionization *n.* 离子化,电离
ionizing electrode 电离电极
iron-clad coil 铁壳线圈
IsaMill *n.* 艾萨磨
isobutyl *n.* 异丁基
isomorphism *n.* 类质同象
isopropyl *n.* 异丙基

J

Jameson cell 詹姆士浮选柱
jamming *n.* 堵塞
jaw 颚
jaw crushers *n.* 颚式破碎机
jig *n.* 跳汰机
jigging *n.* 跳汰
Jones magnetic separator 琼斯型磁选机

K

kerosene *n.* 煤油
Knelson concentrator 尼尔森分选机
Kick's theory 基克定律

kinetic energy 动能
kinetics of flotation 浮选动力学

L

lagoon *n*. 尾矿池
laser diffraction 激光衍射
lattice *n*. 晶格
launder *n*. 溜槽
leach pad 浸出垫
leaching *n*. 浸出
liberation *n*. 解离
lignite *n*. 褐煤
limestone *n*. 石灰岩,石灰石
limonite *n*. 褐铁矿
linoleic acid 亚麻酸
linoleum *n*. 油毯
lip *n*. 突出部分,边缘,前缘
liquefaction *n*. 液化
liquid helium coolant 液氦冷却剂
liquid nitrogen 液氮
lithotypes of coal 煤岩组分
lixiviant solution 浸滤剂
locked particle 连生颗粒
loess *n*. 黄土
low-intensity magnetic separators 弱磁场磁选机
lump *n*. 块
lustre *n*. 光泽

M

machine vision 机器视觉
maghemite *n*. 磁赤铁矿
magma *n*. 岩浆
magnet yoke 磁轭
magnetic circuit 磁回路
magnetic field strength 磁场强度
magnetic flux density 磁通密度
magnetic flux leakage 磁通泄漏,漏磁
magnetic induction 磁感应密度
magnetic permeability 磁导率
magnetic separation 磁选
magnetic separator 磁选机
magnetism *n*. 磁学
magnetite *n*. 磁铁矿
magnetization curve 磁化曲线
magnetization *n*. 磁化
magnetohydrostatics 磁流体静力学
magnetomotive force 磁通势,磁动势
malachite *n*. 孔雀石
manganese *n*. 锰
manganese dioxide 二氧化锰
marble *n*. 大理岩
marl *n*. 泥灰岩
matrix *n*. 矩阵;排列
meagre coal 瘦煤
mechanical classifier *n*. 机械分级机
media cleaning 介质净化
median size 中值粒径
medium recovery circuit 介质回收流程
medium sump 介质桶
mercaptan *n*. 硫醇
mesh of grind 磨矿细度,磨矿粒度
metalliferous ores 金属矿石
metamorphic rock 变质岩
methyl *n*. 甲基
mica *n*. 云母
mica schist *n*. 云母片岩
micron *n*. 微米
microorganisms *n*. 微生物
middling *n*. 中矿
mild steel 低碳钢
Miller process 米勒工艺
mill grate 磨机排矿格子板
milling *n*. 磨,制粉,选矿
mined-out area 采空区
mineral *n*. 矿物
mineral concentration 矿物富集
mineral dressing 矿物拣选,选矿

mineral moisture 结晶水分
mineral separation 矿物分离
mineralisation *n.* 矿化
mineralized froth 矿化的泡沫
misplaced material 错配物
module *n.* 组件,模数,模量
molar ratio 物质的量比
molecule *n.* 分子
molybdenite *n.* 辉钼矿
molybdenum *n.* 钼
monazite *n.* 独居石
muffle furnace 马弗炉
multi-stage grinding 多段磨矿

N

nanosize *n.* 纳米
naphtha *n.* 石脑油
near-gravity material 邻近密度物料
negative polarity 负极性
Newton's law 牛顿定律
niobium *n.* 铌(Nb)
nominal aperture 标称孔径
nonconductor *n.* 非导体
nonferrous metal 有色金属
non-homogeneous magnetic field 非均匀磁场
nonionic collectors 非离子型捕收剂
non-magnetic *adj.* 非磁性的,非磁的
nonmetals *n.* 非金属
non-polar surfaces 非极性表面
nugget *n.* 天然金块

O

obsidian *n.* 黑曜岩
occurrence *n.* 产状
oleic acid 油酸
olivine *n.* 橄榄石
open circuit *n.* 开路
"open-flow" flotation machines 敞开式浮选机
open-pit mining 露天开采
optical microscopy 光学显微镜
ore dressing/sorting 矿石拣选,选矿
orifice *n.* 沉砂口
orthophosphate *n.* 正磷酸盐
overflow lip 溢流口(面)
overflow *n.* 溢流
overflow weir 溢流堰
overgrinding *n.* 过度粉碎,过磨
oversize *n.*; *adj.* 筛上产品;过大的
oxalic acid 草酸
oxhydryl *n.* 羟基
oxidation-reduction potential 氧化还原电势
oxidizing *n.* 氧化(作用)

P

paramagnetic minerals 顺磁性矿物
particle size 颗粒粒度
partition *n.* 分配,分布,分开
partition curve 分配曲线
paste backfilling 膏体充填
paste thickening technology 膏体浓缩技术
peat *n.* 泥煤,泥炭
pebble *n.* 砾石
percolate through 渗入
performance curve 性能曲线
peridotite *n.* 橄榄岩
periphery *n.* 周边,周围
periphery *n.* 分选室,齿板箱
permanent magnetic roll 永磁磁辊
permeable base 有孔(透气)底板
permeability *n.* 渗透性
phenocryst *n.* 斑晶
pH modifier pH 调整剂
phosphate *n.* 磷酸盐
phosphorus *n.* 磷
photometric sorting 光度分选,光电分选
physical bond 物理键

physicochemical property 物化性质
physisorption *n*. 物理吸附
pilot scale 中试规模,试验性规模
pinched sluice 尖缩溜槽
pine oil *n*. 松油
pitchstone *n*. 沥青岩
pitman *n*. 连杆
pivot *n*. 轴,支点
plant *n*. 车间
plastic flow 塑性流动
plate and frame filter press 板框压滤机
plate box 分选室,齿板箱
pneumatic tables 气动摇床
pneumatic machine 充气浮选机
polar surfaces 极性表面
polyamide *n*. 聚酰胺
polyester *n*. 聚酯
polyglycol ether 聚二醇醚
polymorphism *n*. 同质多象
polypropylene *n*. 聚丙烯
porphyry *n*. 斑岩
precipitation *n*. 沉淀
preconcentration *n*. 预选,预富集
pregnant leach solution 浸提液
primary crushing *n*. 粗碎
projected area diameter 投影面积直径
propagation *n*. 传播
proximate analysis 工业分析
pulp density 矿浆浓度
pulp *n*. 矿浆
pulsating current of water 脉动水流
pulsion stroke 脉动冲程
pulverization *n*. 研磨,粉碎
pulverize *v*. 粉碎
pumice *n*. 浮石
pyrite *n*. 黄铁矿
pyrometallurgy *n*. 热冶学
pyroxenite *n*. 辉石岩
pyrrhotite *n*. 磁黄铁矿

Q

quarry *n*.；*v*. 采石场;采石
quartz *n*. 石英
quebracho *n*. 白坚木(南美产漆树科乔木)
quicksand *n*. 流沙,重悬浮体

R

ragging *n*. 人工床层
railway ballast 道砟
rake *n*. 耙子,刮板
ratio *n*. 比值
ratio of concentration 选矿比
reciprocating screens 往复筛
recovery *n*. 回收,回收率
redox reaction 氧化还原反应
reducing agent 还原剂
reduction ratio 破碎比
refuse *n*. 矸石,尾矿
regulator *n*. 调整剂
rehabilitation *n*. 复原
Reichert cone 赖克特圆锥选矿机
relative density 相对密度
remanence *n*. 剩磁
resonance frequency 共振频率
resonance screens 共振筛
retard *n*. 阻滞,妨碍
reverse flotation 反浮选
revolving screen 旋转筛
rheology *n*. 流变学,流变能力
riffle *n*. 格条
rigid surface 刚性表面
ring *n*. 环
rinse *v*.；*n*. 漂洗,冲洗;漂洗
rinse water 冲洗水
roast *v*. 煅烧,焙烧
roasting *n*. 煅烧,焙烧
rock *n*. 岩石
rock salt *n*. 岩盐

rod mill *n.* 棒磨机
roller shaft 滚轴
rotary thermal dryers 回转热力干燥机
rotor-type electrostatic separator 转鼓型静电分选机
rougher *n.* 粗选槽
run-of-mine ore 原矿石
rupture *n.* 破裂;断裂
rutile *n.* 金红石

S

saddle *n.* 马鞍形
Sala-HGMS carousel separator 萨拉型转盘式高梯度磁选机
sandstone *n.* 砂岩
saturation magnetization 饱和磁化强度
scalping *n.* 脱介
scavenger *n.* 扫选槽
scheelite *n.* 白钨矿
schiff bases 席弗碱胺类
schist *n.* 片岩
scour *n.* 冲刷
scouring *n.* 冲刷(洗)
scrap iron 废铁
scraper *n.* 刮刀
screen *n.*; *v.* 筛子,筛分机;筛
screen performance 筛分性能
scroll *n.* 螺旋
secondary crushing *n.* 中碎
sedimentary breccia 沉积角砾岩
sedimentary rock 沉积岩
sedimentation *n.* 沉淀,沉降
seepage *n.* 渗漏
segment *n.* 切片,断片
self-aeration *n.* 自吸气
semi-autogenous mills 半自磨机
semi-logarithmic plot 半对数图
settling cones *n.* 浓密斗,沉降锥
settling rate *n.* 沉降速率
shaft *n.* 矿井
shaking screen 振动筛,振荡筛
shaking table 摇床
shale chamber 沉物室
shale *n.* 页岩
shear force 剪切力
shim *n.* 垫片
shredding *n.* 粉碎
sieve analysis 筛分分析
sieve bend 弧形筛
sieve bend screens 弧形筛
sieve *n.* 筛分
sift out 淘汰
silica *n.* 硅石
sink and float test 浮沉实验
sink product *n.* 沉物
sink-float process 浮沉过程
size analysis 粒度分析
size distribution 粒度分布
size separation 分级
sizing *n.* 筛分
skin flotation 表层浮选
slag-forming fluxes 造渣溶剂
slate *n.* 板岩
sleeve *n.* 偏心套筒
slime *n.* 矿泥,煤泥
slot *n.* 缝,口
slotted polyurethane panels 开槽聚氨酯板
slow-floating *adj.* 难浮的
sludge *n.* 矿泥,泥浆
slump*n.* 坍落度
slurry *n.* 矿浆
smelting fluxes 助熔剂
smithsonite *n.* 菱锌矿
sodium bisulphite 亚硫酸氢钠
sodium borate 硼酸钠
sodium carbonate 碳酸钠
sodium nitrate 硝酸钠
sodium oleate 油酸钠

solenoid *n.* 螺线管
solution *n.* 溶液
solvent extraction 溶剂萃取
sort *v.* 拣选,分拣
sorting column 分级柱
sp. gr. = specific gravity 相对密度
specific energy 比能
specific susceptibility 比磁化率
specular hematite 镜铁矿,镜赤铁矿
sphalerite *n.* 闪锌矿
spider *n.* (圆锥破碎机)横梁
spigot *n.* 沉砂口
spigot product 沉砂产物
spike *n.* 尖状物
spindle *n.* 主轴,心轴,锭子
spiral *n.* 螺旋选矿机
splitter *n.* 分隔板,分样器,分配器
spring *n.* 弹簧,发条
spring roll 弹簧辊
squeegee *n.* 刮刀
stannum *n.* 锡
starch *n.* 淀粉
static thickener 静态浓缩机
stationary screen 固定筛
steel rod 钢棒
steepness *n.* 陡度,坡度
sterling silver 纯银
stirred mills 搅拌磨
stirrer *n.* 搅拌器(棒)
stope *n.* 回采工作面,采矿场
Stokes' equivalent diameter 斯托克斯的等效直径
Stokes' law 斯托克斯定律
strain *n.* 应变,变形,拉紧
stratification *n.* 分层(作用)
stray magnetic field 杂散磁场
stress *n.* 压力,应力
stroke *n.* 打(冲)击,冲程
structural steel 结构钢
sub-aeration machine 底部充气浮选机
subaeration *n.* 底部充气
sub-bituminous coal 亚/次烟煤
submicrometer *n.* 亚微米
suction stroke 吸气冲程
sud *n.* 皂水,肥皂泡
sulfhydryl collector 巯基类捕收剂
sulfonate *n.*; *v.* 磺酸盐;使……磺化
sulphide *n.* 硫化矿,硫化物
sump *n.* 水仓,水坑
supercharging *n.* 增压
superconducting separators 超导磁选机
supergene sulfide 浅成硫化矿
support bearing 支撑轴承
surface moisture 表面水分
surfactant *n.* 表面活性剂
suspension *n.* 悬浮液
swing jaw *n.* 摆动颚,动颚
syenite *n.* 正长岩
synchronization *n.* 同步

T

tailing *n.* 尾矿
tailing disposal 尾矿处理
tailings dams 尾矿坝
talc *n.* 滑石
tank *n.* 槽,箱
tannin *n.* 丹宁酸
taper *n.*; *v.* 锥形(度);缩成锥形,逐渐缩小
tar *n.* 焦油
teeter *n.* 流态化,悬浮态
teeter chamber 悬浮室
tenorite *n.* 黑铜矿
tensile *n.* 张力,拉力
terminal surface energies 终态界面能
terminal velocity 末速
tertiary crushing 三段破碎(细碎)
test sieve 标准筛
tetrabromoethane *n.* 四溴乙烷(TBE)

texture *n*. 构造
thallium formate 甲酸亚铊
thallium malonate 丙二酸铊
the angle of repose 休止角
thermodynamics of flotation 浮选热力学
thermal drying 热力干燥
thermic regeneration 热再生
thickener *n*. 浓密机
thickening *n*. 浓缩
thiocarbonate *n*. 硫代碳酸盐
thiourea *n*. 硫脲
throughput *n*. 处理能力,通过能力
tillite *n*. 冰碛岩
tin *n*. 锡
tip *n*. 顶端,尖端
tipping motion 翻转运动
titanium-bearing *adj*. 含钛的
toggle *n*. 肘板
torque *n*. 转矩,扭矩
total moisture 全水分
trachyte *n*. 粗面岩
traction thickener 周边传动式浓缩机
trap *n*. 夹带
trash *n*. 废弃物
tray *n*. 盘,托盘
trench *n*. 沟渠
triboelectric separation 摩擦电分选
trommel *n*. 滚筒筛
trommel screen 圆筒筛
trough *n*. 沟,槽,池
tumbling mills 滚筒磨
turbulent resistance 紊流阻力
two-stage separation 二段分选

U

ultimate analysis 元素分析
ultra-fine particles 超细颗粒
underflow *n*. 底流
undersize *n*. 筛下产品
uniform magnetic field 均匀磁场
upgrading *n*. 选别,提高品位
upstream tailings dam 上游式尾矿坝
urea *n*. 尿素

V

vacuum filter 真空过滤机
vacuum *n*. 真空
valve *n*. 阀,活门
van der Waals forces 范德瓦耳斯力
velocity *n*. 速度
ventilation *n*. 通风
vertical current 垂直流
vertical stratification 垂直分层
vessel *n*. 分选槽,容器
vibrating screen 振动筛
viscosity *n*. 黏度,黏性
viscous resistance 黏滞阻力
vitrain *n*. 镜煤
voidage *n*. 空隙率
volatile matter 挥发分
volatile *n*.; *adj*. 挥发物;挥发性的
voltage *n*. 电压
volume magnetic susceptibility 体积磁化率
volumetric concentration 体积浓度
Von Rittinger's theory 冯-雷廷格定律
Vorsyl separator 沃西尔重介质旋流器
vortex finder 溢流管
vortex *n*. 涡流,漩涡
vortextractor 涡流排料室

W

washability *n*. 可选性
washability curves 可选性曲线
washed coal 精煤
washing plant 选矿厂,选煤厂
water repellent 疏水性
weakly magnetic 弱磁性的
wear *n*.; *v*. 磨损;耗损

wedge *n.* 楔形物
wetting phenomena 润湿现象
wet sieving 湿筛
wet sizing screen 湿法筛分机
wolframite *n.* 黑钨矿，钨锰铁矿
Wemco cone separator 维姆科圆锥选矿机
work of adhesion 黏附功

X

xanthate *n.* 黄原酸盐，黄药

Y

yield stress 屈服应力

Z

zero vertical velocity 轴向零速度
zinc cementation 锌置换
zinc chloride 氯化锌
zinc precipitate 锌置换金泥
zircon *n.* 锆石

REFERENCES(参考文献)

[1] BURT R. The role of gravity concentration in modern processing plants[J]. Minerals Engineering,1999,12(11):1291-1300.

[2] CHOI W Z, ADEL G T, YOON R H. Estimation of model parameters for liberation and size reduction[J]. Minerals & Metallurgical Processing,1988,5(1):33-39.

[3] CLEARY P W. Predicting charge motion, power draw, segregation and wear in ball mills using discrete element methods[J]. Minerals Engineering, 1998, 11(11): 1061-1080.

[4] ENROL G KELLY, DAVID J SPOLLISWOOD. Introduction to mineral processing [M]. New York: A Wiley-Interscience Publication, 1982.

[5] FARAJZADEH S, CHEHREH CHELGANI S. Gravity separation by falcon concentrator-an over review[J]. Separation Science and Technology, 2022, 57(13): 2145-2164.

[6] FATHI H. A short history of hydrometallurgy[J]. Hydrometallurgy, 2005, 79(1/2): 15-22.

[7] FLAVEL M A. Control of crushing circuits will reduce capital and operating costs [J]. Mining Magazine, 1978: 207.

[8] GREENWOOD N N, EARNSHAW A. Chemistry of the elements[M]. 2nd ed. Oxford: Butterworth-Heinemann, 1997.

[9] GUPTA A, YAN D S. Process control [M]//Mineral Processing Design and Operation. Amsterdam: Elsevier, 2006: 622-671.

[10] HABASHI F. Recent trends in extractive metallurgy[J]. Journal of Mining and Metallurgy, Section B: Metallurgy, 2009, 45(1): 1-13.

[11] KAPOOR A, DINARDO O, GOULD W D, et al. Application of rotating biological contactor technology for mine effluent treatment and metal bioleaching operations [C]//Mining and the Environment Ⅲ Conference, 2003.

[12] KAYA M. Recovery of metals and nonmetals from electronic waste by physical and chemical recycling processes[J]. Waste Management, 2016, 57: 64-90.

[13] LYNCH A J, MORRELL S. The understanding of comminution and classification and its practical application in plant design and operation[J]. Comminution Theory and Practice, AIME, 1992: 405-426.

[14] MATRICARDI L R, DOWNING J. Kirk-othmer encyclopedia of chemical technology[M]. New York: John Wiley & Sons Inc, 2019.

[15] SLOTTEE S, JOHNSON J, CROZIER M. Paste thickening iron ore tailings[C]//XXXV Ironmaking and Raw Materials Seminar, Ⅵ Brazilian Symposium on Iron Ore

Florianópolis, Aug 30 to Sept 2, 2005, Santa Catarina, Brazil.

[16] SVOBODA J. Magnetic techniques for the treatmemt of materials[M]. Dordrecht: Kluwer Academic Pub.

[17] TASKER P A, TONG C C, WESTRA A N. Co-extraction of cations and anions in base metal recovery [J]. Coordination Chemistry Reviews, 2007, 251 (13/14): 1868-1877.

[18] TIWARY C S, KISHORE S, VASIREDDI R, et al. Electronic waste recycling via cryo-milling and nanoparticle beneficiation[J]. Materials Today, 2017, 20(2): 67-73.

[19] UM N. Hydrometallurgical recovery process of rare earth elements from waste: main application of acid leaching with devised τ-T diagram[M]//Rare Earth Element. InTech, 2017.

[20] VERBURG R B M. Use of paste technology for tailings disposal: potential environmental benefits and requirements for geochemical characterization [C]// Proceedings of International Mine Water Association (IMWA) Symposium 2001, Belo Horizonte, Brazil, 2001.

[21] WILLS B A, FINCH J A. Wills' mineral processing technology [M]. Oxford: Elsevier Science & Technology, 2015.

[22] X J WANG. Effective bioleaching of low-grade copper ores: insights from microbial cross experiments[J]. Bioresource Technology, 2020, 308: 123273.

[23] 蔡璋,匡亚莉.选矿工程专业英语[M].北京:煤炭工业出版社,1993.

[24] 吕宪俊.矿物加工专业英语[M].北京:化学工业出版社,2007.